S0-DUT-569

# GEOLOGY AND SOCIETY

## Environmental Resource Management Series

### Consulting Editor

Donald R. Coates
State University of New York, Binghamton

*In the same series:*

Soils and the Environment
G. W. Olson

Field Guide to Soils and the Environment
G. W. Olson

Mineral Resources
John A. Wolfe

Land-Use Planning
Julius Gy. Fabos

# GEOLOGY AND SOCIETY

Donald R. Coates



A Dowden & Culver book

Chapman and Hall
New York London

First published in 1985 by
Chapman and Hall
29 West 35th Street, New York NY 10001

Published in Great Britain by
Chapman and Hall Ltd
11 New Fetter Lane, London EC4P 4EE



Printed in the United States of America

ISBN 0 412 25160 4 (cased edition)
ISBN 0 412 25170 1 (paperback edition)



Library of Congress Cataloging in Publication Data

Coates, Donald Robert, 1922-
Geology and society.
"A Dowden & Culver book."
Bibliography: p.
Includes index.
1. Geology. 2. Environmental protection.
I. Title.
QE33.C664 1984 550 84-12073
ISBN 0-412-25160-4
ISBN 0-412-25170-1 (pbk.)

British Library Cataloguing in Publication Data

Coates, Donald R.
Geology and society.—(Environmental resource management).—(A Dowden & Culver book)
1. Geology
I. Title II. Series
551 QE33

ISBN 0-412-25160-4
ISBN 0-412-25170-1 Pbk

*This book is dedicated in love to*

*Mom*

*Dad*

*Stan, and*

*Eleanor*

# PREFACE

Mankind lives, works, and plays on the earth's surface. The majority of such human activities change in some manner the geological materials and processes of our planet. It is the basis of this book that an understanding of this relationship is significant. Furthermore, it is demonstrated that natural processes and events can in turn greatly affect society. Erosion devastates farms and grazing lands. The spectacular hazards of earthquakes, volcanoes, floods, and landslides can lead to disastrous loss of life and property. Thus, one theme in this book is to provide perspective on the duality of these impacts on the environmental scene. Geology is a key component that can lead to an improved understanding of these changes and influences. A carefull orchestration of geological studies can help soften the deleterious aspects of nature and minimize the harmful byproducts of civilization. This is the central message that is repeated throughout these pages.

This book is one of a series of volumes published and in preparation in the series entitled "Environmental Resources Management." Books already in print include two on soils, and others on mineral resources, and land-use planning. These will be followed by books on surface-water resources, groundwater resources, environmental pollution, energy resources, coastal environments, glacial environments, arid regional environments, and others.

Chapter 1 provides the introduction and sets the stage and tone for the book. Chapters 2, 3, and 4 deal with the resource base of society – minerals, fuels, and water. The geological hazards of volcanic activity, earthquakes, landslides, and floods are addressed in Chapters 5, 6, 7, and 8. Chapters 9 and 10 discuss alteration of features in coastal and in soil terrane. Chapter 11 on geoengineering shows how the involvement of geology can be important in design and construction activities. Biochemical pollution and waste problems are described in Chapter 12. Guidelines for environmental management are indicated in Chapter 13, whereas Chapter 14 shows the legal aspects of environmental decision making.

The purpose of this book is to provide a brief introduction into environmental affairs as perceived from the viewpoint of the geosciences. Thus, it is suitable for students in geology, geography and environmental science and for land-use managers and those people with decication to the proper stewardship of our lands and waters. I wish to acknowledge the careful editing done by Joan La Fleur. Illustrations are fundamental ingredients of environmental books and mine have been drawn from a wide range of people and sources for which I am deeply indebted. Special thanks are due to Earl Olson, John Wolfe, and David Alexander, as well as many others who have contributed their materials.

DONALD R. COATES
Newark Valley, New York

# CONTENTS

CHAPTER 1

# INTRODUCTION

The cities and irrigated farms that once thrived and blossomed more than 4 millennia ago in the "Fertile Crescent" of Mesopotamia are now a wasteland. The thickly forested slopes that once contained the magnificent "cedars of Lebanon" are all denuded, and the soil laid bare to bedrock. The ruins of great cities in North Africa and the adjacent fields that once yielded huge grain quantities have been lost to the sand of the Sahara Desert. Ancient cities that once were important seaports such as Tarsus are now many kilometers from the coast, the result of adjacent soil washing to the sea and forming immense deltas. These and countless other examples could be cited of what happens when society becomes an incompetent steward of the environment.

Mankind lives, works, and plays on the surface of the earth so his activities are always destined to change terrain features and processes in some manner. Furthermore, civilization derives sustenance from the geologic environment — the water, the soil, mineral and energy resources. When these substances are withdrawn, used, stored, or manipulated, additional changes occur that may lead to alteration, deterioration, and destruction of the land–water ecosystem.

Even while mankind is changing the natural environment, Earth processes are operating that greatly affect man, such as the devastation wrought by Mount St. Helens in 1980, and the disastrous 1982 floods and mudslides in the San Francisco region. Indeed, when it is thought that the great technological skills, as exhibited by landings on the moon and other types of space conquest, should place mankind on a special pedestal, he becomes humbled by the losses resulting from volcanic activity, earthquakes, landslides, floods, and hurricanes. Almost daily, the news media chronicle stories of new geological events that affect mankind and the environment. For example, the Binghamton Evening Press for September 13, 1979, carried four such items in headlines on page 1 (Fig. 1.1).

Tioga
20 Cents
NEWSSTAND

# The Evening Press

Thursday, Sept. 13, 1979
Binghamton, N.Y.

## Frederic weaker; rips Gulf Coast

### Hurricane rips 4 states

MOBILE, Ala. (AP) — Hurricane Frederic hurled its fury at the Gulf Coast today from Florida to Louisiana, ripping apart homes, shattering hospital windows and flattening businesses with winds of up to 130 mph before it began to abate, still packing a powerful punch.

Before dawn, the power of the winds had dropped to 80 mph.

There were two confirmed deaths as the storm hit Alabama, Louisiana, Mississippi and the Florida Panhandle, churning up 15-foot tides before it weakened over land and posed a different threat — flooding.

"I would say there is not a dwelling, business or any other building in Jackson County that does not have damage

take the brunt of the storm.

Nearly half a million people fled their homes as Frederic stalked the path deadly Camille took 10 years ago in killing more than 250 persons.

Frederic damaged the roof and ripped a cupola from the historic City Hall in Mobile, where more than 5

### Quake dumps town into sea

JAKARTA, Indonesia (AP) — The most powerful earthquake in two years plunged half an Indonesian town into the sea, officials reported, and heavy casualties were feared.

The quake yesterday afternoon measured 8.0 on the Richter scale. Officials said it devastated Ansus, a town of 8,000 on Yapen Island some 2,300 miles east of Jakarta. Yapen, with a total of 40,000 people, is in Cendrawasih Gulf, on the northern coast of Irian Jaya, the western half of New Guinea.

Half the houses in Ansus were swallowed by the gulf, Interior Department spokesman Faisal Taimin reported. But he said there was no word on casualties yet.

church, a police barracks and other buildings.

Interior Minister Amir Machmud and a team from the office of the governor of Irian Jaya flew to the area.

The quake also rocked Biak, an island of 76,000 people about 70 miles north of Ansus, and an American stav-

### Eruption kills 9 on Mt. Etna

CATANIA, Sicily (AP) — Rescue workers recovered three more bodies near the summit of Mount Etna today raising to nine the number killed in the first fatal eruption in years of Europe's most active volcano, police reported.

The three bodies were not immediately identified. The other six persons killed in the eruption last night were all Italians, including three physicians who were attending a medical congress in Sicily.

Search operations, suspended because of darkness, resumed at dawn today. Police said they feared the death toll could go higher as rescue teams probe a mass of rocks spewed up near the lip of the crater.

condition with a skull fracture. As were the dead and most of the seriously injured, she was hit by one of th[illegible] rocks known here as [illegible] because they [illegible] ing lava.

## Acid rain forecast getting worse

By KATHARINE SEELYE
Gannett News Service

WASHINGTON — Acid rain, blamed for damaging wildlife in the Adirondacks, is going to get worse, according to scientists meeting.

Auto emissions and the expected increase in coal burning will account for some increase of what is now a global problem.

"The Environmental Protection Agency has pushed scrubber technology," Aubrey Altshuller of the agency said at a meeting of the American Chemical Society yesterday. "This will balance out the increases (in coal use) — but not quite.

"We estimate a small increase in sulfur dioxide. But nitrogen oxides will increase quite substantially in the next 10 to 15 years."

These are much more difficult to control than sulfur levels, which have decreased with the burning of low-sulfer rather than high-sulfur coal.

Although the acid rain will get worse, no one knows what the effects will be.

"We don't know yet what levels it will reach," Carl Schofield of Cornell University said in an interview.

He said the worst situation in the world now is probably in Sudbury, Ontario, where the nickel smelters are among the world's largest sources of sulfur pollution.

"There has been a complete denuding of the landscape with destruction of the soil — even the topsoil is gone — so that restoration is impossible," he said.

The biggest damage by acid rain so far in the United States has been in the Adirondacks in upstate New York. Researchers say pollutants there, mainly from coal burning power plants in the industrial Ohio River Valley, have interfered with the reproduction of fish and killed all fish in at least 170 lakes.

He said scientists don't know yet whether "episodic" events such as snow-melts in the spring that release a buildup of the acids have caused the extinctions or whether more subtle, chronic processes are at work.

See ACID, 12A

**FIGURE 1.1** Four environmental geology topics made headlines on page 1 of the Binghamton Evening Press of September 13, 1979. This demonstrates geology is news.

## THE ENVIRONMENT AND THE GEOLOGICAL SCIENCES

The environment embraces the total composition of human surroundings. It includes conditions that influence the character of the natural setting — the weather and climate, water, soils, rocks, minerals, landforms, and organisms. The human and cultural environment are those works that have been constructed on the lands and in the waters. At times, these are so massive that they overwhelm the total landscape, such as the cityscapes of the urban environment.

The geosciences comprise a group of related disciplines whose primary objective is the study of the materials, processes, and forms of the earth. Just as in medicine where there are specialists who deal with different parts of the human body and reactions that occur within it, so are there many specializations in the geosciences, each with a particular emphasis. The work of geoscientists ranges the full spectrum from analyzing elemental particles of matter that determine the radiometric date of minerals or events, to those dealing with the movement of entire mountain ranges, continents, and ocean basins. Each subdiscipline employs its own set of equipment and techniques that are used to unravel the earth's secrets, but they are all united by a common quest to decipher these mysteries, and thereby better understand the behavior of physical and chemical laws as they pertain to the lands, waters, and rocks of the planet, indeed the entire realm of the land–water ecosystem.

Within the environmental field, some of the geoscience disciplines have become more heavily engaged than others. The **economic** or **resource geologist** deals with the earth materials that are beneficial and necessary to man, the mineral and fossil fuel resources. The **engineering geologist** (or geoengineer) evaluates the physical properties of materials and their forces to determine their adaptability for sustaining structures emplaced in or on them by man. **Geomorphologists** analyze the surface and near-surface terrain features, and those processes that alter the landscape such as the rivers, ground waters, oceans, winds, glaciers, and gravity movements. **Geohydrologists** (at various times also called hydrologists and hydrogeologists) are interested in the full spectrum of water resources. **Geophysicists** provide special data in earthquake studies, **volcanologists** study volcanic activity, and **geochemists** are involved in pollution investigations in addition to their work in the emerging field of **geomedicine**.

Environmental geoscience is as old as mankind. Geological resources have been used to sustain or enhance the type and quality of life from the first time that humans used flint to cut their meat or arm their weapons, or used earth pigments to paint themselves or their caves. Environmental geoscience has been at the forefront with each quantum jump in the history of man. The Agricultural Revolution was dependent upon the proper use of the right type of soil, and often required importation of water by some kind of geoengineering structure. The Industrial Revolution required plentiful coal and iron resources to blend the skills of man with the materials to manufacture the new products. The Petroleum Revolution was dependent upon plentiful oil supplies, which allowed for the effective use of the internal combustion engine, giving man unheard of power and mobility.

However, environmental geoscience should be more than just locating new mineral and fossil fuel resources, more than simply cataloging how processes operate or inventorying landforms and hydrographic features. This new dimension must also be a concern to study mankind as a process of change, whereby his activities metamorphose earth materials and features. Therefore, to be an environmental geoscientist, one must become involved with the reciprocal relationship of man and nature. This requires working in the policy, planning, and management arena of environmental affairs when his competencies can be brought to bear on problems in his sector of expertise. This involvement should extend into the realm of stewardship whereby predictions should be determined and programs adopted that reduce manmade changes in Earth systems to the barest minimum. Environmental geoscience in this view is science with a conscience, and conforms to the famous words of Ian McHarg in truly being a "design with nature."

## ENVIRONMENT AND HISTORY

Throughout history man has had an ambivalent attitude toward Nature. Such a dichotomy was common during early Greek and Roman times. On one hand, the Greeks recognized a unity of mankind and Nature and worshipped Nature in the embodiment of the gods who represented different parts of the world. Zeus inhabited mountaintops, and Artemis protected the wild creatures. Even certain natural areas, groves of trees, and other earth features were preserved as being sacred to the gods. However, such sanctuaries were invariably small, and at the same time that these were set apart, the main populace was busy ruining the landscape. Plato was one of the few ancients who recognized the detrimental aspects of man at the time. In his *Critias*, he vividly described the changes that had occurred within the short span of two generations, including references to deforestation, erosion, and the drying up of springs.

Even as Roman poets were extolling the beauty of Nature, other Romans treated the landscape as a province to be conquered. For example, a first-century Roman writer proclaimed, "There is nothing that gives either you or me as much pleasure as the works of Nature." However, at the same time, Columella was writing about how the Roman farmers were ruining their soil heritage. Indeed, Simkhovitch believed that part to the blame for the fall of the Roman Empire must be placed on practices that allowed for highly accelerated increases in soil erosion and the loss of fertility of the croplands (see Coates, 1972).

Much earlier civilizations seem to have declined and fallen because of an inability to solve environmental problems. The Sumerian civilization in Mesopotamia required importation of water from the Tigris and Euphrates rivers. In time, however, the canals became clogged with sediment, and the soils became so saline that they lost productive capacity. According to Barbara Bell, the Dark Ages of Egypt in the second millennium B.C. was caused by the total reliance on Nile floodwaters for irrigation of the fields. Several low rainfall years produced starvation, and the accompanying unrest caused governments to be overthrown. Fields were additionally ruined by advancing sands from the Sahara Desert.

In the centuries prior to 1900, writers and scientists throughout the world were strangely silent about the ill effects produced by mankind on the environment. The first book to chronicle the dimensions of wanton destruction by man was *Man and Nature*, written in 1864 by George Perkins Marsh.

For the first time, a worldwide inventory was made of the many places and the many ways in which Earth's lands and waters were being degraded.

> Man, as we have seen, has done much to revolutionize the solid surface of the globe, and to change the distribution and proportions, if not the essential character, of the organisms which inhabit the land and even the waters...So far as he has increased the erosion or running waters by the destruction of the forest or by other operations which lessen the cohesion of the soil, he has promoted the deposit of solid matter in the sea, thus reducing the depth of marine estuaries, advancing the coast-line, and diminishing the area covered by the waters.

Other early writers who described nineteenth century despoilation include Gilbert (1917) who wrote of the impact of gold mining on the slopes of the western Sierra Nevada. Here, gold miners created so much excess sediment by hydraulic mining that they increased floods in downstream areas, ruined farmlands, and caused sedimentation in San Francisco Bay. In the East, Glenn (1911) recounts the degradation of soil in the southern Appalachian region from the combined effects of agriculture, lumbering, mining, and industry.

### The Twentieth Century

This century has seen the culmination of mankind's despoilation of the environment. These unprecedented changes contain both a different range of problems plus an acceleration of the same age-old afflictions. The same environmental excesses that have always plagued mankind are still with us — soil erosion, flooding, siltation, deforestation, salinization, and a host of others (Figs. 1.2, 1.3, 1.4, 1.5). However, to make matters even worse, not only has the magnitude of these problems taken on greater dimensions, but a totally new series of problems has emerged that was not present prior to 1900. Typical of these newcomers are such vexing problems as subsidence, mountainous wastes, and hazardous materials. The accelerating rate of such changes is in part due to the rapidly expanding population, but other factors are also responsible for such lamentable conditions.

The twentieth century has also witnessed massive changes in the mechanization, industrialization, and mobilization of society. An entirely new array of inventions, equipment, machines, and resources has placed extraordinary stresses on the environment. The automobile has created the need for a dense network of highways, which in turn, has produced large-

FIGURE 1.2 Spring floods and landslides in 1983 ravaged parts of Utah causing damages of $28.9 million (Olson, pers. comm.). Homes were destroyed in Farmington, Utah, where the town is built on the alluvial fan of Rudd Creek. Courtesy of Earl Olson.

FIGURE 1.3 Mudflows from Rudd Creek, Utah, partially buried many structures in Farmington. Courtesy of Earl Olson.

**FIGURE 1.4** On October 28, 1983, a 7.3 Richter magnitude earthquake hit the Lost River Range mountain area in Nevada. It caused more than $500,000 in damages to facilities and buildings, and dislodged boulders that tumbled into Challis, Nev. (Olson, 1983). Courtesy of Earl Olson (see also Figs. 6.4 and 6.5).

**FIGURE 1.5** Two children, who were standing near this doorway in Challis, Nev., were killed by the October 28, 1983, earthquake. Courtesy of Earl Olson.

scale modification of landscapes. Fluid fuels became necessary resources, such as diesel oil, gasoline, and natural gas, and in turn, have led to the multiplication of new types of machines and transport facilities. These new energy sources have made possible the drilling of deep wells. Deep water wells have opened new areas to grazing, settlement, and irrigation. The new machines permit mining of mineral and fuel resources by open-pit methods for coal, copper, and iron, often leading to massive degradation of terrain (Fig. 1.6). The new energy also provides greater mobility into new regions that previously were relatively unpopulated. In these ways, many new problem areas of larger scope have arisen including salinization of coastal aquifers, channelization of rivers, and contamination of the lands, water, and air (Fig. 1.7).

The Urban Revolution is still another mega-change of the twentieth century. Population concentration in congested urbanized areas has created another set of headaches. Cities become so large that they outstrip local water resources and require the importation of water from great distances. The invention of air conditioners and other comforts make living possible in such otherwise inhospitable regions as deserts. To serve such cities, it has become necessary to develop vast reservoir systems that may be far distant,

**FIGURE 1.6** Coal strip mining Oliver County, N. Dak. The overburden consists of 1 m of loess and 3 m of glacial till. The coal strata in the Fort Union Formation range from 3 to 21 m thick. Courtesy of Soil Conservation Service (USDA).

**FIGURE 1.7** Careless acts by mankind can cause widespread devastation. Courtesy of John Conner.

as in California, 1000 km from the metropolitan area. Cities greatly alter the landscape, change the regime of streams, and gobble up needed cropland. Their waste products further degrade adjoining areas, and can influence weather factors.

Although beyond the scope of this book, urbanization has led to a large socio-politico-economic cleavage between the urban and rural inhabitants. Their needs are different as well as their perception of the environment. Such conflicts become a vexing problem in governmental management of environmental affairs.

## Environmental Ethics

By its very fabric, environmental affairs breed conflicts. People simply want different things out of the environment. One ethic might be termed the **business ethic**, or the utilitarian-developmental approach to the natural world. Those that espouse this approach believe that Nature is placed on the planet for man's benefit, and so should be used and commanded in his service. They view the ultimate good as deriving the maximum yield that Nature can be forced to offer. This is akin to the Aesop grasshopper fable of not storing up goods for use tomorrow.

A second behavior pattern is that of the **preservation ethic**. This is the opposite extreme from the business ethic, whereby the belief is held that natural materials, features, and processes have as much right to endure as the human species. Therefore, Nature should be as least molested as possible and preserved down through the ages for "posterity." The ultimate view of this syndrome is exemplified by the abstraction of wilderness areas from development and heavy usage.

The **conservation ethic** occupies a somewhat median position. A conservationist is one who does not repudiate Nature but attempts to reach an accommodation in which the utilization of natural resources is done only after careful design that will assure a continuing supply as long as possible. Furthermore, a variety of measures are undertaken so that the maximum good occurs for the maximum amount of time.

During the twentieth century, the United States was responsible for ushering in three conservation movements. The first in the early 1900s lasted only a few years and was terminated by World War I. Its focus was largely on conservation of mineral and timber resources. The second conservation movement started in the 1930s, and the principal impetus was toward soil conservation and measures to control water flooding and resource problems. Again, it had died out by the beginning of World War II. The third, and present, conservation-environment movement began in the mid-1960s with a totally different emphasis. Although it was very broad-based and covered nearly the complete spectrum of environmental degradation, its greatest impact was the first major development of legislation and a national consciousness about contamination of the environment — its ruination, by the exacerbating problems of pollution and water materials (Fig. 1.8). It de-

**FIGURE 1.8** Modern civilization has often produced a throw-away type of ethic and the creation of unnecessary waste. Courtesy of Soil Conservation Service (USDA).

veloped such momentum that the 1970s became called the "environmental decade." This new movement has reached worldwide scope with the organization of many international environmental groups and agencies. The United Nations has also served as a type of clearing house for some of these groups. Along with the growth in "environmentalism" came the growth in private interest groups whose charters showed a devotion to the saving of all parts of the environment. Although the Sierra Club has a long history, many newcomers came into the fold such as the Friends of the Earth, The Wilderness Society, the National Defense Council, and the Conservation Foundation. They have provided extensive lobbying efforts in governmental legislation, and their members constitute a potent adversary group whenever issues are at stake that differ from their precepts of environmental protection.

## ENVIRONMENTAL GUIDELINES

*Planet Earth is finite.* In the words of one pundit, "There is no free lunch." Mankind cannot afford to have a frontier attitude, that more space and resources are always available. Instead many resources are fast being exhausted, and living space is at a premium as 70 million new mouths to feed are added yearly to the world population. This collision course of people and resources will not only affect the quality of life but may instigate severe international stress and disorder. The trinity of problems – population, food, and resources – cross both social and natural science lines. Their solution must be an integrated one with dual cooperation and management.

*Environmental systems are very complex.* One problem when dealing with environmental affairs is that everything is connected to everything else. It comprises a total system with multiple feedback loops and chain reactions. For example, a dam cannot be constructed in isolation without regard for repercussions elsewhere. Such damming of a river can lead to excessive upstream siltation, accelerated downstream erosion, destruction of wildlife habitats, and even initiation of earthquake activity. In the words of Isaac Newton, "For every action there is a reaction." Geoscientists must learn how to anticipate and predict the type and magnitude of environmental impacts that result when man decides to alter natural systems. Furthermore, it is especially important to calculate when a physical threshold will be reached wherein a major disruption is likely to occur. For example, the determination of the safety factor for potential landslide development, or the discharge of a stream before the onslaught of a disastrous flood.

A third range of environmental principles is that *political realities may force environmental management into an accommodation that is at best a compromise.* It has been said that there are no absolute truths in environmental affairs, only approximations. Because all spectra of society may not agree with, or be served by, a specific action, it is generally futile to hope for 100 percent endorsement for any policy. Instead, conflict is almost the "name of the game." There are pro- and anti-nuclear enthusiasts, pro- and anti-coal productionists, and pro- and anti-stream channelizationists. Scientists even occupy both sides of whether to destroy or save phreatophytes. Conflict and tradeoffs are the rule, and not the exceptions, in many environmental decisions (Fig. 1.9).

**FIGURE 1.9** Environmental affairs often lead to conflicts because different segments of society have contradictory goals. Courtesy of Caterpillar, Inc.

*Environmental stewardship is a never-ending quest.* To establish a sustainable working relationship between mankind and Nature requires the wisdom of Solomon, the patience of Job, and the genius of Einstein. It is hoped that when policy and decisions are made that the long-range benefits to mankind as an inheritor of the earth will outweigh the costs that are involved, both in terms of financial expenditures and damages to the environment.

Prior to the 1960s, the environmental literature was near zero — few books, no journals, and exposure in the media that was about invisible. In the last two decades, the explosion of material on environmental affairs has reached unheard of proportions and has been the fastest-growing segment in all published literature. A few notable exceptions to these generalizations did occur. The most noteworthy included the books *Vanishing Lands* (Jacks and Whyte, 1939), *Our Plundered Planet* (Obsorn, 1948), *Road to Survival* (Vogt, 1948), *Challenge of Man's Future* (Brown, 1954), and *Man's Role in Changing the Face of the Earth* (Thomas, 1956).

In the 1960s and then into the 1970s, a deluge of paperbacks rode in with the new technology for their production, and their content covered a wide spectrum of the environmental scene. The following books are typical of this genre: *Silent Spring* (Carson, 1962), *The Frail Ocean* (Marx, 1967), *Moment in the Sun* (Rienow and Rienow, 1967), *The Population Bomb* (Ehrlich, 1968), *America the Raped* (Marine, 1969), *The Diligent Destroyers* (Laycock, 1970), *Eco-Catastrophe* (Ramparts, 1970), *The Environmental Crises* (Helfrich, 1970), and *Patient Earth* (Harte and Socolow, 1971).

The usage of the term "environmental" did not really catch on until the late 1960s, but was presaged by Dasmann's 1959 book entitled *Environmental Conservation*. The first geoscience book with the title, however, didn't appear until 1970 with the publication of Flawn's *Environmental Geology*. Since then, more than a dozen books have appeared that specialize on the topic. In addition, there are now numerous books that provide the geological aspects of the various materials and processes, and many of these occur in the list of readings at the end of this book.

Other sources of environmental geoscience information are the numerous journals that now contain data and ideas on the topic. There are more than 100 environmental journals, many of which hold great interest to earth scientists. Government publications contain still further vital information of great value to environmentalists. Many of the agencies have their own publication series. For example, the U.S. Geological Survey's Professional Papers, Bulletins, and Water Supply Papers are of special interest. In similar manner, the U.S. Bureau of Mines and the Soil Conservation Service (U.S. Department of Agriculture) provide important data. And the list continues with publications of the government's Department of Energy, Environmental Protection Agency, Corps of Engineers, National Oceanic and Atmospheric Administration, and many others. Of course, other countries have counterparts to many of these types of series.

For those working in environmental affairs the "gray literature" should not be overlooked. These materials are often excellent, but do not form the usual refereed and quality-type assurance review of the typical publication sources. The gray literature includes a wide range of works such as governmental open-file reports, Environmental Impact Statements (EIS) as required by the National Environmental Policy Act (NEPA) of 1970, and a host of consulting reports from industry, engineering and consulting firms, and other agencies.

## THE ENVIRONMENTAL DECADE

The 1970–80 period has been called the "environmental decade." The stage was set with the publication of Rachel Carson's *Silent Spring* in 1962. It awakened complacent lawmakers and environmentalists to the dangers of DDT and other chemicals with which the earth was being poisoned. Although there were some statutes that addressed environmental matters prior to 1970, the real breakthrough came with the passage of the National Environmental Policy Act of 1970, signed on December 31, 1969. Other events of 1969 that greatly increased national environmental awareness include:

1. The Santa Barbara, Calif., oil spill did more to galvanize environmental action than any other single event. This disastrous incident spewed viscous crude oil over 21 km of oceanfront beaches and killed countless birds and wildlife.

2. Federal officials met with representatives of the Everglades Coalition to resolve the Miami-Dade County, Fla., jetport proposal, which if enacted would have placed Everglades National Park in jeopardy of degradation.

3. A consortium of oil companies applied for a permit to construct the 1265 km long Alaskan pipeline.

4. The Sierra Club won a convincing preliminary court judgment against the development of a large resort at Mineral King Valley, Calif.

5. Los Angeles residents were dramatically warned to desist from such outside activities as golf, jogging, and exercise that entailed deep breathing of the smog-laden air.

6. The Township of Ramapo, N.J., passed a zoning ordinance that restricted growth, thereby touching off lengthy legal disputes and drawing nationwide attention from other burgeoning communities.

7. The federal government continued to initiate a program to build the SST (supersonic transport) and asked for $662 million in development funds. There was national concern that such high-level planes would irreparably pollute the atmosphere and contribute to ozone depletion.

The environmental decade produced most of the basic legislation needed for environmental reform and management. These acts range from the revolutionary NEPA to statutes dealing with air and water pollution; solid, hazardous, and toxic waste; land-use and coastal policies; and resource development and procurement of minerals, fuels, energy, and recreation.

## THE PRESENT STATUS

Since 1969, the National Wildlife Federation has published an annual summary that provides an assessment of the relative increase or decrease in levels of environmental degradation (Fig. 1.10). Their report for 1982

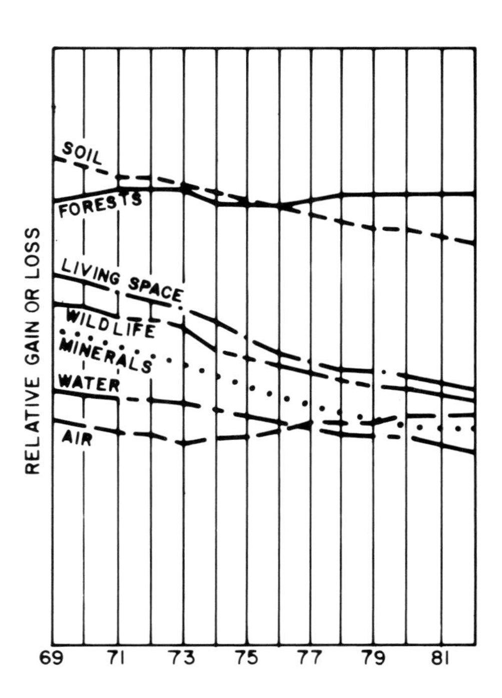


**FIGURE 1.10** Environmental Quality Index for the period 1969–72 as graphed by the National Wildlife Federation. Graph and text courtesy of the National Wildlife Federation.

showed a somewhat balanced treatment of losses and gains. Some of the good news is that industry and homeowners are almost 20 percent more energy efficient than they were at the time of the oil embargo in 1973. Conservation measures had also reduced gasoline consumption to a point that was much lower than in 1973. Clean-air programs, according to the Council on Environmental Quality saved $21 billion in health, property,and other damages and caused 14,000 fewer deaths of people than would have otherwise died from polluted air. Water pollution is a severe problem along with water supply. The General Accounting Office reported that one-third of all new municipal sewage treatment plants do not meet legal water standards. Water-supply systems are so antiquated or inefficient that more than 25 percent of water that flows toward its objective never arrives for its intended purpose. Soil erosion is more acute now than at the height of the "dust bowl" years of the 1930s, with losses ranging from 5 to 9 billion tons a year depending on weather patterns. Conversion of farmland to non-farm uses has reached a staggering 3 million acres (1.2 million ha) per year.

Furthermore, environmental problems are worldwide in scope. Lake Baikal, USSR, is a case in point. This deepest lake in the world prior to 1960 was essentially pure and pristine throughout. Since that time, however, it has undergone a continuing avalanche of pollutants from new industries in the region. Even after 10 years of attempts to reduce pollution, in 1977 the USSR Academy of Science reported (Pryde, 1983, p. 215), "...the danger of Baikal being destroyed had increased rather than decreased, that the entire lake was on the brink of irreversible changes." Environmental problems are monumental in India and become worse with each passing year. The soils are rapidly being degraded by accelerated erosion, salinity, and waterlogging. Deforestation has reduced the number of trees so that less than 10 percent of the country has them. This has helped to accelerate desertification, increased flood levels and damages, and caused excessive silt to form in dams and downstream reaches.

## THE FUTURE

The increase in world population may prevent the goals of environmentalists from being achieved. During the 12 months prior to June 30, 1983, Earth's population experienced a record growth to a total of 4.7 billion. The continuing increase is largely attributed to a slowing in the death rate of children, longer lives, and dramatic birth increases in many Third World nations. If present trends continue, there will be more than 6 billion inhabitants within the next 15 years. It has become a debatable issue on how much additional stress will be placed on the environment, by a continuing population expansion. The Global 2000 Report (1981), which was commissioned by the President, states:

> If present trends continue, the world in the year 2000 will be more crowded, more polluted, less stable ecologically, and more vulnerable to disruption than the world we live in now. Serious stresses involving population, resources, and the environment are clearly visible ahead. Despite greater material output, the world's people will be poorer in many ways than they are today. For hundreds of millions of the desperately poor, the outlook for wood and other necessities of life will be no better, for many it will be worse. Barring revolutionary advances in technology, life for most people on earth will be more precarious than it is now — unless the nations of the world act decisively to alter current trends.

The report also included detailed projections to the year 2000 for population, income, food, fisheries, forests, water, nonfuel minerals, energy, and environmental consequences. It emphasized that:

- Forty percent of the world's tropical forests will disappear.
- Tropical deforestation and other causes will render extinct at least 500,000 living species, some 20 percent of all those on Earth.
- Soil erosion will remove, on the average, several inches of precious topsoil from croplands all over the world.
- Desertification will accelerate.
- Atmospheric concentrations of carbon dioxide and ozone-depleting chemicals will increase at rates that could alter the world climates significantly.

A more cheery scenario has been sounded, however, by Herman Kahn and Julian Simon in their *Global 2000 Revised* (Kahn, 1982). They say, "If present trends continue, the world in 2000 will be less crowded, less polluted, and less vulnerable to resource-supply distribution than the world we live in now." They predict declining scarcities, lowered prices, and increased wealth. Their primary thesis for such a rosy picture is predicated on faith in man's ingenuity and the rate of technological advances. However, their report provides little or no information on environmentally oriented matters.

## POSTLUDE

There are two distinctly different roads that have fed the mainstream of today's environmental thought. One is concerned with the maintenance of human values that depend on reference for the land–water ecosystem, and the other is concerned with the efficient and long-term scientifically organized use of natural resources. It is the challenge of society and the scientist working together as partners to assure that these two sides of environmentalism can be complementary and compatible. The role of the geologist can be substantial toward this objective. He should strive to:

- obtain results as quantitative as possible, which can be integrated into the pluralistic study of systems where nature is complex, yet interdependent;
- anticipate and eliminate short- and long-term damage when human forces intervene with natural processes;
- collaborate with the full range of lawyers, economists, and planners who are involved in the total equation for environmental management; and
- provide information and documentation in public forums that show integrity and responsibility in the stewardship of natural materials, resources, and processes.

CHAPTER 2

# MINERAL RESOURCES

Minerals are inextricably linked to the rise of mankind and civilization. Their importance has been repeatedly documented by archeologists who have named the Ages of Mankind on the basis of the materials used by society – the Stone Age, the Copper Age, the Bronze Age, and the Iron Age. The more recent times also carry names of other geologic resources, such as the Coal Age, the Petroleum Age, and the Atomic Age. The strength and power of early nations often stemmed from mining activities and the amount of such metals as gold and silver in the treasury. Minerals have also played a significant role in exploration of the world, colonization of foreign lands, and the trade, commerce, and industrialization of civilized societies. The health of the United States economy is mineral-related (Fig. 2.1). For example, in 1972, domestic raw materials valued at $32 billion were converted into processed materials and energy that exceeded a value of $150 billion and formed the basis for a gross national product of $1.1 trillion. A decade later, these values had doubled (Table 2.1).

Earth is a closed system where matter cannot be created. There are finite limits to such geologic resources as minerals and fossil fuels. These materials differ from other resources, such as the living resources of timber and crops, because there is no second growth – when mined out, they cannot be replaced. Thus, mineral resources, as contrasted with other products of the earth, are non-renewable, must be mined in place, have extraordinarily high discovery costs, and development costs may increase with time.

Use of mineral products continues to expand among industrialized lands but especially with those nations that are just now moving into the industrial stage. According to the U.S. Bureau of Mines, mineral consumption throughout the world will more than double in the 1975–2000 period. The use of metals and rocks is so vast that it permeates nearly all human activities (Fig. 2.2). The endless list includes all sectors of individual and organized society – health and welfare, tools and machines, buildings, transportation and commerce, currency, recreational equipment, etc. This universal appetite for minerals has fostered wars (one reason given for World War II was to define it in terms of the "have" or Allied Nations vs.

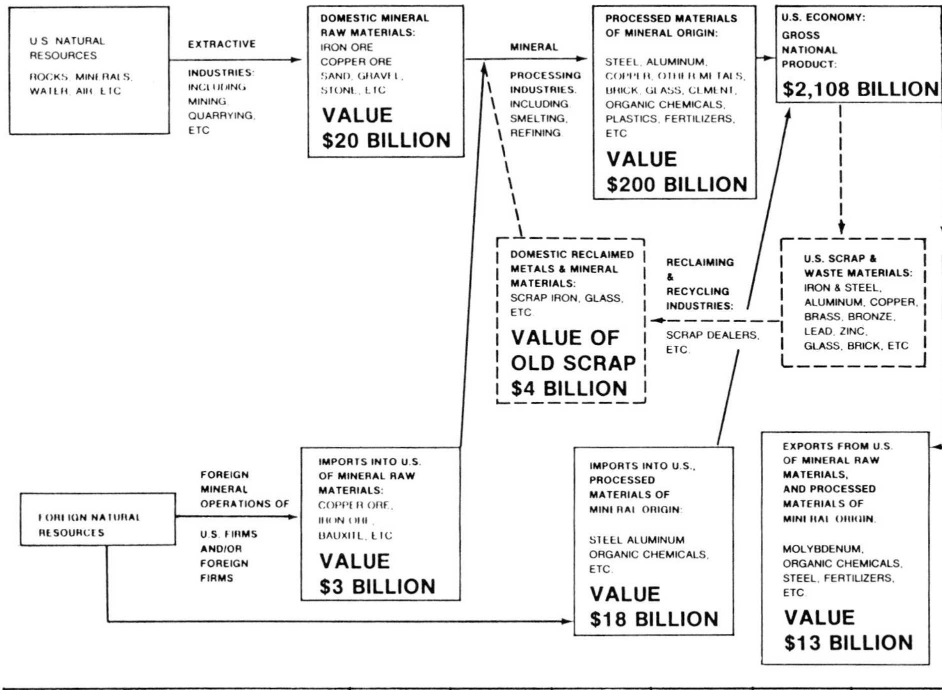


| ECONOMIC ACTIVITY | 1973 | 1974 | 1975 | 1976 | 1977 | 1978 |
|---|---|---|---|---|---|---|
| DOMESTIC MINERALS: RAW MATERIALS | 12 | 14 | 15 | 17 | 18 | 20 |
| RECLAIMED MATERIALS | 4 | 4 | 3 | 4 | 3 | 4 |
| IMPORTED MINERALS*: RAW MATERIALS | 2 | 3 | 3 | 3 | 3 | 3 |
| PROCESSED MATERIALS | 7 | 12 | 10 | 12 | 14 | 18 |
| DOMESTIC PROCESSED MATERIAL OF MINERAL ORIGIN | 114 | 146 | 135 | 153 | 171 | 200 |
| U.S. GROSS NATIONAL PRODUCT | 1307 | 1413 | 1529 | 1700 | 1887 | 2108 |
| EXPORTS OF MINERAL RAW MATERIALS AND PROCESSED MATERIALS | 7 | 11 | 11 | 10 | 10 | 13 |

ESTIMATED IN BILLIONS OF DOLLARS
*REVISED

BUREAU OF MINES, U.S. DEPARTMENT OF INTERIOR
(based in part on U.S. Department of Commerce data)

FIGURE 2.1 The role of non-fuel minerals in the U.S. economy. Estimated values for 1978. U.S. Department of Interior, Bureau of Mines.

**TABLE 2.1** World Production of Mineral Commodities in millions of metric tons

| Rank | Commodity | Tonnage (1979) | Status* |
|---|---|---|---|
| 1 | Aggregate | 12,000 | S |
| 2 | Portland Cement | 957 | S |
| 3 | Iron Ore | 887 | A |
| 4 | Clay | 306 | S |
| 5 | Silica | 200 est | S |
| 6 | Salt | 185 | S |
| 7 | Phosphate | 128 | S |
| 8 | Gypsum | 82 | A |
| 9 | Sulphur | 55 | S |
| 10 | Manganese | 27 | C |
| 11 | Potash | 26 | A |
| 12 | Trona | 18 | S |
| 13 | Aluminum | 16 | D |
| 14 | Magnesite | 11 | A |
| 15 | Chromium | 11 | C |
| 16 | Barite | 7.6 | A |
| 17 | Copper | 7.6 | D |
| 18 | Zinc | 6.0 | A |
| 19 | Fluorite | 5.4 | D |
| 20 | Asbestos | 5.3 | A |
| 21 | Lead | 3.5 | A |
| 22 | Borates | 3.0 | S |
| 23 | Nickel | 0.8 | A |
| 24 | Zircon | 0.8 | D |
| 25 | Graphite | 0.6 | D |
| 26 | Magnesium | 0.3 | S |
| 27 | Tin | 0.26 | C |
| 28 | Molybdenum | 0.103 | S |
| 29 | Titanium | 0.100 | A |
| 30 | Antimony | 0.077 | D |
| 31 | Tungsten | 0.046 | D |
| 32 | Vanadium | 0.041 | A |
| 33 | Uranium | 0.035 | S |
| 34 | Cobalt | 0.028 | C |
| 35 | Silver | 0.011 | A |
| 36 | Gold | 0.0012 | A |
| 37 | Platinum | 0.0002 | C |
| 38 | Diamond | 0.0000096 | C |

*North American status, including U.S. and Canada; S = surplus, A = adequate, D = deficient, C = critical (Wolfe, 1984, personal communication).

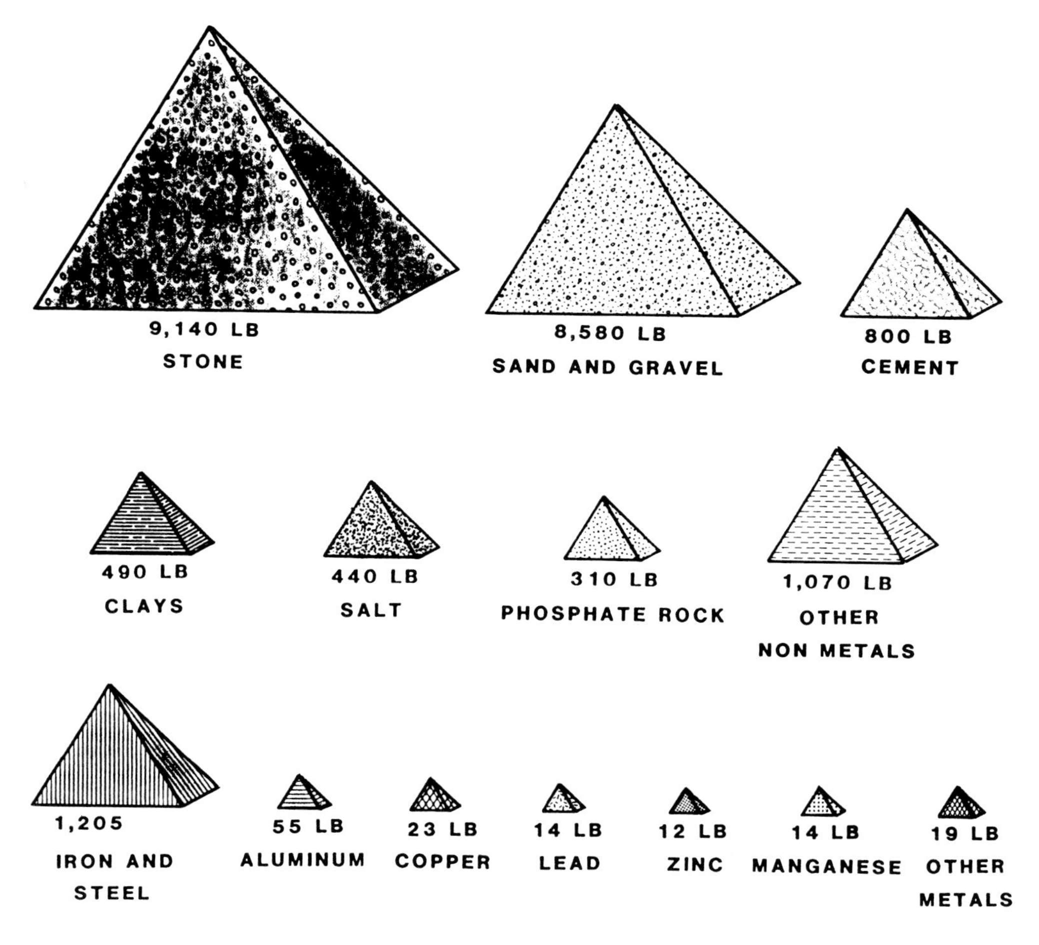


FIGURE 2.2 Over 22,000 pounds of non-fuel mineral materials are now required annually for each U.S. citizen. U.S. Department of Interior Bureau of Mines.

the havenots or Axis Nations), but problems that are now becoming evermore apparent have received little attention on either the national or international levels.

## CLASSIFICATION

Minerals are those naturally occurring in organic earth materials that have definite physical and chemical characteristics. They are composed of elements which, except for construction rocks, are the substances being used by man. Rocks are composed of minerals, and when the metallic minerals are useful to man and can be mined at a profit, the material is referred to as "ore." In this chapter, we will discuss only mineral resources, distinct from the fossil-fuel resources that are described in Chapter 3.

Mineral resources can be classified in several ways. These materials are generally grouped as to their origin, their physical composition or characteristics, and their probability for use and discovery.

### Origin of Mineral Deposits

Minerals occur in rocks, and one very broad classification scheme is to describe them as forming in the three fundamental type of rocks – igneous, sedimentary, and metamorphic. Because silicon and oxygen comprise a large percentage of most rocks, in order for an ore to occur rather unusual processes are required to concentrate it sufficiently for profitable extraction. For example, those deposits that formed as a result of igneous activity can be concentrated at three locations. Some form within the central magma chamber by processes that differentiate such metals as chromium, magnetite, and nickel. Another group of deposits may occur in a zone of alteration where the magma intrudes into the original "country rock." This type of contact metamorphism yields such metals as lead, zinc, and some iron. However, minerals that form from the hot solutions and gases arising from the magma into fissures of the surrounding rocks are by far the most numerous and best known of metalliferous deposits. There is a wide range in size, shape, and composition of these vein deposits. Such characteristics depend on the pressure, temperature, and type of magma, as well as the distance travelled and constitution of the enclosing materials. These metals include silver, gold, tin, tungsten, mercury, antimony, and others too numerous to mention.

Among the deposits formed by sedimentary processes, at least three deserve brief mention. One set of minerals, such as salt, boron, nitrates, potassium, gypsum, and phosphates, forms by deposition from solution in lakes and shallow seas. Weathering and groundwater can also produce concentrations of metals by the chemical removal of valueless constituents. Many copper deposits are formed in this manner, and aluminum and iron ores also are selectively created. Mechanical concentration in placer deposits provides a third mechanism for forming rich deposits where the mineral being concentrated has certain unusual qualities as to weight or durability. Thus, gold, tin, platinum, and diamonds may occur as segregated lag deposits in streams (occasionally even in beach sands) after the water currents have washed away the lighter and more easily broken materials. The gold found at Sutter's Mill was a placer deposit that led to the California gold rush of 1849. Diamonds in South Africa, platinum in the Ural Mountains of the Soviet Union, and even some gold in Australia and Mexico (these being formed from the winnowing action of wind) are examples of rich placer deposits.

### Mineral Properties

This classification scheme groups minerals into metallic or non-metallic categories. Metals are those elements that are good conductors of electricity and heat, are opaque and often have a shiny type of luster. The non-metalliferous minerals are commonly grouped into subclasses that represent such factors as their use or the degree of change of the resource by man. Among the subsets of non-metallic materials are such classes as crushed stone, dimension stone, sand and gravel, abrasives, gem stones, fertilizers, and ceramics.

### Reserves and Resources

Mineral deposits may be described as being a reserve or a resource. Figure 2.3 shows the classification scheme that has been adopted by the U.S. Bureau of Mines and the U.S. Geological Survey. The primary distinction between reserves and resources is dependent on the exactness of the geologic interpretation, the technology, and the economic situation at the time of evaluation. Therefore, **reserves** are already identified and assessed deposits that are known and can be profitably mined under the existing economic conditions. Total **resources** include both known reserves and other inferred deposits that are judged to exist because of extrapolated information but all of which may not be profitably recoverable under the existing conditions. As an illustration, reserves are those funds that are already in a person's bank account, and resources are anticipated monies that have not yet materialized. The status between reserves and resources is constantly changing because of economic, political, technological, and geologic assessments. In 1973, after the oil embargo by OPEC (Organization of Petroleum Exporting Countries) the world's mineral reserve wealth was expanded by extrapolation to 100 million tons of copper, 40 million tons of nickel,

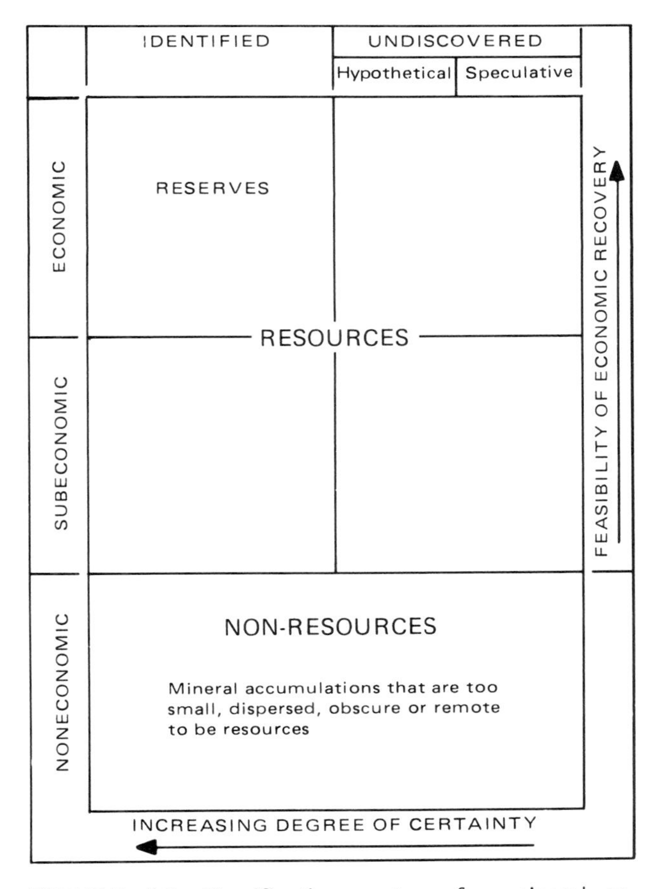


**FIGURE 2.3** Classification system for mineral resources and reserves. After U.S. Geological Survey.

38 million tons of zinc, 21 million tons of lead, and several billion tons of iron and aluminum ores.

## OCCURRENCE AND DISTRIBUTION OF ORE DEPOSITS

The metalliferous deposits are unevenly distributed among nations and geographical locations. A rather unique set of circumstances is necessary to concentrate metals before they are present in sufficient quantities to mine. Table 2.2 shows the degree of enrichment that is necessary in the important metals before they can be considered an ore.

Some minerals occur at or near the surface so they can be mined by surface-mining methods. Others are deep underground, as in South Africa

**TABLE 2.2** Comparison of minimum-grade ore and common rock that must be mined to yield 1 ton of selected metals

| Metal | (1) Crustal abundance (parts per million) | (2) Tons of average continental rock required | (3) Tons of minimum-grade ore required | (4) Concentration above background required (2 ÷ 3) |
|---|---|---|---|---|
| Aluminum | 83,000 | 12 | 4 | 3 |
| Iron | 48,000 | 20 | 3 | 6 |
| Titanium | 5,300 | 190 | 100 | 2 |
| Manganese | 1,000 | 1,000 | 3 | 330 |
| Vanadium | 120 | 8,300 | 100 | 83 |
| Zinc | 81 | 12,000 | 40 | 300 |
| Chromium | 77 | 13,000 | 3 | 4,300 |
| Nickel | 61 | 16,000 | 100 | 160 |
| Copper | 50 | 20,000 | 200 | 100 |
| Lead | 13 | 77,000 | 30 | 2,570 |
| Uranium | 2.2 | 455,000 | 670 | 680 |
| Tin | 1.6 | 625,000 | 6,000 | 104 |
| Tungsten | 1.2 | 830,000 | 200 | 4,150 |
| Molybdenum | 1.1 | 910,000 | 400 | 2,275 |
| Mercury | 0.08 | 12,500,000 | 200 | 62,500 |
| Silver | 0.065 | 15,000,000 | 10,000 | 1,500 |
| Platinum | 0.028 | 36,000,000 | 330,000 | 110 |
| Gold | 0.0035 | 285,000,000 | 125,000 | 2,280 |

*Source:* Mining and minerals policy, 1977, Annual Report of the Secretary of the Interior.

and India, where deep shafts extend 3300 m below the ground so that gold can be extracted. Although mining is now almost entirely on land, many studies have indicated that localities under the seas and oceans contain valuable metals. Many metals are associated with mountain and deformed areas that either in the distant geological past, or more recently, as in the Andes and the North American Cordillera, have been sites of collision and subduction by the massive crustal plates of the earth. Such locales are also positions where volcanism and earthquake activity are most prominent.

Historically, the United States has had very rich metal deposits. However, with long sustained use and the rise in production by other countries, many of our supplies have been mined out, or the total production is much less than that of many other countries. The Soviet Union and Canada are especially rich in a wide range of ores. Some metals are mostly concentrated in single countries. For example, the majority of the world's tin comes from southeast Asia and Bolivia, and Canada has the largest working mines for nickel.

The seas and oceans are being increasingly viewed as areas for future mineral development. The Pacific Ocean is especially rich in deep-sea nodules that contain mostly manganese and iron oxides but are also relatively rich in nickel, copper, cobalt, and molybdenum. One prime area known as the Clarion–Clipperton zone far to the west of Mexico contains about 2.1 billion metric tons of potentially recoverable nodules that average 25% manganese, 1.3% nickel, 1% copper, 0.22% cobalt, and 0.05% molybdenum. Other possible sites occur where hydrothermal systems along ocean ridges or at ocean floor spreading centers transport metals upward and into the adjacent rocks and sediments. The East Pacific Rise materials contain as much as 29% zinc, 6% copper, and 30 ppm (parts per million) silver around the hydrothermal vents. There are other adjacent sites that contain high concentrations of iron, manganese, cobalt, molybdenum, and vanadium. At a site known as Atlantis II along the central rift valley of the Red Sea, the top 10 m of sediments contains 29% iron, 3.4% zinc, 1.3% copper, 0.1% lead, 54 ppm silver, and 0.5 ppm gold. Such potential sources may indeed become the location for future mining when all the political, legal, and technological problems have been resolved.

## EXPLORATION AND DISCOVERY

Until about 40 years ago, the vast majority of mineral deposits were found because of exposures of the ore on the ground. Such minerals were diagnostic to the trained eye, and since prospectors were widely travelled, most of the large deposits were staked out years ago. The hunt for new deposits, therefore, has concentrated on discovery of lower grade deposits, on those deposits at depths greater than the usual shallow prospecting, on newer metals that had little value previously (such as titanium, tantalum), and in areas that were largely unsearched in the past.

The modern exploration for new deposits utilizes not only the traditional methods of geological mapping, but also a wide range of new techniques that have evolved through the disciplines of geochemistry and geophysics, along with satellite imagery. The common thread in all these methods is the detection of subtle contrasts in physical properties of earth materials, which may provide tell-tale fingerprints of important minerals. Geochemical prospecting utilizes the systematic sampling and measurement of soil, rock, and vegetation differences in an attempt to discover anomalies that indicate a metalliferous deposit. For example, characteristic chemical halos that appear in soil may provide evidence for metals at depth. The presence of helium, sulfur gases, carbon dioxide, and mercury may signal important ores below. Geophysical methods may be used in the air, on the ground, or in boreholes to determine variations in such physical properties in rocks as density, gravity, electrical conductivity, heat conductivity, magnetism, seismic behavior, and radioactive intensity. Instruments have been developed that measure and record each of these characteristics. The data obtained are then calibrated with the relationship of such readings to known standards so that judgment can be made about the potential and probability of significant contents of extractable resources. For example, the neutron activation probe that is used in boreholes is highly sensitive in detecting many metals such as copper and nickel.

Satellite imagery by high orbiting mechanical laboratories such as the original ERTS (Earth Resources Technology Satellite) and its replacement system known as LANDSAT are especially useful in mapping vast areas. Data produced by these instruments have located not only promising new structures capable of containing mineral deposits, but can also detect vegetation differences, which might be mineral indicators. For example, some trees have leaves that are sensitive to such metals as nickel, copper, and zinc. Such elements in the soil can retard chlorophyll production, which can be detected by infrared bands that measure light intensity differences.

A special program for cataloging mineral resources of the United States was started in 1978 by the U.S. Geological Survey and entitled the "Coterminous United States Mineral Resources Assessment Program." By 1982, 106,000 km$^2$ had been mapped in 12 states. The program's goal is to complete the mapping of 1.6 million km$^2$ by the year 2000 for those areas that either have a favorable potential for the occurrence of strategic minerals or those minerals that are present on federal lands in significant amounts. This is the first systematic regional study of mineral resources in the country and typical discoveries that have been made include:

1. Three areas near Rolla, Mo., with high potential for lead, zinc, silver, copper, nickel, and cobalt with a combined value of $3 billion.
2. Location of a previously unknown molybdenum source at Dillon, Mont.
3. Siting of new gold occurrences in Tertiary gravels of the Sapphire Range near Butte, Mont. Here, gold is associated with sapphire throughout a thickness of 10–30 m of gravel.

Computers have added a new dimension in prospecting for minerals. A system called PROSPECTOR has been developed by Campbell and coworkers (1982), which is programmed with available knowledge with a set of rules called inference networks. In an evaluation of information at Mount Tolman, Wash., the computer-assisted program helped identify high-grade ore mineralization patterns in a prophyry molybdenum deposit. When the prospect was drilled, the area rated favorable by PROSPECTOR was completely overlain by the known deposit.

## MINING AND DEVELOPMENT

There are numerous considerations that enter the decision to mine a deposit — economic, marketing, engineering, and geologic. Large amounts of capital are necessary to prepare the way for a successful operation. There is often a 15-20 year or more time lag between the initial discovery of a new ore deposit and the date the mine actually starts producing. For example, in 1929, it was known that a large copper deposit occurred in the White Pine area of Upper Michigan. However, not until 1955 at a cost of $61.7 million was production finally consummated. During the 1955-69 period, exploration costs in the western United Stated amounted to about $1 billion. The Henderson molybdenum mine in Colorado took many years of geologic investigation and required another eight years of development and $250 million before a single ton of ore was produced.

There have been some encouraging mineral assessments in the United States in recent years. There is the likelihood of new porphyry copper deposits in the Cascade Mountains, of asbestos in California, and zinc in fractures of coal depsoits in Illinois. A new lead belt occurs at depths greater than 300 m in southeast Missouri. Also, platinum and palladium were discovered in Montana in 1974, but one of the chief finds was the location of gold as disseminated deposits at Carlin, Nev., in 1965, which has turned out to be the most important gold discovery in 50 years.

The international situation can provide significant leverage on the metals market. When the price of the precious metals started their phenomenal rise in the late 1970s to prices 10 times the previous value, abandoned mines suddenly became profitable, and a greatly expanded series of explorations for new mines was begun. In addition, new sophisticated methods were employed for separation of the metal from its bondage with the native rock. For example, the Empire Mine near Sierra City, Calif., has recently been opened with reserves of 10 years at present-day prices. Mines at Tuscarora, Nev., have become profitable with the installation of the "heap leaching" process. This method for metal separation involves an impermeable pad of clay or vinyl upon which the broken ore is placed and then sprinkled with a cyanide and water solution. The cyanide dissolves the gold, which combines with zinc to form a bland sludge. This is then baked, vaporizing the zinc and leaving the gold as residue. Other typical old mines that for a while were again in operation are at Cripple Creek, Colo., which during its boom years produced more than $500 million in gold.

## SUPPLY AND DEMAND

The 1908 White House Conference of Governors meeting was convened to discuss the state of the nation's resources, and was the first of its kind ever held (Blanchard, 1908). Theodore Roosevelt expressed the concern of the convenors, "...what will happen when the forests are gone, when the coal, the iron, the oil, and the gas are exhausted?" There are two ends of a broad spectrum among those who debate the supply of minerals. On one end are those who argue that doomsday is around the corner, and we are approaching an irreversible exhaustion of mineral resources. Only by the adoption of the most drastic conservation and economic measures can the situation be partly salvaged. The opposing school of thought holds that minerals are simply part of rock, and because there is an ample amount of rocks, there are sufficient minerals for the foreseeable future.

The actual status of future mineral supplies occupies a position between the extremists. The alarmists probably do not sufficiently allow for market and price adjustments, new technologies, and man's adaptive and innovative qualities to discover new deposits and alternative materials. However, the developmentalists may be overly optimistic in their appraisal of human resourcefulness. Although technology has often been able to rise to the occasion, each solution has its social, economic, and environmental costs and disruptions (Fig. 2.4). Furthermore, disregard of such consequences can lead to apathy and neglect of the realities of the length of time required to orchestrate the full range of discovery and procurement so vital in assuring continuous mineral supplies.

World War II was a watershed for the United States in terms of supply and demand for mineral resources. Prior to 1941, the nation mined and consumed nearly half of all world mineral resources. However, by 1971, the United States share had dropped to about 27 percent. World consumption of 18 basic mineral resources increased nearly six times, but U.S. consumption had not quite doubled. However, American production of mineral resources lagged behind our use of them, requiring vast foreign imports (Fig. 2.5). By 1972, consumption for 15 critical industrial materials amounted to 60 percent of total consumption, and the situation has continued to worsen when now the majority of crucial minerals must be imported. Such trends are unhealthy as viewed from both national defense and economic prosperity objectives.

There is a wave of increasing nationalism on the part of underdeveloped countries. They are becoming unwilling to export only raw materials, and instead, want either much higher prices for mineral commodities or to sell materials in a partly finished stage. Such LDC (Less Developed Countries)

**FIGURE 2.4** Mining sites can lead to conflict between industry and preservation-type environmentalists. Courtesy of Caterpillar, Inc.

now contain much of the mineral wealth of the world and have learned the skills of driving "hard bargains" for their products. It is becoming obvious that new accommodations and liaisons will be necessary in order to implement effective policies that will maintain a continuous flow of minerals so vital to the highly complex appetites of the industrialized nations.

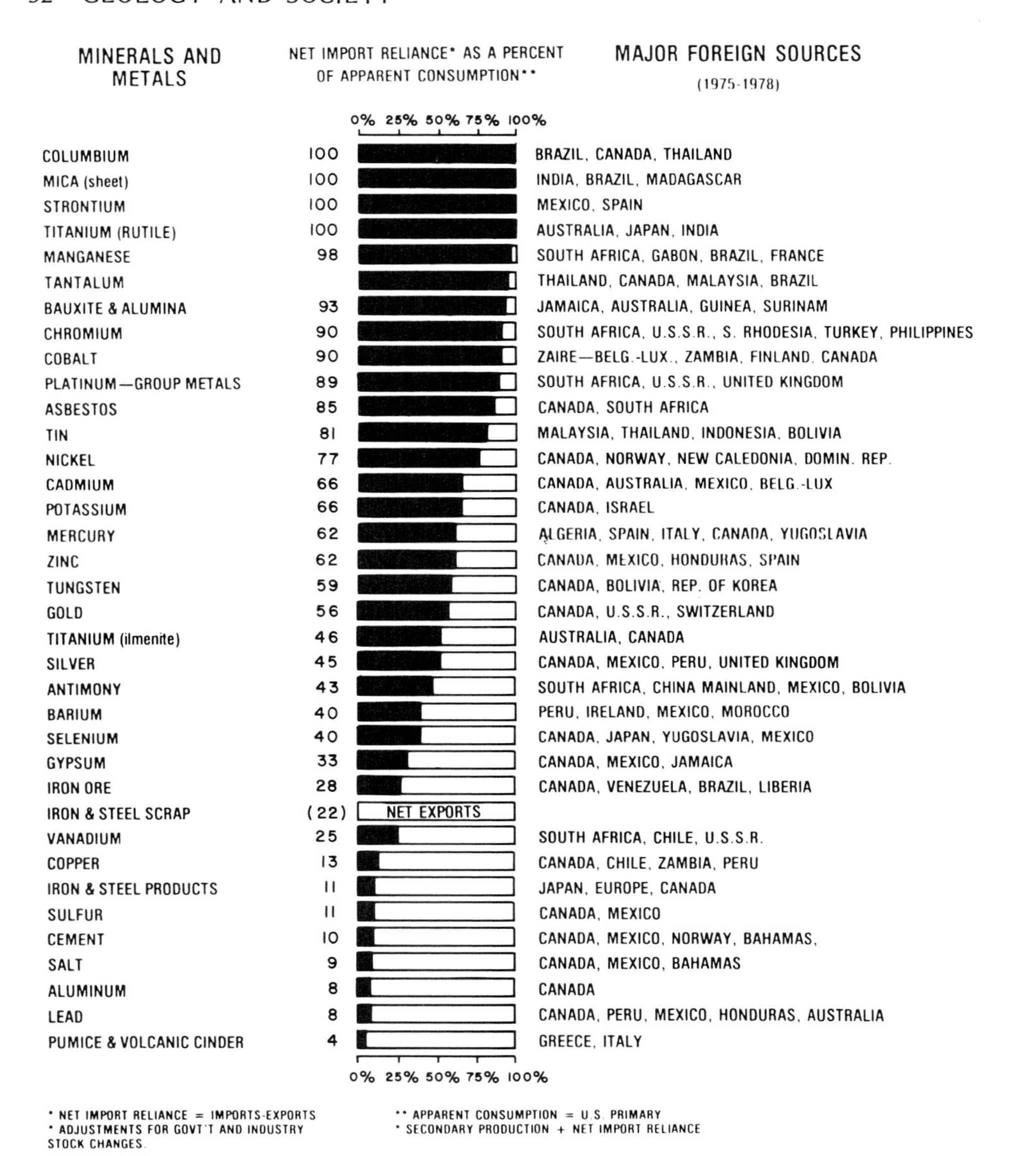


FIGURE 2.5 Import reliance of mineral resources by the United States. U.S. Department of Interior, Bureau of Mines.

## NON-METALLIC MINERALS

The previous discussion was primarily related to the metallic mineral deposits – the ores. However, in many countries, and even in many states of the United States, the non-metallic minerals (including rocks) have total values that exceed monies derived from the metals (Fig. 2.2). There is an even broader range of mineral types in this grouping than those occurring in the metalliferous class. Thus, the scope of such deposits is much more varied in terms of use, occurrence, physical properties, and value. For example, the high-bulk materials such as sand and gravel have the lowest value per unit (measured in cubic yards, cubic meters, or tons), whereas the precious gems such as diamonds, emeralds, and rubies have the highest value per unit (measured in carats — 1 carat equals 200 milligrams).

### Construction and Building Resources

Materials used in construction of the myriad building enterprises created by humans constitute the largest tonnage of mineral resources. The mining and use of rock aggregate (sand and gravel, and crushed stone) comprise the major percentage of this use. The best sand and gravel deposits are those that are clean and well sorted by water, as in floodplains, deltas, beaches, or glacial meltwater deposits such as those occurring as outwash, kames, and eskers. Since 1970, sand and gravel have always produced revenues in excess of $1 billion, and in some states such as New York, it is the most valuable of all mineral resources. Dimension stone is rock that is shaped, cut, or sawed to special sizes — a process known as "dressing." A great variety of rocks, which may be igneous, sedimentary, or metamorphic, are used depending on their field setting and physical properties. Such rock should be sufficiently uniform so that it is easily quarried. The value depends on the use, but most rock should be strong, be able to sustain dressing without fragmentation, and have certain aesthetic qualities such as color and attractive patterns. Because of the great bulk of construction rocks, the quarry should be as close as possible to the marketing area. A factor known as "zero demand distance," the distance of transportation when the transportation cost equals the mining and sales cost at the plant, plays an important role in a successful mining operation. Granite, limestone, sandstone, and to lesser degrees marble and slate, are the most common dimension-stone source rocks.

Other indispensable materials used in the building trades include glass (silica), clay, cement (calcium carbonate), plaster (gypsum), asbestos, and a host of others.

### Fertilizer Materials

This group comprises another range of mineral resources that are very bulky and have large tonnages of consumption. The elements that are in most demand are phosphorus, calcium, potassium, nitrogen, and sulfur. The use of mineral fertilizers expands each year, and crop yields are invariably linked to the successful emplacement of them on the soil.

### Other Uses

Non-metallic minerals are used in a vast number of ways throughout society, industry, and commerce. Health and pharmaceutical needs require an unusual array of minerals, including such elements as boron. Abrasives are vital to many industrial, drilling, cutting, and smoothing operations and include diamonds, corundum, garnet, and quartz. Gemstones are increasing in intrinsic value as people and nations search for a currency that has international stature. Such countries as South Africa and Brazil have always been noted for their large diamond production, and Bolivia, for emeralds. Recently however, Sri Lanka, a country the size of West Virginia, has the greatest variety of gemstones of any similar-sized area on Earth. Although diamonds, emeralds, opals, and turquois are absent, there are more than two dozen other types of gems, including ruby, sapphire, alexandrite (worth more than $10,000 a carat), and unusual types of beryl, aquamarine, garnet, spinel, zircon, tourmaline, topaz, and moonstone.

## IMPACTS CAUSED BY MINING

Most of the deleterious environmental effects of mining operations that will now be discussed are also inherent in the mining of fossil fuels (Figs. 2.6, 2.7). Society often judges active mines to be ugly, dirty, and noisy, and undesirable operations because they depreciate the value of surrounding properties and cause health problems (Fig. 2.8). Indeed, blasting, dangerous excavation and shafts, movement of heavy equipment, dust and gases in underground workings, caveins, and explosions have taken their toll of workers and, in some cases, people in adjacent areas (Fig. 2.9).

Even abandoned mines can pose problems and hazards in terms of physical dangers from unsealed shafts, precipitous rock cuts, unstable debris piles, and acid contaminated water drainage. The U.S. Bureau of Mines in a 1978 study identified 22 potentially health-hazardous uranium tailings sites, and estimated $130 million would be required to make such areas safe. Similar tailings were actually used in construction of homesites at such places as Grand Junction, Colo., and Salt Lake City, Utah. Millions of dollars have been spent in attempts to remove these hazardous radioactive building materials.

**FIGURE 2.6** Massive mine tailings, Kalgoorlie gold fields, Western Australia. Large scale terrain derangement from the waste products of mining. Courtesy of Karl H. Wyrwoll.

**FIGURE 2.7** Clay pits in Laramie Formation at Golden, Colo. This air photograph has north at the bottom and shows a 0.75 km distance. Courtesy of U.S. Geological Survey.

By 1980, more than 1.7 million ha (hectares) had been disturbed throughout the United States by mining activities (the size of Connecticut and Rhode Island combined). Coal mining accounts for nearly half the total, followed by sand and gravel, stone, and then various amounts of iron, phosphate, and others.

Mining effects can also be felt at distances away from the disturbed lands. If not properly reclaimed, tailings and dredging piles become sites for excessive erosion, and sediments are transported by water into adjacent streams and water bodies. Such siltation clogs drainages, increases flooding risks, damages water-supply systems, and causes deterioration of aquatic life, thus upsetting ecosystems (Fig. 2.10). The gold hydraulic mining methods used on the western slopes of the Sierra Nevadas provide vivid

**FIGURE 2.8** Hudson Highlands, N.Y. View looking west across the Hudson River near Bear Mountain. The quarry excavation operation was abandoned due to action by the Scenic Hudson Preservation Society.

**FIGURE 2.9** Dangerous open excavation at Kalgoorlie, Western Australia. These gold mines were originally opened in the nineteenth century and have largely been mined out. Courtesy of Karl H. Wyrwoll.

**FIGURE 2.10** Debris washed into the Bued River, Phillipines, from the mining operations upstream. The channel has become so clogged that major rehabilitation is required to reduce flood damages. Picture taken April 1979. Courtesy of John Wolfe.

testimony to this chain of events. Towns along the streams sustained increased water losses, and parts of San Francisco Bay were smothered with the large sediment load. Even navigation suffered because of new shoals that were created.

The refining and processing of mineral sources at plants can be another source of environmental damage. For example, the smelter operations at Sudbury, Ont., Ducktown, Tenn. (Fig. 12.5), and Swansea, Wales, have devastated soils and vegetation for kilometers around the plants and ushered in new cycles of erosion on the land. Thus, the range of mining impacts can be large unless proper measures are applied to mitigate the possible effects.

Measures that need to be taken to minimize damages caused by mining activities include continuous reclamation, whenever possible, as mining progresses; construction of effective holding ponds to cause siltation to occur onsite; efficient antipollution installations for all smokestacks; and the return to as natural conditions as soon as possible when mining is completed. There is some encouragement along these directions. For example, of the 1.5 million ha that were disturbed by mining activities during the 1930–71 period, more than 590,000 ha were reclaimed or 40 percent of the total.

### Case Histories

*Far West Rand Mining District, South Africa.* Damages that result from mining activities can take a variety of forms. The disasters at this site near

Johannesburg were caused by dewatering operations, which were begun in 1957. By 1960, the program was in full operation, and sinkholes and other ground displacements were occurring. The overlying bedrock of the gold-producing strata consists of dolomite and chert sedimentary rocks. The carbonates throughout a long period of weathering and erosion have developed deep caverns and a subsurface pinnacle-type of topography in which the intervening terrain is filled with unconsolidated sediment. Dewatering was undertaken to lower the water table so that more ore was accessible by dry mining methods. Subsidence has consisted of compaction of unconsolidated material and sinkhole development when arches collapse above the water table (Figs. 2.11, 2.12). The pumping operation has caused water levels to decline about 100 m in most places and as much as 550 m at some sites (Foose, 1967; Bezuidenhout and Enslin, 1970). Between December 1962 and February 1966, eight sinkholes formed larger than 50 m in diameter and deeper than 30 m. The most catastrophic event occurred in 1962 when a 30 m deep sinkhole killed 29 people, and five others were killed at another site in 1964.

*Lake Peigneur, Louisiana.* This near-tragedy occurred on November 20, 1980, because of lack of proper communication between Texaco, which was exploring for petroleum, and owners of the Diamond Crystal underground salt mine. Underground, the salt mine had rooms 24 m high and corridors

**FIGURE 2.11** This is a rejuvenated paleosinkhole created by dewatering operations of the underground gold mines at Kaolin Street, Carletonville, South Africa. Courtesy of R. J. Kleywegt, Director, South Africa Geological Survey.

**FIGURE 2.12** This sinkhole destroyed several houses at Blyvooruitzicht, South Africa. It was caused by the gold mine dewatering operation. Courtesy of R. J. Kleywegt.

as wide as a four-lane highway. Most of the mine was under the lake at a depth of 390 m (Martinez et al. 1981).

At 5:00 A.M., the drilling pipe of the oil rig on the lake became stuck when it encountered the roof of the salt mine. The drilling platform began listing, and by 6:00 A.M., the crew abandoned the rig. At 7:30 A.M., the drilling rig and platform toppled over into the lake. At 8:00 A.M. in another part of the salt mine, 48 miners and three visitors had descended and discovered a leak. By 9:00 A.M., they had all been safely evacuated to the surface. At 11:00 A.M., a giant whirlpool developed at the site where the drilling rig had vanished. Most of the damage occurred from 11:30 A.M. to 1:00 P.M. (Figs. 2.13, 2.14, 2.15). During this period, the whirlpool opened to a diameter of almost 1 km and swallowed 11 barges, one tugboat, and demolished three buildings, 26 ha of lake frontage, and ruptured an underwater gas well, which caught on fire. At the southern tip of Lake Peigneur, a 45 m waterfall was created which reversed the flow of a small stream. Miraculously, no lives were lost in the disaster.

These events set in motion a series of lawsuits and counterlawsuits between Texaco and Diamond Crystal for more than $55 million in damages. In all, a total of more than seven lawsuits were filed, some by the 200 miners who lost their jobs as the aftermath of the destruction. The lake has now recovered and covers as much land as before the debacle.

**FIGURE 2.13** Jefferson Island, Lake Peigneur, La. This is a salt dome with underground salt mining 400 m below the surface. This aerial photograph was taken prior to the 1981 catastrophic collapse with a view north-northwest. Courtesy of Joseph Martinez (see also Martinez, et al., 1981).

FIGURE 2.14 Lake Peigneur drained catastrophically and a newly incised channel developed with steep gradiant into the barren lake. Note barges and docks that foundered into the sediment of the lake floor. Courtesy of Joseph Martinez.

FIGURE 2.15 Landsliding by rotation slump blocks from failure within the Pleistocene strata across the collapsed area at Lake Peigneur, La. Courtesy of Joseph Martinez.

## MINERAL POLICY AND PLANNING

The United States, and indeed many nations of the industrialized world, are at a crossroad in terms of mineral availability and policies to solve the attendant problems of diminishing resources. Increased importation of mineral resources poses both national security and economic threats to society. At the same time, there are continuing controversies between arch-preservationists and mining development entrepreneurs. For example, the preservationists argue that the Mining Act of 1872 is antiquated and vests too much power with mining companies, whereas the mineral producers claim the Wilderness Act of 1964 is too restrictive and that the environmental wave is choking off important resources that are necessary, as in Alaska. Indeed, many argue environmental laws are seriously stiffling mineral exploration and production.

The Mining Act of 1872 is an anachronism when compared to more recent laws governing other types of resource production. It contains no provisions for a balanced resource allocation, for environmental protection or reclamation, for multiple terrain use, or for assignment of government fees. Instead, it acts as a government subsidy to mining with a guarantee of "free access" to public lands. It only requires $100 of work annually on the property to maintain a claim. Thus, a company can mine and sell minerals from governmental land without payment for such privilege. This is in great contrast to other resources such as timber where a fee is based on the amount of lumber involved. Resources such as coal, oil, gas, and oil shale fall within the jurisdiction of the leasing system established by the Mineral Leasing Act of 1920. Sand, gravel, and stone are covered under the Multiple Surface Use Act of 1955. When mining for any of these materials, the individuals or companies must pay to prospect for them on public lands.

Although the land being mined in the United States is generally owned by the mining company, important resources do occur on federal or Indian lands. When such lands are mined through leasing agreements with the government, a certain share of the revenues must be paid to the federal treasury. For example, the following solid minerals have provided a government royalty of more than $800 million during the 1920–82 year period: copper, fluorspar, lead, limestone, phosphate, potash, sand–gravel, sodium, uranium, and zinc (U.S. Department of the Interior, 1983).

Revenues from mining and fossil-fuel operations in 1980 for use on federal and Indian lands reached an all time high, exceeding $2.6 billion. This was $700 million more than the previous high in 1979. The increased income was largely from higher oil and gas prices. Royalties on oil and gas production of the Outer Continental Shelf accounted for 73 percent of the total revenue. By 1982, the figures jumped to total revenue of $9.3 billion, with $3.987 billion for outer shelf lease bonuses.

The Wilderness Act of 1964 allowed hard rock mining exploration in some designated areas until 1993, but in general, such regions have not supported mining operations. Many wilderness areas have high potential for mineral and fossil-fuel resources. The White Mountain Wilderness in New Mexico has molybdenum; the Charles M. Russell National Wildlife Refuge in Montana probably has oil, gas, coal, and bentonite; and the Gila Wilderness and Gila Primitive Area, N. Mex., have good possibilities for gold, silver, copper, molybdenum, lead, zinc, and fluorspar. Whether to open such sites to thorough study and possible mineral exploitation is becoming a heated debate.

The need to establish a coherent and comprehensive mineral policy in the United States is evident when actions of such consortiums as the Organization of Petroleum Exporting Countries (OPEC) become united. If world market prices for petroleum can be fixed, as they were in 1973 and several times subsequently, there is nothing to prevent other nations from trying to secure a near-monopoly on certain strategic minerals. This is especially true for those many metals in which the United States is now deficient. For example, three countries in southeast Asia, Malayasia, Indonesia, and Thailand, account for more than two-thirds of the world's tin production. Since 1956, they, and other producers, have had 5-year agreements with consuming nations to prevent price fluctuations. In July 1981, the consensus deteriorated, and several nations declined to sign a new 5-year agreement. Prices on the London Metal Exchange began rising when tin production and purchases increased at an accelerated pace. Malaysia thereupon initiated meetings with the goal to develop TINPEC, a cartel similar to OPEC, for the purpose of controlling world supplies and prices of tin. Fortunately, the United States had already stockpiled 200,000 metric tons of tin in its strategic reserve for emergency use.

## Policy Recommendations

There are a variety of programs and methods that are necessary to alleviate the inadequate way that mineral resources are managed and perceived. The following points might provide the basis for a policy that would sustain and assure a more stable flow of resources and their availability in this technological age:

1. Develop educational programs that stress the importance of mineral resources. This would set the proper receptive climate for closer liaison

among people, government, and industry. Only through a broad-based support system can appropriate mandates be developed.

2. Review laws and ordinances pertaining to mineral resources and update them in accordance with consistent regulations and insights that will allow multiple land use of areas necessary for mineral extraction.

3. Provide incentives for improved recycling programs and the elimination of waste. Each year more than $2 billion worth of metals is destroyed in waste areas, which might be recoverable.

4. Encourage the use of metal substitutes whenever possible. Wood and glass can be used for many articles now made of metals. Use of plastics (although they depend on petroleum) can also conserve metals.

5. New incentives should be given for exploration along with higher depletion allowances for the early years of production from mining operations. Thus, it should be realized that discovery of new deposits is becoming increasingly difficult and costly.

6. The encouragement of additional research aimed at providing new technologies for ore recovery and improvement of more efficient mining and production strategies. The problem of "high grading" needs careful examination. This is the technique of mining out only the highest ranking ores first. Although this produces large immediate profits, making stockholders happy, it may make the mining of lower grade ores even more difficult and costly in the future. Of course with lower grade ores, there is the possibility of greater environmental impacts, and they cost more per unit to mine because more rock must be extracted and disposed.

7. A crash program to develop the technology for mining of the seas, the deep seas, and the continental shelves. Such research must address not only the methods to be used for the extraction of the ores, but also how this can be accomplished to minimize ecological harm to marine oganisms.

## RECLAMATION

Federal laws such as the Surface Mining Control and Reclamation Act of 1977, and state laws such as New York's Mined Land Reclamation Act of 1975 have done much to salvage lands that would otherwise become derelict terrain. The following steps should be taken in surface-mining operations to lessen any potential for causing environmental damage:

- Preplanning. Land restoration needs to be an integral part of the mining development plan. Both interim and final land use reclamation should be implemented.

- Soil and slope stabilization. During mining, erosion and siltation-control methods need to be instituted. Spoil piles should be covered, and areas planted as soon as possible. Mining areas should be minimized to expose the least rocks to weathering and potential for acid buildup. Slopes should be graded to reduce landslide risk. Catchment basins can be constructed to trap washed sediment. Haul and access roads should be sited for minimum lengths and composed of material to reduce fugitive dust.
- Water control and storage. Surface runoff must be controlled and diverted from site with proper engineering methods. Water seeping from the site should be detained on site and treated to prevent toxic substances from contaminating downstream hydrologic systems. Settling ponds and lakes can aid in containing excess water from causing downstream erosion.
- Regrading and revegetation of the site should be adopted both during and after exhaustion of mining operations. Slope recontouring should be consistent with the original character of the terrain, and may even enhance the natural beauty and aesthetics of the environment.

## POSTLUDE

Many events, both national and international, have caused grave concern over the health of the American mineral industry. The years 1982 and 1983 proved especially difficult due to sluggish mineral demand, continuing worldwide surpluses, and low prices. More than 20 percent of the industry's exploration and production capacity was dormant. The Commonwealth of Pennsylvania is a case in point. For the first time in 250 years, there was not a single operating metallic mineral mine. In September 1983, the New Jersey Zinc Company mine at Friedensville closed the last operation. It had been continuous since the 1840s. As recently as 1981, this was the second largest zinc producing mine in the country, and its yield placed Pennsylvania as the sixth largest zinc producer in the United States. The two major factors that led to the closing were the enormous increase in electricity costs to run the pumps that kept the mine dewatered and the lowered demand for zinc because of the recession. Other recent closings have included the Grace Mine at Morgantown, Pa., which shut down iron ore production, although large reserves still remain along with modern mining and milling facilities, and the world-renowned Cornwall iron mine, which had a continuous mining history back to 1742.

Thus, it is ironic that today, Pennsylvania, which during the past two centuries was an important producer of such ores as iron, manganese, chromium, lead, silver, nickel, copper, cobalt, graphite, magnesium, zinc, uranium, and vanadium, now has no metallic mining. Fortunately, such trends are not always irreversible. Commodity price changes, new technology, discovery of even higher grade ores, of needs to become more self-sufficient can cause renewed mineral production.

CHAPTER 3

# ENERGY RESOURCES

Mankind has always required some form of external energy supply for his very existence. His first source was food, which supplied body heat, but in cold climates, fire was necessary to supplement clothing for bodily warmth. In addition, cooked food tasted better when heated. The domestication of animals supplied an increased energy power source when used in lieu of human muscle power. However, the real change in energy came with the invention of the waterwheel in the first century B.C., which again multiplied the power available to man. Harnessing wind power by development of the windmill occurred in the twelfth century.

The most significant revolution finally was consummated by the perfection of the steam engine, which dominated a new era, the Industrial Revolution. Unfortunately, this new method of generating power required abundant sources of fuel. Wood was the first principal fuel but was finally overtaken by coal, so that by 1900, only 20 percent of the world's energy was supplied by wood. Although wood is the principal fuel source for half the world's population, it constitutes less than 1 percent of the total energy production. Meanwhile in the twentieth century, a new source of energy was gradually coming into its own, petroleum, and by 1975, two-thirds of the world's energy was being created by the wonder fuels of oil and natural gas.

The human spirit is never complacent with present conditions, so with the explosion of the first atomic bomb at Alamogordo, N. Mex., in 1945, another new era was ushered in — the Atomic Age, or Nuclear Revolution. However, nuclear energy never quite lived up to its advanced billing, and today, it is only an "also ran" in total energy production. Fortunately, there are still other energy options available.

Perhaps the final chapter in mankind's quest will be the efficient development of renewable resources — wind, water, oceans, and solar types of energy. However, in this chapter we are going to emphasize the geologic energy resources — those that are non-renewable. The use of these resources has many problems because they all produce some type of environmental degradation. This means there is a constant battle between environmental advocates and those who champion industrial development. Not only do tradeoffs occur when these conflicts are resolved, but because all nations do not have equal energy resources, this leads to commerce readjustments and international problems. Thus, geologic materials and energy resources become focal points in environmental matters and international affairs.

## AN OVERVIEW

Energy resources are unevenly distributed throughout the world. Much of south Asia, central Africa, and South America, areas teeming with rapidly expanding populations, are fossil-fuel deserts. Their use of forests, which are not being replaced, is leading to the destruction of vital soil resources on a massive scale. Many of the industrialized regions to the north are rich in coal and uranium, and some are comparatively rich in petroleum. For example, the Soviet Union is the world's largest oil producer, and the United States is third, after Saudi Arabia. There are exceptions, however. Western Europe imports more than half its oil, and Japan imports 90 percent of all its energy resources (Table 3.1).

Fuels account for nearly 94 percent of the world's energy production of 297 quadrillion Btu's (or 297 quads). Coal constitutes 27 percent of the total global energy budget. At the present rate of use, the known reserves will last into the twenty-second century. However, the supplies of oil will be exhausted in 30 years, if new fields are not discovered. Although new gas fields continue to be discovered, such additional resources will probably be offset by production decline in currently used fields. Nuclear power has been so plagued with economic, environmental, and political problems that its future is uncertain. Its development seems to be losing momentum, and the total generating capacity may even be in decline by the end of the century. Figure 3.1 shows total energy supply in the United States.

Hopefully, the eventual transition from non-renewable to renewable resources will endow the global economy with greater stability, fairness, and permanence than has occurred in societies that currently use fossil

**TABLE 3.1** World oil supply (millions of barrels per day)

| | 1973 | 1983 (estimated) | 2000 (projected) |
|---|---|---|---|
| OPEC | 31.0 | 17.2 | 26.6 |
| Non-OPEC | 6.6 | 13.2 | 15.3 |
| United States | 9.2 | 8.6 | 7.5 |
| Other | 0.8 | 5.4 | 5.1 |

*Sources:* U.S. Department of Energy, Oil and Gas Journal averages.

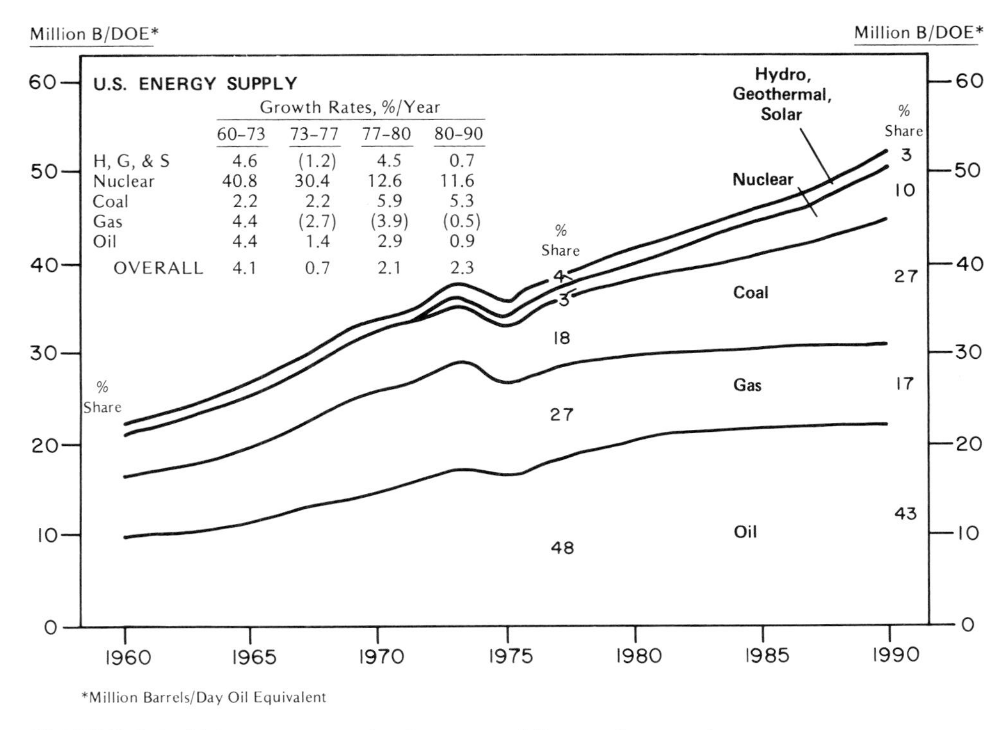


**FIGURE 3.1** U.S. energy supply. Courtesy of Exxon Corporation.

fuels. To achieve such a sustainable society will require huge monetary investments, a greatly expanded engineering technology, and the political drive of governments and industry to work together to cause the new millennium – the Renewable Energy Revolution.

## OIL

Oil, plus natural gas, oil shale, tar sands, and coal, comprise that family of resources known as the fossil fuels. When coal is omitted, these materials constitute the petroleum industry (the word petroleum is derived from the Latin, *petra*, rock, and *oleum*, oil). Few other resources in the history of mankind have prompted such vast societal changes and problems of international unrest as now occur with the production, marketing, and use of oil. It takes great capital investment and scientific and technical skills to locate and obtain this valuable resource. Since its discovery in 1859, near Titusville, Pa., world oil consumption has nearly doubled with each decade. Is it any wonder that oil has become not only the glamour resource of the industrialized world, but also the product that has vastly metamorphosed countries and upset the balance of national wealth?

## The Geology of Oil

Oil is of biogenic origin, developing from animal and vegetal material in a sedimentary rock environment. It occurs in special marine settings, and the oil is entombed in rocks that are now millions of years old. There are two stages in the burial of the organic material. The animal/plant debris settles out of shallow brackish marine waters along with fine-grained sediment of clay and silt. This mixture of matter preserves the organic material under anaerobic, or oxygen-free, conditions. The second burial stage occurs with deposition of additional sediment that covers the entire deposit with thicknesses of materials that range from 100 to many 1000s of meters. The tremendous pressure of the covering sediment, along with greatly increased temperatures, leads to a 100 percent or more compaction of the materials, and oil migration and ultimate accumulation at other localities.

When first-formed, hydrocarbons are present in only small quantities. It is therefore necessary that several processes and conditions operate that sufficiently concentrate them to permit economic recovery. This requires the migration of the oil and gas from the fine-grained source rock to coarser-grained reservoir rocks and then its accumulation into a "trap" where it is held in bondage until released by well production.

To act as a reservoir rock, the host materials must have reasonable porosity for adequate storage of the migrated fluid, as well as permeability, which is the ability of the rock to permit the easy flow of materials through it. This requires the interconnection of the rock openings, which then serve as avenues for fluid movement. Migration of the hydrocarbons is facilitated by water and gas, which provide fluid pressure in the system.

The final resting place of oil in its journey occurs where its migration is impeded by a barrier or natural dam, called a trap. The requisites for such a trap are that it provide an impervious seal to permit accumulation of hydrocarbons and prohibit escape of fluids. Figure 3.2 shows some typical traps that fulfill these conditions. The oil of the Mideast is found commonly in anticlines, whereas faults form traps in California, salt domes, in the Louisiana region, and unconformities, in Pennsylvania and New York.

## Production Methods

For oil to migrate, or even move into the bore of a well, it must squeeze through rock and grain openings that are minute. In addition, the viscosity

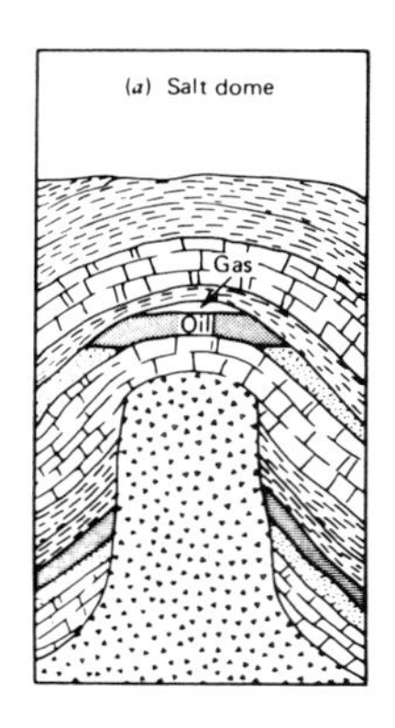


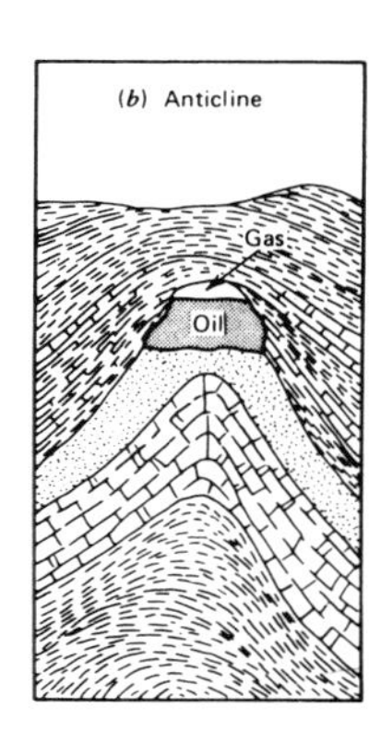


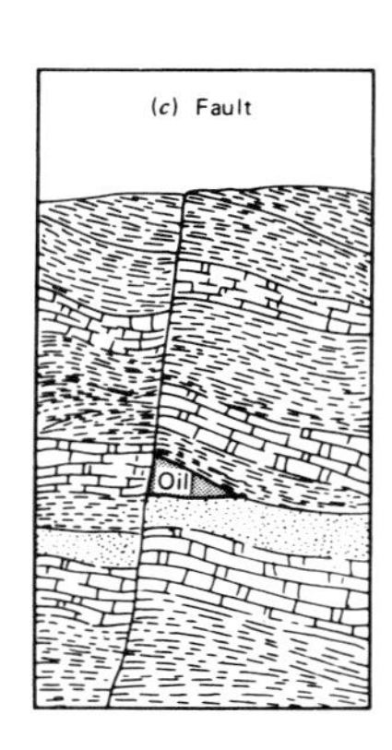


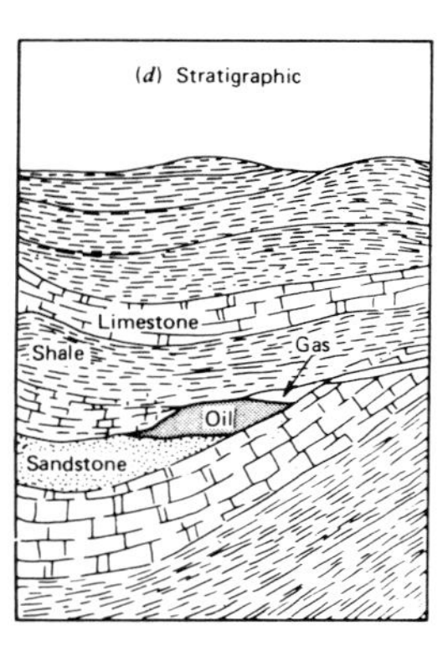


**FIGURE 3.2** Types of oil traps. Courtesy of U.S. Geological Survey.

of oil and forces of adhesion make it cling to the rock. Thus, considerable force is needed for movement. The chief propulsive force is either water pressure or gas pressure, or both. Fluids move from points of high pressure to those of lower pressure, and the lighter hydrocarbons rise to higher points than the denser water. A well that is pumped too fast can create a local area of low pressure within the reservoir, or oil pool, permitting water to rush in. This causes a loss in well productivity. Even when the well is pumped under efficient procedures, much of the oil remains within the reservoir, more than 60 percent.

To retrieve more oil after there has been too much lowering of the water and gas drive, secondary recovery methods may be initiated. These include artificial injection of gas, air, or water to induce sufficient pressure to once again force the oil to become mobile. Such rejuvenation under the proper conditions may permit extraction of an additional 15 percent or more from the reservoir. Even tertiary recovery methods may be employed under certain conditions. *In situ* combustion by fire may create sufficient heat to lower the viscosity of the oil and permit it to flow once again. Hydraulic fracturing is still another technique that can produce increased yield. This consists of reversed pumping of air or fluids into the rocks at such high levels that the rocks are fractured and create larger openings through which the hydrocarbons can move more freely.

Man's thirst for oil, and the exhaustion of old wells, requires enormous expenditures for exploration methods and drilling of new, and often much deeper, wells. Most shallow oil fields have been discovered, so the search becomes ever more difficult and expensive. The older methods of strict geologic investigations from the surface have been largely supplanted by geophysical prospecting methods. This requires highly sophisticated equipment including the most commonly used techniques of seismic prospecting. Dynamite charges are exploded in the ground, which cause shock waves to reverberate through the rocks. The complex signals are recorded by a seismograph. The results are fed into a computer that aids in solving the geologic problems. For example, such methods were the first to reveal the presence of hydrocarbons in the overthrust belt of the West – an area that has proven especially fruitful for new natural gas discoveries. Such remote sensing methods are the only ones that can be used to look for possible fossil fuels hidden deep under marine waters, as in the Gulf of Mexico and the North Sea.

At the present rate of American oil production, we have only nine years of remaining oil reserves. To equal this depletion rate would require the discovery of three fields the size of Prudhoe Bay, Alaska, during this decade. More than 2.5 million wells have already been drilled in the United States, which is four times the total for the rest of the world. In 1980, a record 60,000 wells were drilled with a total investment of $20 billion. With only a few exceptions, the outlook seems bleak for major new discoveries.

## Oil Consumption

During 1980, the United States used more than a fourth of the total world production of 60 million barrels of oil per day. Of this total, only 9 billion barrels were of domestic origin. The rest, about 8 million barrels, had to be imported largely from OPEC (Organization of Petroleum Exporting Countries), which produced half the world supply. Total cost of the imports – more than $50 billion. No wonder the United States had a severe trade deficit! Today, however, the situation has changed. The United States imports less oil from OPEC than from other countries, and less monies are being spent in exploration for new fields.

## Environmental Effects

As with the extraction of all materials from within the earth, oil production also creates deleterious impacts. Tanker spills, pipeline leaks, and uncapped wells can run amuk, such as the well off the coast of Mexico in 1980 (known as the Ixtoc I oil spill). They can severely damage wildlife and nearshore ecosystems, cause costly cleanup, and lessen revenues at recreation and developed sites. Land subsidence can occur when hydrocarbons are withdrawn only 100s of meters below ground surface. The first recognition of this relationship was in 1925 in the Goose Creek Oil Field, Tex. Here, a surface area of about 25 $km^2$ was lowered more than 1 m. The most spectacular case, however, is the Wilmington Oil Field, at Long Beach, Calif. The first important subsidence occurred in 1940 and amounted to 0.4 m. By 1945, maximum subsidence was 1.4 m, but continued to become greater at a rate of about 0.6 m per year (Figs. 3.3, 3.4, 3.5). When the severity became obvious and damages had reached the millions of dollars, a program of water injection was initiated to arrest the process. Although this program eventually became effective, before the trend was reversed, maximum subsidence of nearly 9 m had occurred, and the total subsidence bowl covered a 76 km$^2$ area. Other hydrocarbon extraction has also caused subsidence such as the production of methane gas in the Po Delta, Italy, and at Niigata, Japan.

The burning of fossil fuels is now of such large proportions and the gases and particulates emitted of such magnitude that it is raising both national and international repercussions (see section on acid precipitation, Chapter 12).

## NATURAL GAS

Petroleum gas is called natural gas to distinguish it from manufactured gases and gases derived from synfuel processes. Natural gas and crude oil are both biogenic, are commonly found together, and form under the same geologic conditions. Chemically, natural gas is mostly methane, which amounts to more than 75 percent of the volume. Gas, being a compressible fluid, generally occupies a relatively small volume of the reservoir. Because of its low density, it occurs at the highest part of the trap. Gas expands when

**FIGURE 3.3** Land subsidence at Long Beach, Calif., prior to 1970. The subsidence is largely from petroleum withdrawal. In recent years, water injection into the subsurface strata has stopped the subsidence. Courtesy of Raymond F. Berbower.

**FIGURE 3.4** Horizontal displacement and buckling of railroad tracks, Long Beach, Calif., caused by subsidence resulting from extensive petroleum withdrawal. Photograph taken by Long Beach Harbor Department at Pier A on July 31, 1958. Courtesy of Raymond. F. Berbower.

**FIGURE 3.5** Sheared and failed bridge columns at the north end of Heim Bridge, Long Beach, Calif. Photograph taken by Long Beach Harbor Department on September 16, 1955. Courtesy of Raymond F. Berbower.

pressure is released, and even when oil is being pumped, gas creates an extra drive to make oil more mobile. Unfortunately, in the early days of oil drilling, gas was considered a useless product and was allowed to escape unused or was burned off. Such practices diminished the recovery of oil in such fields, because the important driving force of gas had been dissipated.

Today, the 165,000 gas wells throughout the United States supply 26 percent of the nation's total energy needs. Each year, more than 20 trillion cubic feet (TCF) are used, but in spite of recent spectacular new finds, discovery of new natural gas resources still lags behind usage (Fig. 3.6). For example, with the gradual deregulation of natural gas as established by the Natural Gas Policy Act of 1978, new large fields have started to come into production, such as the overthrust belt of the West and the Tuscaloosa Trend in the south Gulf coast of Louisiana. However, proved reserves in 1979 still showed that production exceeded discoveries by 39 percent.

Several options are available that may help reverse the natural gas deficit. Expanded exploration programs may discover new fields. Fortunately, gas can occur at deeper depths within the earth than oil. For example, economic recovery of oil is now generally impossible below depths of 6000 m, whereas some producing natural gas wells are more than 7500 m deep. Natural gas also occurs on the north slope of Alaska, and if a trans-Canadian pipeline is built, such a source could bring new supplies to the United States. Another long-range possibility is continued development of technologies that will

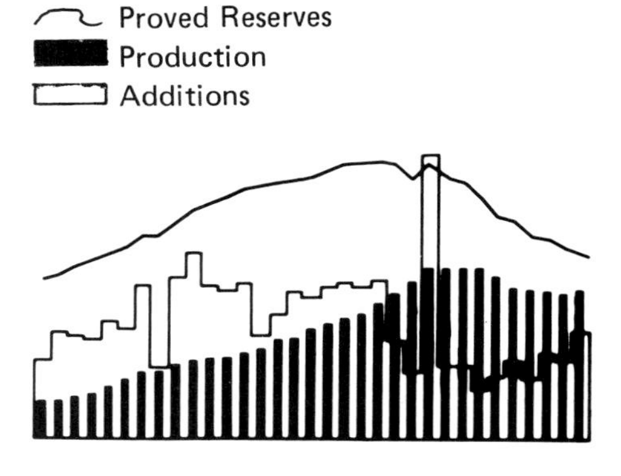


**FIGURE 3.6** Proved reserves of natural gas reached a peak in 1967 and began declining with the exception of 1970 when Prudhoe Bay reserves were added. The solid bars represent annual withdrawal of reserves, and other columns represent yearly additions to reserves. Courtesy of American Gas Association.

permit tapping of unconventional gas deposits. Such deposits, which differ from the present conventionally used resource, occur more tightly bonded to rocks, which do not readily release the gas. Enormous volumes are locked into such rock as the Devonian shales of the Appalachians, within coal seams of many states, the brine-filled deep waters of the Texas–Louisiana Gulf Coast region, and the tight sands of many geological formations. Even today, industry is producing about one TCF annually from such tight sands, and some geologists believe that potential deposits may aggregate as much as 800 TCF.

## COAL

Coal has been called a variety of terms, but most appropriate in this day and age are such phrases as "black diamonds," and the "bridge to the future." Coal is by far the most common and abundant fossil fuel, and was responsible for ushering in the Industrial Revolution. Coal mining in the United States began in the mid–1700s, but did not overtake wood as a major fuel source until the 1880s. As recently as 1925, it was still supplying 70 percent of our energy, but by the 1940s, it was overtaken by both oil and natural gas because they were cheaper, cleaner, and more easily transported.

Because of the huge reserves of coal in the United States, we might be called the "Saudi Arabia of coal." At present however, it provides only a fifth of the energy. In 1979, 680 million tons were burned, and 77 percent of this use was by the electric utilities. Indeed, due to the deleterious impacts produced by burning coal, some adversaries ask the question, "Can we afford to burn coal?"

### Geology of Coal

Coal can be viewed as metamorphosed plants. It is the compressed and altered residue of plants that grew in ancient fresh and brackish wetlands. As plant remains accumulated, they were transformed first into peat and later altered chemically and physically into the more solid forms of coal. The formation of peat requires a very humid to subtropical climate to support extraordinary luxurient vegetation. Furthermore, the topography needs to be in flat low-lying areas, even near marine waters. Such a terrain contains a high water table that permits prolonged accumulation of plant material in a reducing environment. Under such conditions, there were often cycles of deposition alternating with non-organic processes forming the more usual shales, sandstones, and even limestone. Furthermore, these were settings with slowly subsiding basins whereby the rate at which land was sinking could accommodate the materials being washed in or growing in the area. These circumstances were repeated numerous times in many places throughout the Appalachian coal basins. For example, one sequence in West Virginia is more than 1000 m thick and contains 117 different coal beds.

Transformation of plant material to coal occurs in two distinct phases. During the biochemical phase, the organic materials lose oxygen and hydrogen, which results in increased fixed carbon. During this degradation period, cellulose and proteins are changed more rapidly than lignin, and the action of microorganisms is a significant process in this decomposition. The geochemical phase consists of those processes that produce dynamic changes in the material. These are generally characterized by elevated pressure and temperature caused by conditions of deeper burial.

Coal is classified by rank according to the percent of fixed carbon and heat content as calculated on a mineral-matter-free basis. Fixed carbon increases from lignite, to bituminous and then anthracite. There are also intermediate ranks such as sub-bituminous. These changes result from a combination of factors including depth of burial, heat, compaction, time, and structural deformation. **Rank** is simply the way of expressing the type of progressive metamorphism that has been achieved. The **grade** of coal, which is an expression of its quality, is independent of rank. Quality depends upon the amount of such deleterious materials as ash, sulfur, and water in the coal. Sulfur is especially undesirable because it contributes to erosion–corrosion of materials, forms boiler deposits, reduces the quality of coke, and is the principal culprit in acid mine drainage, air pollution, and acid rain (precipitation). Sulfur content of U.S. coals ranges from 0.2 to 7 percent, averaging 1 to 2 percent. It is highest in the bituminous coals of Pennsylvanian age in the Appalachians, and lowest in deposits of the West.

### Mining Techniques

The type of method used for coal extraction is a function of such variables as the thickness of the coal seam(s), their depth and type of overburden, and their topographic and geographic setting. The principal decision is whether to use surface or underground mining techniques to obtain the coal. Each has its own particular set of advantages and disadvantages.

In **surface mining**, a type of sub-method called **area mining** is used when the topography is especially flat, and the coal seam is conformable to the surface. The overburden is bulldozed or clam-shelled into long windrows, so coal can be removed along parallel cuts (Fig. 3.7). After coal is mined out along that line, the windrow material is shoved back into the elongated

FIGURE 3.7 Large coal scoop for surface mining. The bucket can hold several cars. Courtesy of U.S. Bureau of Mines.

trench. A new cut and windrow is made, and the process is repeated throughout the operation (Fig. 3.8). **Open pit mining** is employed when the coal seam is particularly thick. The overburden is shuffled to different sites as coal is removed and eventually smoothed out when the mine is exhausted. When the terrain is hilly with much sloping topography, and coal seams are essentially horizontal, **contour strip mining** methods are used (Fig. 3.9). The overburden is removed above the coal, starting at the exposed part of the hillside and extending at the same level along the slope. This method is continued until the ratio of overburden to coal becomes excessive and uneconomic. **Auger mining** can be used to obtain additional coal by boring mechanically into the exposed seam.

**Subsurface mining** consists of constructing tunnels into the coal deposit so that it can be extracted and brought to the surface. A **room and pillar method** is one approach whereby a network of corridors are excavated and interconnected in the seam. Coal left between the corridors provides support to the roof to prevent collapse. When conditions permit, a **longwall system** is used, designed to produce especially long underground faces of coal.

**FIGURE 3.8** Strip mining of coal using the windrow method. Aerial view of Colstrip, Mont., looking southeast. The orphaned windrow banks show in the background. Courtesy of U.S. Bureau of Reclamation.

After an area is mined out, collapse is allowed to take place, thereby permitting a larger coal volume to be removed.

## Geography and Resources

Coal occurs in many countries, but like oil, is not equally distributed. Three countries, the United States, the Soviet Union, and China, possess nearly two-thirds of known reserves. Such reserves will last more than two centuries at the current rate of consumption, but there undoubtedly are

**FIGURE 3.9** Contour mining of coal in the Appalachians. U.S. Bureau of Mines.

many additional reserves that have not been calculated. Although the United States has nearly one-fourth of the total reserves, both the Soviet Union and China currently produce about as much coal as the United States. In 1974, the U.S. Geological Survey estimated total coal resources in the United States at about 4 trillion tons, of which 1731 billion were identified as occurring at depths less than 1000 m and 2237 billion were probable but unmapped deposits.

Use of coal has expanded in the United States from 560 million tons in 1950 to 820 million tons in 1982. Surface mining accounts for 59 percent and underground mining produces 41 percent of the total. In 1978, there were 250,000 miners, 6000 mines, and 26 states producing coal. If coal is to become an even more important part of the energy picture, a number of problems will have to be solved. Some of these are in the socio-economic sector, such as shortages of capital and high interest rates, manpower problems with potential strikes and declining productivity, and transportation problems involving inadequate rail and terminal facilities. Perhaps even more pressing are the environmental concerns that must be addressed.

## Environmental Effects

Some people ask the question, "Is coal savior or demon?" Another comment making the rounds is like the goods news, bad news jokes. The good news is that the United States has abundant coal resources that can be used to eliminate dependence on foreign oil. The bad news is that we will probably have to use it. Indeed, few other geological resources have reaped such a harvest of environmental issues.

The extensive use of coal produces two categories of mining-related problems – on-site damages and those impacts created during the burning of coal. Coal is dangerous to produce, difficult to transport, unsightly to store, and fouls the air when burned.

Coal mining is the most hazardous of all major occupations. Twentieth century mining in the United States has caused more than 100,000 deaths during mining operations and permanent disability to 1 million others. The worst long-range health hazard is black lung disease, caused by coal inhalation, and this costs the federal government more than $1 billion each year for compensation payments to victims.

Surface mining devastates the landscape, but fortunately, state and federal laws (such as the national Surface Mining Control and Reclamation Act of 1977) have helped reduce such damages (Figs. 3.10, 3.11, 3.12). However, the hydrologic damages are still an undesirable legacy. Streams have silted up from the debris washed from spoil piles and increased the threat of flooding. Acid mine drainage has afflicted thousands of kilometers of streams, especially in Appalachia, and to clean them would cost more than $10 billion. In the West, the groundwater table has been seriously lowered near many surface mining operations, such as Decker, Mont., where the 15 m of water table decline has dried up wells in the area.

**FIGURE 3.10** Surface spoils from haphazard coal mining pose severe acid drainage contamination problems. This site is 5 km from Utica, Ill. Courtesy of Soil Conservation Service (USDA).

FIGURE 3.11 A hydro-seeding operation in the Appalachians to induce vegetative regeneration of a coal-mined area. Courtesy of National Coal Association.

FIGURE 3.12 This was the site of a former coal mine at Lumberport, W. Va. The area has now been completely reclaimed and planted with alfalfa. Courtesy of National Coal Association.

Subsurface mining has led to fires in some coal seams, and the fires have produced noxious fumes, which have despoiled surrounding land and killed wildlife. Land subsidence is still another problem when mines are shallow without adequate roof support (Figs. 3.13, 3.14, 3.15). Here, the ground surface deforms with open cracks, small-scale faults, troughs, depressions, and pits. In the United States, more than 800,000 ha of land have already subsided, and additional thousands are in jeopardy. Of course this also occurs in other nations, such as Australia where many of the coal mines in New South Wales have caused extensive damage from subsidence (Fig. 3.16).

Off-site problems connected with coal production and its use are discussed later in Chapter 12 under the title of acid precipitation.

## SYNFUELS

Synfuels is a new term that has sprung up in the jargon of energy production. These are fuels, either liquid or gas, that have been manufactured from other materials. They can be synthesized from other fossil fuels, or even

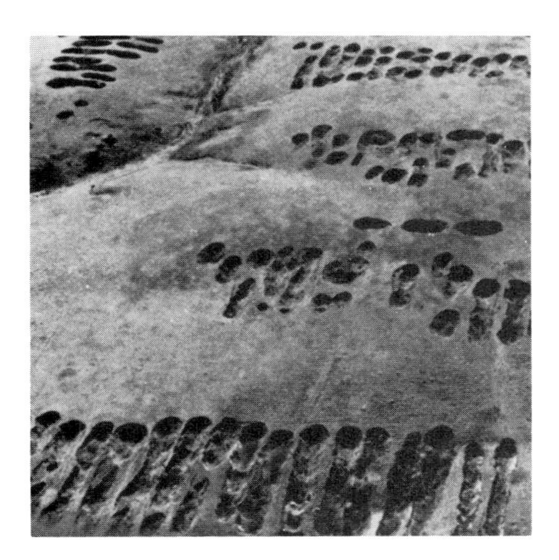

**FIGURE 3.13** Sinkholes and troughs caused by subsidence from mining of underground lignite coal near Beulah, N. Dak. Coal was mined here during the 1918-52 period with 15 m of overburden above the rooms and pillars of the coal seams. Photograph taken October 1976 by C. Richard Dunrud. Courtesy of U.S. Geological Survey.

**FIGURE 3.14** Damage to house in Pittsburgh, Pa., region from underground coal mining. Courtesy of Jesse Craft.

**FIGURE 3.15** This school building had to be abandoned because of severe structural damage caused by subsidence connected to the underground coal mining. Courtesy of Jesse Craft.

**FIGURE 3.16** Sinkholes and ground subsidence from underground coal mining operations in New South Wales, Australia.

developed from such materials as waste, or biomass conversion from forest and crop products. Their future status is uncertain because, again, there are many attendant problems with their development and use. The geologic materials that qualify as potential synfuel sources are coal, oil shale, and tar sands.

Congress passed the Energy Security Act in 1980 with the purpose of developing a totally new industry: synfuels. It authorized the formation of the Synthetic Fuels Corporation (quasi-governmental agency in the fashion of the Tennessee Valley Authority), which was to develop programs and expend $88 billion by 1992 for the production of synfuels. Only small amounts of money have thus far been allocated, and its future is clouded according to recent governmental projections.

### Oil Shale

Oil shale is a fine-grained sedimentary rock with sufficient organic content to yield substantial amounts of oil when heated. Although found in many other countries, the United States is reported to have nearly 40 percent of the total world reserves, estimated at 5 trillion barrels. These deposits are largely concentrated in the Green River Formation that underlies parts of

Colorado, Wyoming, and Utah. Although these vast reserves are probably 15–20 times greater than all probable oil production in the country, the amount that can be ultimately and economically obtained is a hotly debated question.

Throughout history, about 400 million barrels of oil have been produced from oil shale, mostly from such places as Scotland, Estonia, and Manchuria. Deposits also occur in France, Germany, Spain, South Africa, Australia, and Brazil. Since 1944, the U.S. Bureau of Mines has operated pilot plants in an attempt to determine the feasibility, economic possibility, and technical aspects of a successful industrial operation. Within the past few years, many new schemes have been proposed for mining and manufacturing of the geologic synfuels, and several commercial corporations have invested funds for its development.

The question that is still unresolved is how best to mine and produce fuel from oil shale – above ground by a system of retorts that crack and break down the rock thus releasing the oil, or by the creation of underground burning with an *in situ* operation of mining and recovery at the same time. Surface mining methods seem to have special deterrants that include the use of large amounts of water, an especially precious commodity in the arid West, and the large-scale destruction of enormous tracts of land. Furthermore, release of contaminants by burning, such as the trace elements, might yield disastrous results to the soil, water, and wildlife throughout the region.

### Tar Sands

Tar sands are geologic formations that contain a mixture of about 85 percent sand, 4 percent water, and 11 percent bitumen. The bitumen is an organic material, which is a dense, sticky, semi-solid liquid that is resistant to flow at normal temperatures. Tar sands are different in occurrence from oil because they do not need a structural trap, but instead require stratigraphic control. They need several prerequisites for their formation: (1) generally a paleo-delta system with rich source beds, (2) an overburden of sediments to restrain migration of fluids, (3) a gradual dip to the rocks, and (4) subsequent biodegradation. This last process is important because a flushing action occurs when the original crude oil is washed and in contact with oxygen and bacteria. This removes the more water-soluble light carbons and leaves as residue the heavy tar-like substances.

Total 1977 tar sand reserves were estimated at 2 trillion barrels for 16 major deposits, of which seven contain 98 percent of the total. These are located mostly in Canada and Venezuela. In the United States, 550 deposits have been identified in such states as California, Kentucky, New Mexico, and

Texas, but Utah has about 80 percent of the estimated 29 billion barrels. The first major effort to mine tar sands began in 1967 with the consortium called Great Canada Oil Sands Ltd. This plant in Alberta synthesizes the extensive deposits known as the Athabasca sands – the single most important deposit in the world. The plant produces about 45,000 barrels of oil per day at a profit. A second major operation known as Syncrude has spent $3 billion on a project to extract oil from tar sands at a 20,000 ha site north of Fort McMurray in northern Alberta. It hopes to produce 125,000 barrels per day when fully operational.

## OTHER GEOLOGIC ENERGY SOURCES

Geologists are involved in many aspects of the fossil fuel industry – exploration and location of deposits, drilling and mining methods for extraction, design of procedures to minimize environmental damage, and development of strategies to reclaim or rehabilitate the resource areas. Geologists also do the same types of jobs with the other geologic energy sources.

### Uranium

The element uranium is locked into minerals that primarily occur in sedimentary and igneous rocks, and is mined in the same fashion as any mineral deposit. Nearly one-fourth of the world's proved reserves occur in the United States, and another one-eighth are found in Canada. Nearly all American uranium occurs in New Mexico and Wyoming, and the reserves throughout the West are sufficient to power all nuclear plants and those under construction into the twenty-first century. In 1980, 300 mines in the United States produced 20 million tons of ore.

The nuclear energy bound up within uranium contains 3 million times the energy of an equal weight of coal. Unfortunately, radioactive materials require extraordinary precautions and a precise technology to assure safety in all aspects of the nuclear industry – mining, processing, using, and finally disposing of the waste products. Such factors, along with the exacting requirements for proper location of nuclear plants and determination of their environmental safety from earthquakes, hurricanes, tornadoes, and other hazards, have caused much higher building and development costs than plants which burn fossil fuels.

During the nuclear reactor boom in the early 1970s, U.S. utilities ordered 100 new plants in a 3-year period, and some "experts" even predicted there would be 1500 by the start of the twenty-first century. However, the combination of rising inflation, escalating plant costs, potential health hazards, less electric demand than forecast, and environmental concerns have caused hard times in the nuclear industry. Mines have closed, miners have lost jobs, and few new plants have been built. Nationwide, 82 nuclear plants account for 12 percent of the generation of electricity in the United States. Yet in some parts of the country, reactors are especially important. They produce 80 percent of the electricity in Vermont, 60 percent in Maine, and 50 percent in Connecticut and Nebraska. Throughout the world, 21 nations now have a total of 166 reactors, with heaviest concentration in Europe. For example, the nuclear share of electricity generated in 1982 according to countries and percent was: Finland – 40%, France – 39%, Sweden – 39%, Belgium – 30%, Switzerland – 28%, Japan – 20%, West Germany – 17%, and the United Kingdom – 16%.

A special role that geologists fill is the determination of plant safety from earthquake hazards. The NRC (Nuclear Regulatory Commission) has very strict guidelines as to where nuclear plants can be built. For example, the site for development must be proved to be safe from earthquake activity that could jeopardize the 40-year life for which plants are designed. To determine this, geologists must make detailed studies of the specific locality and the surrounding region to make sure there is no "capable" fault present – the possibility of an earth rupture that would be sufficiently seismogenic to damage the plant. A capable fault is defined as one that has had movement within 35,000 years, or more than two movements within 500,000 years, or is linked to a major fault system within several miles of the plant site.

The environmental problems associated with nuclear energy are unique. Radioactive materials are the most deadly of all matter. The great scare that arose in 1979 from the near-disaster with the Three Mile Island Nuclear Plant in Pennsylvania galvanized public opinion to the great potential threat that such plants can have to the safety of people throughout a very wide area. Furthermore, the problem of ultimate disposal of radioactive waste has not yet been solved. Such materials continue to be placed into temporary storage until a final solution has been reached. Thus, it appears that nuclear energy is not the panacea it was once championed as being.

In the United States, not a single new nuclear plant has been ordered since 1978. After the near disaster at Three Mile Island, Pa., cancellations of previously designed plants accelerated. Work was halted on 18 reactors in 1982, bringing the total cancellations since 1972 to 97. At the present time, there are 57 federally approved plants in various stages of construction. However, a 9–0 ruling by the U.S. Supreme Court in April 1983 temporarily doomed further planning for new nuclear plant sites. The Court upheld a California moratorium enacted in 1976 that bans certification of new facilities until the federal government finds a way to dispose permanently of the highly radioactive waste products. The Nuclear Waste Policy Act of 1982 provides a strict timetable for the selection of two underground repositories, and mandates that the first one must be in operation by 1998. At the present time, most geologists who have worked on the repository problem favor deep burial in salt formations, which are predicted to act as a filter and absorb any dangerous radioactive ions.

### Geothermal

Geothermal energy is the natural heat of the earth. It results from the radioactive decay of rocks and the action of magma (molten rock) within the earth's crust. Most of this heat is too deep to be tapped by man. However, wells are now currently being drilled to 7.5 km, and with increased technology, will ultimately reach twice that depth. Most geothermal heat is too diffuse to be recovered economically, just like the ocean, which is a veritable storehouse of energy, but concentrating it is the real trick.

The thermal energy is locked in solid rock and in water and steam that fill the pores and fractures. Most geothermal reservoirs occur in the upwelling parts of major water convection systems. These are hot spots of larger regions where the flow of heat is 1.5–5 times the average. Such regions are commonly also zones of young volcanoes and mountain building localized along the margins of major crustal plates. Geothermal sources are either of hot water or dominated by hot vapor (dry steam). In those with hot water, the fluid is produced by boiling and moves into a recovery well as a mixture of both water and steam. The water must be removed before the steam is fed into the turbine for electric generation. Vapor-dominated systems also consist of water and steam, but the rock dries the fluid as it moves toward the surface so that only the superheated steam remains, which is then piped directly into the turbine. Hot-water systems are 20 times more common than those with vapor.

The first use of geothermal heat to generate electricity was at Lardello, Italy, in 1904. The most common surface assurance of geothermal energy is hot water or small fumaroles. Reykjavik, Iceland, heats nearly all buildings from such energy, and hot springs at Boise, Idaho, have heated homes since the 1890s. Wairakei, New Zealand, has extensive geothermal heat energy. The largest producing area in the United States is called The Geysers in California. Here, more than 100 wells bring hot steam to the surface and produce sufficient electricity to account for one-half the consumption of San Francisco.

There is increased interest in the development of geothermal energy in many parts of the United States. More than 900,000 ha of federal lands have been leased for exploration and possible development. New drilling in 1979 for potential geothermal wells was 25 percent more than in 1978. Other possible sources and technologies are also being considered. In places, hot dry rock occurs within drilling range. Under such conditions, water could be injected, and the resulting steam used, as in natural systems. The geo-pressured sands of the American Gulf Coast region may offer another possible source. Here, the rate of sedimentation has been so fast as to trap residual heat within the rock at temperatures that are much higher than normal – 150°-180°C. The deep hot water could be pumped to the surface and used as a source of heating.

Environmental impacts can even occur from the development of geo-thermal energy, and the type of influence depends upon the extraction method. Ground subsidence is occurring at The Geysers and Wairakei (Fig. 3.17). Since geothermal effluents are warm and are commonly mineralized, they pose a threat to surface and groundwater quality. The mineral matter also clogs pipes and wells, which require constant maintenance and eventual replacement. Gaseous emissions may contain sulfur dioxide, hydrogen sulfide, and radioisotopes, all of which are harmful and require special treatment and control.

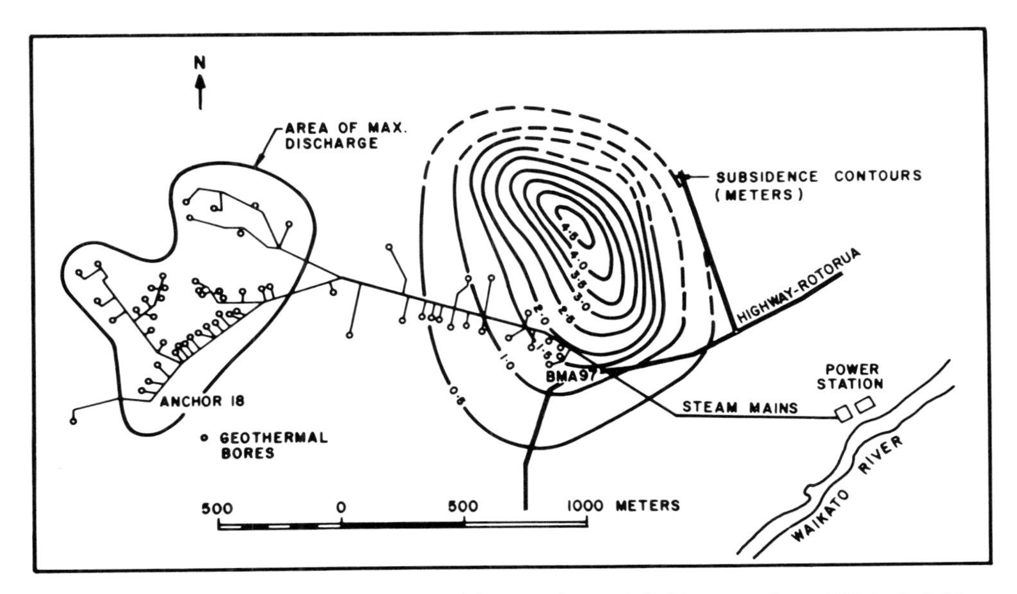


**FIGURE 3.17** Total subsidence caused by geothermal fluid extraction at Wairakei, New Zealand, 1964-74. Courtesy of W. B. Stillwell.

### Hydropower

The generation of hydroelectricity requires a considerable distance of fall for the on-rushing water to turn the turbines, which create the electrical energy. This situation can occur in natural settings such as Niagara Falls, or by a dam where water is higher in the reservoir pool than at the outlet gate below. An advantage of this power is it can easily be regulated by opening or closing control valves. However, droughts place a disadvantage, especially when there are other needs for the water. This is particularly true for dams in the western United States where multi-purpose dams and reservoirs are the pattern, and the region is subject to drought conditions.

The first U.S. hydropower facility began operating at Appleton, Wis., in 1822. Rapid expansion of dams in the early decades of the twentieth century coupled with use from waterfalls, helped to supply hydropower that accounted for 40 percent of the world's electricity. There is now more than 15 times as much hydropower generation worldwide, but it represents only 23 percent of the total electric generation.

Although hydropower generation is pollution-free, the dams can have a multitude of environmental impacts (see Chapter 11). However, in man's never ending quest for new energy sources, in this case electrical energy, the 50,000 dams that exist in the United States may help the country. Studies are now being undertaken to determine the feasibility of which ones might be economically turned into electric generating stations. We have perhaps delayed too long the local creation of electricity, which apparently has been so successful in the thousands of small generators on dams now in existence in China.

## ENERGY POLICY

The United States has had a Department of Energy for many years, but there is not a comprehensive policy that governs many of the ramifications of energy production and use. There are a series of individual statutes that address various components, but some of these are in a constant state of flux. The oil embargo in 1973 by the OPEC cartel sent shock waves throughout the industrialized nations and caused a dramatic reassessment of their energy policies. In the United States, Congress established the Strategic Petroleum Reserve, a $30 billion project designed to store about 750 million barrels of crude oil in mammoth salt caverns along the Gulf Coast in Texas and Louisiana. In 1983, it was only about one-half full.

Conservation practices and smaller cars have greatly reduced oil consumption in the United States during the past decade. By 1983, the country was importing 27 percent less oil than in 1973, and much less is purchased from OPEC. For example, only 2.3 percent is supplied by those countries that embargoed oil. Worldwide consumption is also down in the non-communist nations. Although OPEC would like to produce an additional 15 million barrels of oil per day, there is no current market for this quantity.

New York State developed a State Energy Master Plan that outlines strategies for maintaining adequate, safe, and economic supplies of energy. Since its creation, the State has cut energy use 15 percent, and from 1978 to 1983, a total of 395 million fewer barrels of oil were used, representing a $12 billion savings.

Despite a decade's growth in population, domestic oil consumption is less today than in 1973. Petroleum, gas, and coal account for 90 percent of the nation's energy needs, as compared to 94 percent in 1973. Coal production has climbed 30 percent, but natural gas has declined 18 percent. The average American household has sliced the total use of energy by 20 percent in the 1973–83 decade.

The Mineral Leasing Act of 1920, with amendments, provides that states whose boundaries encompass federal mineral leases will share the revenues with the government. Accordingly, in 1982, 23 states shared $609 million in disbursements of bonds, rental, and royalty revenues. The revenue generated by mineral leasing of federal and Indian lands is one of the largest non-tax sources of income to the federal government (Table 3.2).

## POSTLUDE

Profound changes are occurring in the world's use and choice of energy resources. Political unrest and conflict in the Middle East pose increasing

**TABLE 3.2** Royalties and revenue from federal and Indian mineral leases in the United States, 1920–82

| | Volume | Value | Royalty |
|---|---|---|---|
| Oil production millions of barrels | 13,874 | $90,684 | $13,673 |
| Gas production millions of thousand cubic feet | 88,284 | $66,726 | $10,362 |
| Coal production millions of short tons (2,000 lb) | 1,110 | $ 9,319 | $ 302 |
| Other production all other solid and fluid minerals | N/A | $13,602 | $ 803 |

*Note:* All dollars are in millions.
*Source:* U.S. Department of Interior (1983)

problems for stability of oil production. Energy prices are rising more rapidly than general inflation rates, which work special hardships on the LDCs. The present oil glut will diminish, and the petroleum industry will again be hard-pressed to meet increasing demands by the 1990s. Coal and nuclear energy have serious and unresolved problems in production and use. Increasing world stress will occur because one-fourth of the world's population consumes three-fourths of the energy, and there will be pressure by the LDCs to reduce income disparity posed by the industrialized nations. Thus, the continuing risk of an energy crisis must be faced. More effective public action to conserve supplies is needed. Governmental action may be necessary to allocate resources more equitably, prevent needless waste, inaugurate new incentives to use energy more effectively, and promote renewable energy resources wherever possible.

CHAPTER 4

# WATER RESOURCES

Water, a combination of hydrogen and oxygen, is a compound that makes our Earth unique when compared with other planets. In spite of this bizarre occurrence, it is so common that 70 percent of the earth's surface is covered by this aqueous cloak. Indeed, if all continents were planed off and dumped in the oceans, there would still be sufficient water to cover all former lands to a depth of 3.2 km! Thus, whoever perpetrated the name of our planet, calling it "Earth," should have more appropriately named it "Water." Furthermore, there is a long list of unusual properties that water possesses, which are fundamental to our health, welfare, and survival. It is basic to all life forms, plant, animal, and human. It constitutes about 70 percent of the human body, and is so necessary that a person can rarely survive more than seven days without it. Water has unusually high capacity for absorbing heat; has low viscosity, thus permitting easy transportation; has surface tension that permits capillarity, which aids growth of plants and provides wetting to surfaces; and with slight temperature change, can appear in three physical states — liquid, solid, or gas.

Unfortunately, water has the appearance of being excessively common and is especially often taken for granted in industrialized nations or in the humid, temperate lands. Its low cost, when compared with other commodities such as petroleum or iron, also provides a deceptive index that its importance is less than other natural resources. However, Karl Wittfogel (1956) in calling attention to the significance of water in human history shows how it was a unifying force in the development of the "hydraulic civilizations." His thesis was that many of the ancient civilizations would not have grown so rapidly if they didn't have water as the commodity around which all social structure could be organized.

Just as water has been important in the rise of civilization so also has the inability to cope with water problems aided their decline. Thus, the Dark Ages of Egypt in the second millennium B.C. have been attributed to governmental ineptitude to deal with lowered water levels in the Nile during drought years. Mesopotamian civilizations were largely ruined when canals became silted and fields became salt encrusted so that the agricultural base of the economy was destroyed.

The American West has been plagued by many severe droughts throughout its history. One of the early tragedies was the drought that started in 1884 and lasted for 12 years. Plentiful rains before these dates had encouraged the immigration of many settlers, but the drought saw numerous crop failures, led to great suffering, and caused widespread emigration. The "dust bowl" period in the 1930s also produced terrible trauma, and the losses and heartaches were popularized by John Steinbeck's classic work *The Grapes of Wrath* (1939).

Water is a fundamental resource for people, industry, and irrigation. However, its prodigious use, and abuse, causes many areas to live on the edge of water bankruptcy. Severe water problems arise because it is often not distributed properly for human use – there is too much, which creates floods; there is too little, which causes droughts; or the quality is polluted, which causes disease and ruins crops. Thus, water can be likened to a type of schizophrenic personality – a Dr. Jekyll and Mr. Hyde. When understood and used properly, water can be a true and worthy servant of mankind, but when neglected and abused, it can be a fiendish tyrant and destroyer. It is ours to choose the proper path. Enlightened environmental management is required for proper stewardship of this vital resource. The alternative is written in the sands of time wherein lack of knowledge and concern have led to destruction of manmade habitats and many natural environments.

## OCCURRENCE

The world's water resources are locked into the hydrologic cycle whereby evaporation of ocean water becomes entrained into the atmosphere. Precipitation from the sky provides water to lakes, streams, and wetlands. In addition, some water may sink below the land surface and become part of the groundwater. Ultimately, all these continental waters flow back to the ocean and become recycled once again (Fig. 4.1). As can be seen in Table 4.1, the oceans, with 97 percent of all waters, dominate the hydrosphere. And since the glaciers contain a little more than 2 percent of potential waters, there is less than 1 percent that occurs on the land and possibly available to mankind.

In 1983, about 40,000 bgd (billion gallons per day) of water passed over the 48 conterminous states as water vapor. About 10 percent of 4200 bgd fell to the ground as precipitation, but 2765 bgd was also evaporated. Consumptive use by humans and crops caused loss of 106 bgd, and the remainder flowed to the oceans and Canada. For example, 325 bgd moved into the Pacific Ocean, 995 bgd to the Atlantic Ocean and Gulf of Mexico, and 6 bgd to Canada. Out of the total of 1450 bgd of water that was not evaporated, only about 675 bgd were considered to be available for human uses.

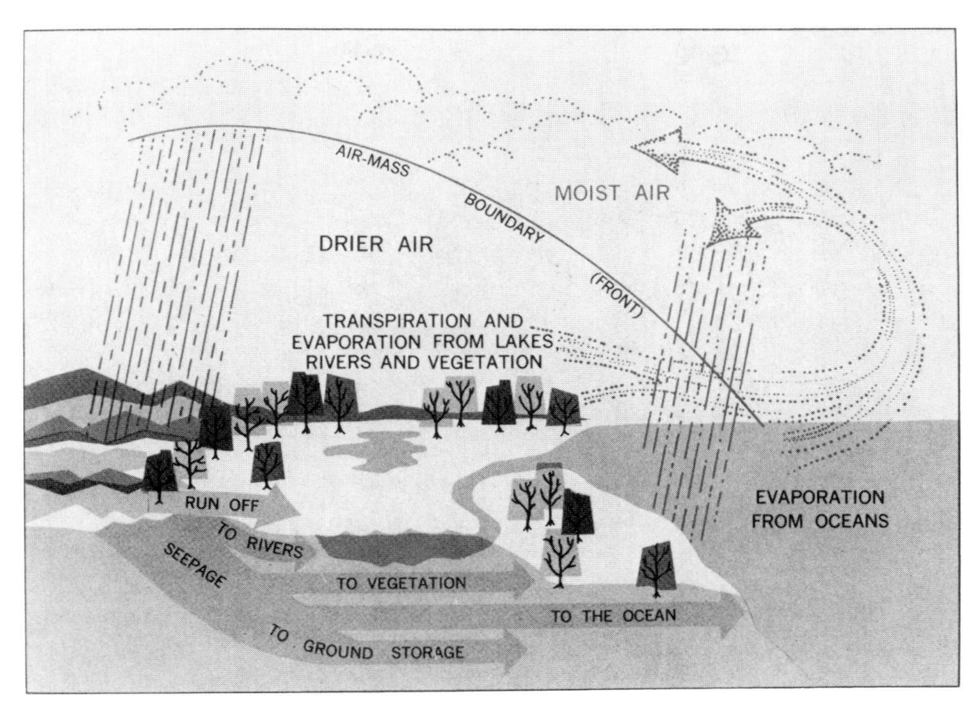


**FIGURE 4.1** The hydrologic cycle. Courtesy of Water Resources Council.

## Surface Water

Surface waters include those in streams (rivers), lakes, wetlands, and reservoirs. Streamflow is highly variable and changes from region to region, season to season, and year to year. Such variations cause headaches to water planners who must arrange adequate supplies for their constituencies regardless of conditions. For example, the effects of drought can be especially severe in those regions that use a high proportion of the runoff or where storage and distribution facilities do not allow sufficient retention for emergencies and extended dry periods. The most costly droughts in recent times occurred in the Northeast in 1955, and 1961–66; and in the West, the droughts of 1976, 1977, and 1980 each created billions of dollars of losses in farmlands and crops.

There are nearly 50,000 reservoirs in the United States with dams 7.5 m high or higher. Reservoir storage exceeds 450 million acre-feet (146,800

**TABLE 4.1** Distribution of world's estimated water supply

| Location | Surface area (square miles) | Water volume (cubic miles)[a] | Percentage of total water |
|---|---|---|---|
| Surface water | | | |
| Freshwater lakes | 330,000 | 30,000 | 0.009 |
| Saline lakes and inland seas | 270,000 | 25,000 | 0.008 |
| Average in stream channels | — | 300 | 0.0001 |
| Subsurface water | | | |
| Vadose water (includes soil moisture) | 50,000,000 | 16,000 | 0.005 |
| Groundwater within depth of half a mile | | 1,000,000 | 0.31 |
| Groundwater — deep lying | | 1,000,000 | 0.31 |
| Other water locations | | | |
| Icecaps and glaciers | 6,900,000 | 7,000,000 | 2.15 |
| Atmosphere (at sea level) | 197,000,000 | 3,100 | 0.001 |
| World ocean | 139,500,000 | 317,000,000 | 97.2 |
| Totals (rounded) | — | 326,000,000 | 100 |

[a] One cubic mile of water equals 1.1 trillion gallons

*Source:* U.S. Geological Survey, news release, August 13, 1972.

billion gallons, see Appendix A), and 41 percent of all stored water is held by the 31 largest reservoirs, which each have more than 2 million acre-feet. In addition, there are about 2 million small reservoirs and farm ponds that are locally important for water supply and recreation. It is especially noteworthy that the Great Lakes contain about 18 percent of all freshwater on the planet.

Only within the last 20 years have wetlands taken on the stature of being a water resource. Because they were generally associated as being undesirable swamps, bogs, and marshes, they were invariably dredged, drained, or filled for developmental purposes. Belatedly, after tens of millions of hectares were destroyed, they have finally been recognized as important as a buffer against floods, storage against droughts, and vital in the food chain of living organisms. Most states now have laws for their protection and preservation.

### Groundwater

The water table represents the top of the groundwater zone. Below this demarcation, all openings in earth materials are water-saturated, whereas above the water table, the rocks and soil openings also contain air and comprise the vadose zone. Unlike surface water, the movement of groundwater is exceptionally slow, generally only several meters per year. In the development of groundwater resources, it is important to locate wells in aquifers (those water-bearing materials capable of yielding water in sufficient quality and quantity for economic use) (Fig. 4.2). Most groundwater withdrawal is by wells that pump the water to the ground surface. Although most aquifers drain water into wells by gravity flow, occasionally the water is confined in an artesian system where water pressure is sufficiently great to cause movement of water through the well to heights above the water table. Springs can also be a source of water, and these occur where there is movement of groundwater onto the land at a position where the water table intersects the topography. And finally, groundwater can act as an underground reservoir where the storage may discharge into streams, thus causing permanent streamflow even during times of no rain or drought conditions. Indeed, about 30 percent of all streamflow occurs from the seepage of groundwater.

The depth to the water table is largely a function of climate — it is near the surface in humid terranes (generally less than 10 m), and it may be more than 100 m below the ground in arid regions. Aquifer conditions are also highly variable and depend upon the type and structure of earth materials. Sand and gravel make the best unconsolidated aquifers, and sandstones and

**FIGURE 4.2** Prospecting for water using electrical resistivity methods on a pediment surface, southern Arizona.

sometimes limestone offer the best rock aquifers. It is rare to develop productive water wells at depths greater than 900 m.

One of the major aquifers of the United States that is in jeopardy due to extensive groundwater mining is the Ogallala Aquifer. This vast underground reservoir is the size of California and extends south of Nebraska to Texas in the Great Plains. It has been responsible for transforming the area into one of the richest agricultural regions in the world with agricultural products that exceed $15 billion a year. Although wells first tapped the aquifer in the 1930s, it became increasingly exploited after World War II with the development of high-capacity pumps. The 150,000 wells that now draw from the aquifer produce an overdraft that about equals the yearly flow of the Colorado River. Water tables are especially low in Kansas and Texas where declines of 10–15 m have occurred within the past 15 years. If current trends continue, the states of Kansas, Texas, Colorado, New Mexico, and Oklahoma may lose as many as 2 million ha of presently irrigated lands within the next two decades.

A major problem occurs with the development of the Ogallala Aquifer because there is negative incentive to conserve. It is the "tragedy of the commons"; only this time, it is groundwater and not pasture. Noting the prosperity of the large irrigation users, others also have increased their use of the common resource with a long-term consequence that can be ruinous. The federal government has even been party to the rapid depletion through subsidization by price supports for such commodities as cotton, crop disaster payments, cost-share soil conservation programs, and low interest loans. In addition, farmers are granted a depletion allowance on pumped groundwater thereby enjoying a tax break similar to that the oil industry had for many years and the mineral industry currently enjoys. Thus, the more water that is pumped and consumed, the less tax that must be paid.

The amount of groundwater and its availability for use are governed by a variety of factors that include the porosity and permeability of soil and rock materials, their topographic setting, and geologic environment. These factors determine rates of movement, seepage and infiltration, evaporation, and other features in the hydrologic budget. In years of low streamflow, all surface flow in the channel may originate from groundwater recharge. Thus, groundwater is also important to surface water conditions and flow continuity of river systems, especially in humid temperate climates and tropical regions.

In the United States, the quantity of groundwater storage in the zone 800 m below the land surface is 49 trillion $m^3$ (180 billion acre-feet). This is more than four times the volume of water stored in all the Great Lakes, or equal to all of the water the Mississippi River has discharged into the Gulf of Mexico during the last 250 years. Groundwater has been increasingly tapped for use during the past three decades. More than 100 million Americans depend on groundwater for drinking, and 25 percent of all water used in the nation comes from this source. As we shall see, such large usage is leading to an increasing number of problems that must be resolved lest the situation become even more acute.

## WATER USE

Water use in the United States and throughout the world continues to expand rapidly. Americans use twice as much water today as they did 25 years ago. Total usage in the country is now 1.8 billion $m^3$ per day (480 billion gpd), or 2000 gallons per person per day! Although record amounts of surface and groundwater are being withdrawn, the rate increased only 8 percent during the 1975–80 period, compared to 12 percent during 1970–75 (Table 4.2). California uses 210 million $m^3$ per day, ranking it first among all states. Four states — California, Florida, Texas, and Idaho — withdraw 25 percent of the nation's total, largely to satisfy irrigation needs. Water withdrawals in the United States are surprisingly similar when those states west of the 100th meridian are compared with eastern states, 62.6 million $m^3$ per day compared with 64.3 million $m^3$. However, the type of usage of the water is very different (Table 4.2). Another difference is the relative amounts that are not returned directly to the water source. For example, in the West, 48 percent is used and returned whereas 88 percent is returned in the East. The

**TABLE 4.2** Total water withdrawals in the United States

| | West | East |
|---|---|---|
| Irrigated agriculture | 552.0 | 41.4 |
| Domestic and commercial | 28.9 | 77.0 |
| Manufacturing | 17.2 | 175.0 |
| Energy | 8.73 | 327.0 |
| Minerals | 8.49 | 18.0 |
| Other uses | 10.5 | 3.7 |
| Total withdrawal | 625.82 | 642.10 |

West are those states west of the 100th meridian, the drier part of the country where precipitation is usually 50.8 cm or less. East contains those states that are in the more humid climate of the country. All figures are in millions $m^3$. Source: U.S. Geological Survey.

remaining percentages are classed as "not returned," which is water lost by evaporation, transpiration, incorporation into products or crops, or consumed by humans and livestock.

### Patterns of Use

Steam powered generation of electricity accounts for 95 percent of all freshwater withdrawals in the United States. However, advances in cooling technology by the year 2000 are predicted to decrease such usage by 11 percent. Although large amounts of water are required for electricity, the consumptive use is only 2 percent of all freshwater withdrawal. Conversely, irrigation consumes 83 percent of the water applied to the land.

In the United States, groundwater use has expanded at a rate of 3.8 percent per year during the last two decades, whereas freshwater use has increased at a rate of only 2 percent. California is the biggest groundwater user where 48 percent of freshwater usage is from wells. The importance of such use is represented by water in the San Joaquin Valley. This basin encompasses 7.3 million ha with valley floor and foothills of 4 million ha. Annual income from the irrigated farm crops exceeds $5 billion – only the states of Iowa, Texas, and Illinois have more farm income. The rich soil, 8-month growing season, and Mediterranean-type climate with dry summers and moist winters are ideal for farming. Unfortunately, because of such heavy groundwater withdrawals, there is a yearly overdraft of 1.85 billion $m^3$. The total overdraft of groundwater in the United States is 79 million $m^3$ per day, which is about 25 percent of the total withdrawn from storage.

Tucson, Ariz., is the largest American city that relies entirely upon groundwater. Here, water is pumped five times faster than replenished by natural recharge so that in the last 10 years, the water table has declined more than 40 m. By 1983, 6000 ha of nearby farmland had to be retired because of high pumping costs and need for the city to husband the water resources. Because of the dwindling water availability in such cities as Tucson and Phoenix, and the perceived need to maintain a large water supply for income-producing farmland and industry, Arizona fought and won a 70-year fight to import water from the Colorado River via the newly constructed Central Arizona Project. This will annually divert 146 million $m^3$ of water that will be lifted 600 m and transported 400 km in a pipeline to metropolitan and irrigation sites. The project will cost $2 billion before completion with additional interest and expenditures of $4 billion during the next 20 years.

Water use is commonly divided into the following categories:

1. Off-channel use includes public-supply (domestic, commercial, and industrial), rural (domestic and livestock), irrigation, and self-supplied industrial and mineral production. Although many of these uses are non-consumptive, when water is returned into the hydrologic system, it commonly is in a deteriorated condition. Most consumptive losses occur with irrigation where 83 percent of the water is utilized.

2. In-channel use includes hydropower generation, navigation, and recreation. U.S. water commerce traffic constitutes 15 percent of all freight transported. Such recreational activities as swimming, boating, fishing, and ice skating account for 25 percent of all outdoor recreation.

*Human use.* In temperate climates, humans require 2 liters of water daily to replace body losses. Citizens in industrial nations use about 570 liters per day (150 gpd) for typical domestic home use — bath or shower, washing dishes, washing clothes, flushing toilets, etc. (Fig. 4.3).

*Industrial use.* Prior to 1950, irrigation was the greatest water use in the United States, but soon after, the full-range of industrial, commercial, and mining uses became greater. Water use in thermoelectric power plants comprises about 80 percent of all industrial use. Although hydroelectric power production returns the water almost immediately to the river, the 12.5 billion $m^3$ per day that pass through the turbines is 275 percent more than the runoff from streams of the 48 contiguous states. Processing industries require large amounts during manufacturing. For example, to make 1 ton, 65,000 gal (246 $m^3$) are needed for steel, 120,000 gal (454 $m^3$) for paper, and 300,000 gal (1130 $m^3$) for aluminum. It even takes 57 gal (215 liters) for all associated activities to produce one hen's egg.

*Irrigation use.* Water is used in large amounts to raise irrigated crops because control of supply has several advantages. Crops can be grown in dry climates; water can be applied as needed to produce higher yields than rainfall farming; and it is possible to harvest more than a single crop per year. Irrigated lands have increased in the United States from 13 million ha in 1955 to more than 21 million ha today. One out of every 7 ha throughout the country are now irrigated. It is so intensive in states like California that 60 percent of all energy supply is related to irrigation farming. Although total daily use of irrigated water is in excess of 360 million $m^3$, it has been estimated that 25–50 percent is wasted by leaking pipelines, infiltration into non-crop sites, absorption by weeds, and excessive water spreading beyond the needs of crops.

**FIGURE 4.3** Chicago's Central Water Filtration Plant on Lake Michigan. It contains the world's largest filtration system capable of chemically treating and filtering 1 million gal/min. Built in 1964 at a cost of $1 billion, it serves a population of more than 5 million in a 1100 $km^2$ area, including Chicago and over 70 suburbs. Courtesy of Chicago Water Department.

To make irrigation more efficient and to use new lands unsuited to conventional furrow methods (Fig. 4.4), new methods have been adopted in some areas within the past three decades. Such innovations as sprinklers (Fig. 4.5) are now in use in many areas. In moderately arid regions, these methods have the advantage of using less water because there is greater accuracy when applied, water can be distributed more uniformly, and it is adaptable to irregular terrain and sandy soils. Each year in the United States, about 250,000 ha are added to lands irrigated by sprinklers. Another method now employed for such crops as citrus trees is drip irrigation (Fig. 4.6). Water is applied directly to the roots of plants by low water pressure systems at frequent intervals. Whereas water efficiency for other methods ranges from 30 to 75 percent, efficiency for drip irrigation is 90 percent. Thus, it has the advantage of water conservation, it produces higher yields because plants grow without stress, and labor and fertilizer costs are reduced. Today, there are 150,000 ha throughout the world under drip irrigation; 55,000 ha are in the United States.

## WATER PROBLEMS

Water is so pervasive and problems associated with it so numerous and widespread that they are discussed in several chapters in this book. Erosion

**FIGURE 4.4** Lateral ditches from the All American Canal irrigate this bean field in the Coachella Valley near Indo, Calif. Courtesy of U.S. Bureau of Reclamation.

**FIGURE 4.5** The concentric circles of this wheel move the sprinkler over this field in the Columbia Basin Project, Wash. Courtesy of U.S. Bureau of Reclamation.

and siltation are part of Chapter 10, flooding in Chapter 8, engineering aspects in Chapter 11, and quality in Chapter 12. Although not publicized with the same fervor as the "energy crises," some nations, and parts of many others, are periodically faced with a "water crisis." The drought in the 1930s that created the Dust Bowl in the United States has been described as the single worst natural disaster in the nation's history in terms of property loss and human suffering. Water shortages and improper farming, timbering, and grazing practices have aided the process of desertification (see also Chapter 10) and placed many lands in jeopardy of destruction. Droughts in the Ganga Plain of India in 1957 led to starvation of 25 million people, and 1 million people migrated away from northeast Brazil during the drought of

**FIGURE 4.6** This young almond orchard is being irrigated by a drip irrigation system near Mendota, Calif. Each emitter at each tree releases about 5 liters of water per hour. This is part of the Central Valley Project that uses imported water from the San Luis System. Courtesy of U.S. Bureau of Reclamation.

1958. Each year, more than 100 million people throughout the world suffer severely from drought conditions. However, the list of water-related problems is a long one and environmental deterioration occurs in many other ways. Three of these are briefly discussed below.

## Groundwater Mining

A condition known as **groundwater mining** is produced when excessive pumping rates contribute to a lowering of the water table. Thus, discharge exceeds recharge, and earth materials are dewatered. Such a condition can lead to a tighter packing arrangement of sediments with resulting subsidence and fissuring of the land surface (Figs. 4.7, 4.8). Subsidence has occurred in many cities throughout the world from overdraft withdrawals of groundwater, such as in London, Venice, Osaka, and Bangkok (Fig. 4.9). Perhaps Mexico City has experienced the world's maximum water-induced subsidence, 8 m, and the combined withdrawal of water and petroleum has caused nearly 10 m of subsidence at Long Beach, Calif.

The largest land subsidence areas are caused by pumping for irrigation. For example, 13,500 $km^2$ have been affected in California's San Joaquin

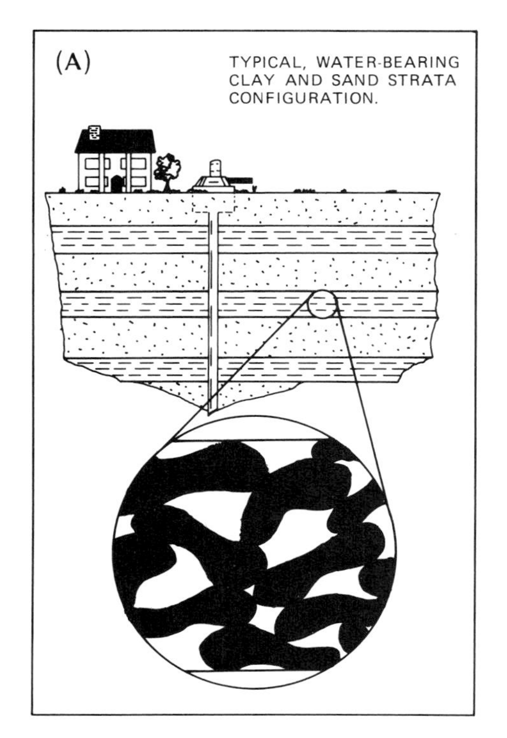


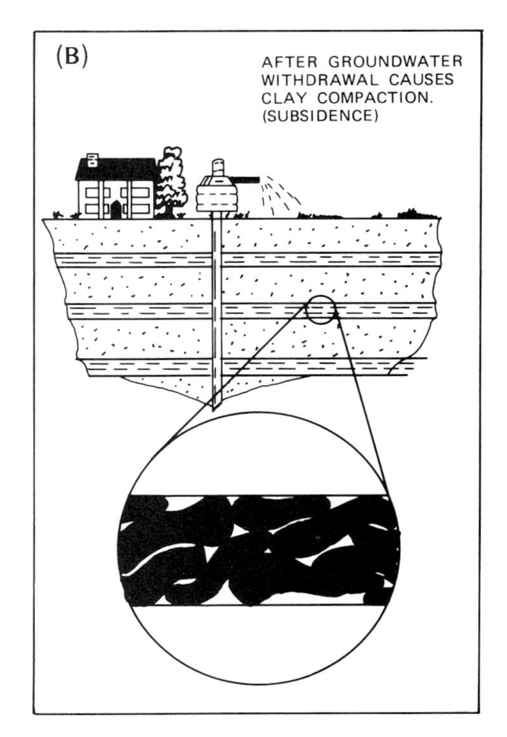


**FIGURE 4.7** What causes subsidence? (A) Typical water-bearing strata with interbeds of silt and clay. (B) After extensive groundwater withdrawal, the silt–clay strata undergo compaction, with a diminution volume that produces subsidence of overlying materials and structures. Courtesy of Harris–Galveston Coastal Subsidence District.

**FIGURE 4.8** Earth fissure Eloy District, Ariz. This fissure extends more than 1 km and in places is nearly 3 m deep. Courtesy of Thomas Holzer.

**FIGURE 4.9** Subsidence is a growing problem at Bangkok, Thailand. The lowered ground surface, caused by excessive groundwater pumping, is causing surface water to flood this slum site. Courtesy of Gerald W. Olson.

Valley, where fields have dropped 1–10 m (Fig. 4.10). In the Santa Clara Valley, 650 $km^2$ have also subsided with a total volume of sediments of 350 billion $m^3$ being involved. Of course, many damages occur to roads, pipelines, buildings, and farming fields running to many millions of dollars per year (Figs. 4.11, 4.12).

## Soil Salinization

Encrustation of salts in soil horizons plagues many drylands irrigation projects. The problem is worldwide and has occurred throughout history whenever slightly saline waters are applied to the land. Even purer waters when continually applied in arid regions can mobilize saline-type ions in the soil, which clog the soils and lead to reduced crop productivity.

The Wellton–Mohawk Irrigation District in Arizona was developed by the U.S. Bureau of Reclamation in 1952. The project diverts water from the Colorado River near Yuma and pumps it 50 km east to irrigate 25,000 ha of

**FIGURE 4.10** This telephone pole in the San Joaquin Valley, Calif., signals there has been nearly 10 m of land subsidence during the 1925–77 period from huge withdrawals of groundwater. Courtesy of Joseph Poland.

desert terrane. The area receives only 9 cm of rain per year and is the sunniest spot in the United States. An impervious substratum prevents downward drainage so that water is held close to the surface. Irrigation water that does not transpire or evaporate soaks into the ground and raises the water table. The resultant accumulation of high water and salt caused the district to install drainage wells in 1961 that would discharge to the Colorado River. At that time, the drainage water had a salinity of 6000 ppm, and the Colorado River intake was only 800 ppm. The water with increased salinity flowed into Mexico, and by 1962, salinity there had increased to 1500 ppm.

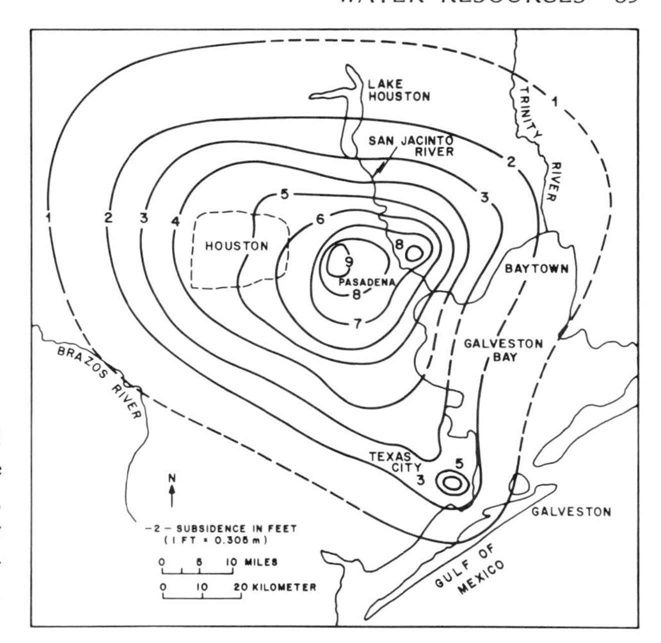


FIGURE 4.11 Total land subsidence by 1980 in the Houston–Galveston area, Texas, caused primarily by groundwater pumping. Courtesy Harris–Galveston Coastal Subsidence District.

FIGURE 4.12 One effect of land subsidence in the San Joaquin Valley, Calif. The protruding casing under the pump shows this site has subsided about 1.5 m. Courtesy of Joseph Poland.

Also in 1961, streamflow of the Colorado River was sharply reduced by completion of the Glen Canyon Dam and reservoir storage into Lake Powell.

These events prompted strenuous objections by Mexico, so the United States spent $12 million for remedial measures, but only marginal improvement had occurred by 1971 when river salinity was 1240 ppm. In a new agreement with Mexico in 1973, the United States agreed to release water with average salinity of 115 ppm or less over the base salinity at the Imperial Dam, which averaged about 800 ppm. In addition, Congress implemented the agreement by passage of the Colorado River Salinity Control Act of 1974. This measure provided funds to begin construction of a desalinization plant near Yuma. It was anticipated that the cleanup effort would cost $333 million. To help reduce the problem, the government purchased and thereby retired from production 2000 ha of citrus trees where salinization was especially severe.

Excessively heavy application of water on irrigated land and inadequate subdrainage cause salinization. Salt can also buildup when water application is too minimal to filter down and flush out the salt. Water logging and salinization are often twin problems in the irrigated lands of drylands. Such problems exist in each of the 30 countries that have more than 400,000 ha under irrigation. Although India has 40 million irrigated hectares, 15 million have been rendered useless because of such problems. Irrigation waters supplied by the Snowy Mountain Scheme of interior Australia have now produced severe problems in that region. In the San Joaquin Valley, more than 26,000 km of tile have had to be emplaced below ground to drain away the deleterious salts during the flushing operation, which causes their mobilization through the soil (Fig. 4.13). Only 40 percent of present farms have on-farm drainage systems. Salt water drains to a lower sink where it can be pumped to another drain site. Each year about 300 million $m^3$ of salty drainage water is produced. To remove it from the region would cost $500 million for canals and pumping systems. Those farms that became affected 20 years ago initially had subsurface drainage tiles that were 60–100 m apart. However, because the problem has become aggravated with continued salt buildup, the tiles are now being placed only 15–30 m apart. Some of the short-term solutions that are being used include the conversion of fields to salt-tolerant crops and plants that have shallower root systems. Some of the highly salted areas have had to be abandoned, and others are used as sumps for adjacent lands.

**FIGURE 4.13** More than 200,000 ha are irrigated from the All American Canal System that brings water from the Colorado River to the Coachella Valley, Calif. This aerial photograph shows a field in the leaching process to flush out salts into the drainage lines below ground. Courtesy of U.S. Bureau of Reclamation.

## Wetlands

Wetlands are lands that are transitional between terrestrial and aquatic systems where the water table is at or near the surface. Wetlands generally support hydrophytes (water-loving plants), contain a substrate that is predominantly an undrained hydric soil, and have saturated ground during most of the year. They can be classified into the following groups: (1) marine intertidal wetlands where the substrate is exposed and flooded by tides, (2) estuarine intertidal wetlands with semi-closed land but still subject to flooding by tides, (3) lacustrine wetlands in topographic depressions or clogged river channels, (4) riverine wetlands that are contained within a channel, and (5) palustrine (inland) wetlands that include other land-locked systems.

Some of the newer reasons cited for the preservation of wetlands include their ability to purify water by removing such contaminants as phosphates and nitrates. Thus, they can recycle nutrients and serve as chemical sinks. Wetlands provide protective habitat for up to one-third of all endangered animal species. Unfortunately, they are constantly besieged and confronted with a wide scope of development schemes, dams, mine operations, urban and industrial expansion, ports, canals, oil and gas production, and agricultural reclamation – all of which take their toll.

By the 1950s, the United States had lost 18 million ha or about one-third of all wetlands in the lower 48 states. The U.S. Department of Interior reported that in the 1950s there were 43.8 million ha of wetlands, but by the mid-1970s, the number had been reduced to 40 million ha, an average annual loss of 185,000 ha.

Unfortunately, this trend has not been reversed. Louisiana still loses an average of 30,000 ha each year, and even by 1982, the national loss was a dismal 186,000 ha (Fig. 4.14). Most losses are palustrine wetlands converted to agricultural lands, but even the total acreage loss to urban development exceeded the size of Rhode Island.

## Transportation

Proper distribution and allocation are some of the vexing problems associated with water supply (Figs. 4.15, 11.8). Growing cities and urban areas such as Phoenix and southern California have developed at sites far removed from major water sources, so water delivery systems are a vital component in patterns of water use. Chapter 11 further explores water engineering projects and construction of dams and canals along with their environmental impacts.

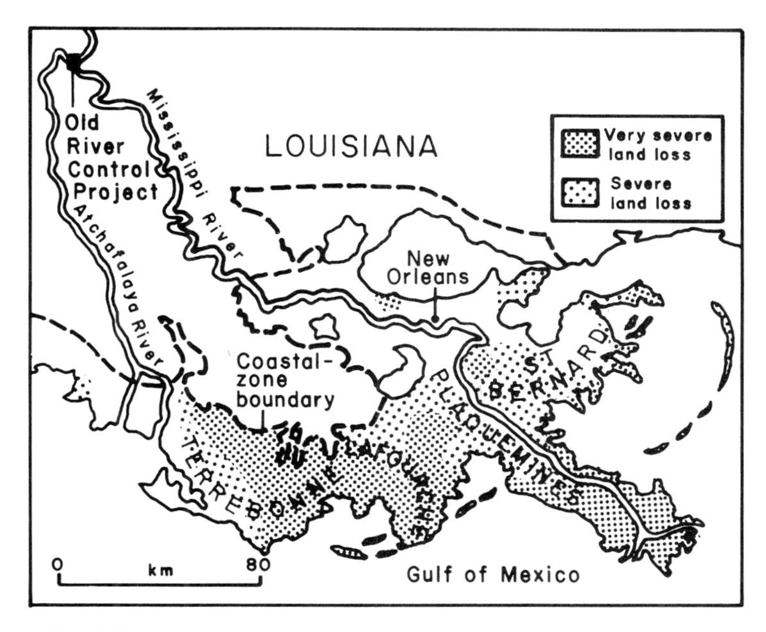


FIGURE 4.14 Map showing loss of land and wetlands in southern Louisiana. Such losses are caused by human intervention and changes in Mississippi River flow and sediment regimes.

## Pollution

Contamination of water, both surface and ground waters, is a problem that has reached epidemic proportions. There are 1.2 billion people who cannot count on safe drinking water, and 80 percent of all world disease is caused by some aspect of water pollution. More than half the rivers in the United States have been officially designated as polluted; that is, their waters violate federal standards for drinking or recreational purposes. A 1975 government survey showed that 17,000 of the 26,000 industries that were investigated discharged wastes directly into waterways. Estimates to clean up the rivers call for expenditures of $600 billion!

Pollution of groundwater has become increasingly serious, because nearly half the American population derive their drinking water from groundwater. Once an aquifer is polluted, its recovery is an exceptionally slow process. Pollution sources range from septic tank and toxic waste disposal, to farming, mining, and construction activities and a variety of non-point sources. Salt intrusion into the freshwater aquifers of coastal cities has forced abandonment of well fields in such localities as Miami, Pensacola, Daytona Beach, Brooklyn, and Los Angeles. This occurs when aquifers are pumped too

**FIGURE 4.15** Aerial view of Ririe Dam, Idaho. The dam is part of the Minidoka Project on Willow Creek, a tributary of the Snake River. The fields in the middle of the photograph bear testimony to the success of this irrigation project. Courtesy of U.S. Bureau of Reclamation.

heavily, thus permitting landward movement of ocean saline water as replacement for the overdraft of freshwater.

## WATER MANAGEMENT

Environmental managers who deal with water resources must address three basic questions: (1) How much total water is available for use? (2) How should the available water be allocated? (3) How can the water supply be increased or demand reduced? There is a growing disenchantment with traditional technological projects, such as dams, reservoirs, and canals, because instead of being part of the solution they have often become part of the problem. There are three categories of aid that can help relieve some of the water problems in particular regions and locations, and these are discussed in the following pages.

### Water Augmentation

Several techniques are in use to increase groundwater supplies. On Long Island, N.Y., there are more than 2500 pits that have been dug into glacial materials, which serve as collection areas for storm runoff so as to recharge underground waters instead of allowing runoff to flow to the ocean (Fig. 4.16). California has several water spreading projects (as in the Santa Ana and San Gabriel River basins) where surface water is diverted into holding basins and ditches so it can infiltrate below land surface instead of being lost to evaporation or flowage to the Pacific.

Weather modification by such processes as cloud seeding is becoming increasingly used in attempts to avert water shortages. This is a highly controversial issue and has led to lawsuits over "who owns the sky." Advocates claim both experimental and practical success with increased precipitation from seeding operations in Australia, South Dakota, Colorado, and Florida. Limited data in the West indicate that cloud seeding in winter has fewer hazards than in summer and increases snowpack amounts in limited areas. Rainmaking can have its unpredictable side. The governor of Idaho accused the State of Washington of "cloud rustling" in causing precipitation that otherwise would have fallen in his state. Seeding operations have been blamed on creating the abnormal cloudbursts that led to the flooding disaster at Rapid City, S. Dak., on June 9, 1972.

Other programs for increase in freshwater supply range from small-scale desalinization plants, such as are necessary for Kuwait and Saudi Arabia, to large-scale interbasin transfers such as the California Water Plan

**FIGURE 4.16** Recharge pit, Syosset, N.Y. Drainage from storm sewers discharge water to this manmade surface pond, whose waters infiltrate to recharge the groundwater zone.

for diverting water from the northern to the southern part of the state, and the project for tunnelling water from the Delaware River basin to the New York City metropolitan area. Study is currently underway evaluating the feasibility for towing Antarctic icebergs to water deficient areas in Australia and the Mideast. Most cost estimates indicate that with appropriate technology, icebergs for such areas could provide cheaper freshwater than the present methods of desalinization.

## Water Remedial Measures

A second category for enhancing water production is the implementation of those engineering projects which will salvage present water supplies, thus increasing efficiency and amounts of usable water. By installing liners to canals, reservoirs (Fig. 4.17), and water catchment areas, and by fixing water leaks, it has been estimated that water availability can be increased 25 percent. Such techniques are especially important in saving irrigation water. Other methods to save this water would include improved applications procedures such as night-time irrigation, the use of the newer spray irrigation, and drip irrigation technologies. Industry can save additional water by instituting more effective recycling methods. Water reuse and the release of

FIGURE 4.17 To prevent seepage in this reservoir, an accordian-style plastic membrane liner is emplaced from a carton on the lowboy trailer. Courtesy of Watersaver Company.

less polluted waters into hydrologic systems can greatly enhance water supplies in highly populated and industrialized regions.

In the southwestern United States, what to do with phreatophytes (plants that use groundwater) is the source of a heated debate. Hydrologists argue for their destruction as a mechanism for saving 25 million acre-feet of water. However, wildlife enthusiasts urge their retention as being vital for refuge and feeding areas so necessary to the survival of many animals and birds. Thus, the controversy provides an interesting sidelight on environmental affairs, shows constraints that operate in "real-world politics," and emphasizes problems of conflict of interest whereby two groups of conservationists have opposite goals.

## Non-structural Conservation

Educational programs with followup by legal procedures and implementation provide a range of measures that can help alert the populace to the importance of water and results that may follow if greater restraints are not employed in its use. For example, one-third of water use in urban areas of the United States is devoted to car washing, lawn and garden watering, and swimming pools. Some urban areas, such as New York City, do not even have water meters. Thus, there is no incentive for its conservation. Water rationing and the establishment of different allocation systems and priorities can also assure wiser resource management. New pricing strategies, such as being argued for energy use, are also applicable for water resources. For example, the rate structure should be organized so it progressively penalizes those with demonstrated higher consumption rates. Incentives in terms of water costs should be provided to those users who document low use with institution of conservational measures, whereas high use without accompanying programs for minimal consumption should be charged higher rates. Only when both the general public and the lawmakers understand the vital significance for cherishing and prudently utilizing this resource will water take its place on the pedestal of important resources.

Total water resource management involves a multitude of environmental affairs that include water supply, flood control, pollution abatement, watershed care, and pricing allocation systems. Thus, evaluation of this precious commodity crosscuts many disciplines and interests of society. Only by comprehensive efforts can true understanding and balanced programs be instituted that will place this fundamental resource into its proper and needed perspective.

## CASE STUDIES IN WATER MANAGEMENT

### South Florida

The water of south Florida comes from a chain of lakes that regularly overflow during the rainy season into sloughs and streams from Orlando to Lake Kissimmee. South of the lake the water moves onto the lower Kissimmee River floodplain then into Lake Okeechobee. Drainage from the south shore of the lake distributes water over the vast 160 km long peat zone south of the lake and onto the 60 km long saw grass plain of the Everglades. To develop agricultural lands, a substantial area south of Lake Okeechobee had been drained by the 1920s. But the muck soil is extremely vulnerable to aerobic decay and desiccation, drought, and wildfire. It has been disappearing at a rate of 0.3 m per decade, in direct proportion to the lowering of the water table. As an aid to the farming system, a series of water retention levees were constructed. However, a severe hurricane demolished a section of the levee, and the ensuing flood killed 2400 people. This led to federal financing of a much larger levee that was finished in 1938 and became the start of large federal expenditures in south Florida. Everglades National

Park was established in 1947 and became the nation's third largest national park, encompassing 5200 $km^2$, and containing unique habitats for flora and fauna. Ironically, while one part of the federal government was attempting to save the Everglades, another part was preparing plans that would place them in severe jeopardy. In 1948, Congress created the Central and Southern Florida Flood Control Project that authorized the U.S. Army Corps of Engineers to build a network of 1280 km of new or improved levees and 800 km of canals. The canals would drain wetlands of the Kissimmee Basin, would regulate water in Lake Okeechobee, and would drain and irrigate the Everglades Agricultural Area south of the lake. In addition, the levees would provide protection for other new development, assuring future growth of the burgeoning "gold coast" cities. The benefit/cost ratio for such construction was also enhanced by arguing that the three lake-like conservation areas that were created would maintain the hydrostatic head of groundwater that was needed to recharge the Biscayne Aquifer from Palm Beach to Miami, thereby preventing saltwater intrusion from the ocean into the gold coast aquifers.

By 1970, it was apparent to all that the Everglades were undergoing debilitating changes so the Corps finally agreed in 1970 to release a minimum of 385 million $m^3$ of water annually to the Park. Then, a new threat appeared when Dade County planned to develop an international jetport in the Big Cypress wetland just north of the Park. A coalition of 21 environmental groups and 2 unions were able to prevent the development by initiation of lawsuits, which led to the Big Cypress National Preserve Land Acquisition bill of 1974. Furthermore, by 1972, Florida had passed four landmark land and water conservation acts and reestablished the South Florida Flood Control District into the South Florida Water Management District with a mandate to protect the Everglades.

Unfortunately, the Everglades continue to be starved for water. The population in the region has grown in 40 years from 2 million to 10 million. In addition, most of the 30 million visitors visit during the dry winter months and use the precious water. All these factors compound the water problem and cause continuing decline of the water table. Each year, sea water pushes farther inland into the canals and aquifers. The channelization of the Kissimmee River has devastated its immediate environment, and the vast change in hydrology is now being cited as responsible for inception of drought years (Fig. 4.18). It has been argued that the surface and near-surface water evaporation and transpiration by the original lush vegetation provided moisture that fell on nearby lands. Because of the lowered water table, the desiccated soils no longer have that capacity and instead, undergo degradation and loss. Some areas have experienced as much as 2 m of subsidence in the last 30 years.

FIGURE 4.18 Channelization of the Kissimmee River in southern Florida. Light area shows the extent of soil degradation as a result of the construction. Note previous meander scrolls in foreground. Courtesy of South Florida Water Management District.

## Colorado River Basin

The Colorado River is 2316 km long and drains 632,000 km$^2$. John Wesley Powell's expeditions starting in 1867 helped provide important information about the geology of the upper basin and River. Even earlier, after an 1857 trip to the lower basin, Lt. Joseph Christmas Ives wrote his superiors in the War Department, "Ours was the first and doubtless will be the last party of whites to visit this profitless locality." This forecast proved to be very wrong. Although by 1900, there were only four people for every 10 km, the same population density as Alaska today, by 1980, that figure had increased 10 times. Throughout the history of the region, however, many controversies have raged.

One of the earliest cases was based on the Winters Doctrine, a U.S. Supreme Court ruling of 1908, which assured the Indians that they could use as much water as they needed. However, in 1963, this was refined by the Court to mean as much water as necessary to serve all "practicably irrigable acres" on a reservation. The first attempt at a comprehensive management plan was signed by all the basin states in 1922 and known as the Colorado River Compact. The compact decrees that the upper basin states of Colorado, New Mexico, Utah, and Wyoming are to receive forever 7.5 million acre-feet of water annually (9.2 billion $m^3$). The lower basin states of Arizona, California, and Nevada can withdraw a similar amount. In 1944, a treaty with Mexico guaranteed that country 1.5 million acre-feet (1.8 billion $m^3$). Unfortunately, these allocations were based on an early estimate that the annual flow of the River was nearly 15 million acre-feet, whereas the average flow amounts to only 13.5 million acre-feet. Thus, if all parties were to use their allocation, there would be a water deficit of about 3 million acre-feet.

A new wrinkle was added when the Supreme Court ruled in 1964 in the case of Arizona v. California that Arizona was entitled to some of the water, which formerly had been used by California. This led to the Central Arizona Project, a $2 billion series of aqueducts, tunnels, and pumping systems, that is designed to deliver water to central and southern Arizona. The Project is set for completion in 1985.

## POSTLUDE

No other segment of environmental affairs is fraught with so many problems as the management of water resources. This basic human and societal commodity is besieged on all fronts with conflicts that involve its quantity, quality, and distribution. Unlike minerals and fossil fuels, water, fortunately, is a renewable resource.

Chief among the water problems that must be solved are the numerous battles that arise over conflicting use patterns among domestic, industrial, agricultural, and mining requirements. Even the very fluidity of the resource allows long travel distance for the spread of pathogens, disease, and toxicity.

In the United States, the administration of water resources resides with 16 federal agencies, but there is still no comprehensive national water policy. Instead, there are a series of statutes that are subject to change, enforcement, and funding allocations. Numerous inequities have been allowed to remain. For example, Idaho receives money for water projects that equals $75 for each resident whereas the per capita share in New York is only $0.15. Most funds for federal projects go for multipurpose uses that combine irrigation, flood control, and recreation.

Although historically, water resource policy has been built around concrete methods to use, store, and distribute water, such as dams and canals, there are now several ways to achieve more efficient water use. These include metering, rationing, mandated conservation practices, water rights reform, and pricing strategies. Thus, there is growing disenchantment with technological fixes that attempt to solve all water problems. Instead, many experts are stressing shifting usage patterns, reuse of water, and reduction in waste. Yes, water is a basic requirement for society. There should be, and must be, an increased public awareness of its preciousness to prevent a "water crisis" that will make an "energy crisis" seem like a minor event.

CHAPTER 5

# VOLCANIC ACTIVITY

"Vancouver! Vancouver! This is it!" were the last words spoken by David Johnson before being overwhelmed by Mount St. Helens' fury on May 18, 1980. Johnson, a volcanologist with the U.S. Geological Survey had been meticulously studying the volcano for weeks, but his 13 km distant observation station was still too close when "all hell broke loose" that fateful Sunday morning. The eruption of this formerly beautiful and majestic mountain in the Cascade Range of Washington is but the latest vestige of the power of natural forces and the devastation that can be wrought.

Volcanoes derive their name from Vulcan, the Roman god of fire, who forged the thunderbolts for Jove and the weapons for Cupid. Homer wrote of the one-eyed Polyphemus hurling missiles at Ulysses, and probably got the idea from Mount Etna's numerous displays. In the Middle Ages, volcanoes were considered the gateway to hell, and Polynesian legends spoke of their control by the fickle goddess Pele who could appear in human form as a beautiful young maiden or an aged hag. Indeed, volcanoes have both beneficial and harmful attributes. Volcanic activity is responsible for life on Earth because throughout the millenniums of time, it has brought water to the surface and produces the gases that constitute the atmosphere – the basic prerequisites for living things. However, modern eruptions and their potential for damage are viewed as hazards or disasters because they place in jeopardy human safety and property.

Throughout human history, more than 200,000 people have lost their lives because of volcanoes (Appendix B). Today, millions of people live in high-risk zones near potentially dangerous volcanoes. Thus, the benefits these terrains produce outweigh the probable destruction they could produce.

Volcanoes comprise some of Earth's most majestic landscapes – Fujiyama, Mt. Mayon, Mt. Rainier, and Yellowstone National Park. It is perhaps the combinations of beauty, but yet the ever-present threat of doom, that make volcanic areas so awesome and spectacular.

## DISTRIBUTION

Although volcanic mountains comprise only a small part of the earth, landscapes of volcanic origin constitute 2 percent of the total land surface of continents and 4 percent of North America. The largest volcanic terranes are the basalt plateaus of the Columbia Plateau and the Deccan Plateau in India where hundreds of thousands of square kilometers are covered with lava flows that may range up to 3000 m thick.

Present-day volcanic activity is largely restricted to several important areas of the world. About 80 percent of the world's 850 active volcanoes are associated with the "ring of fire," a zone of combined earthquakes and volcanism that largely encircles the Pacific Ocean (Fig. 5.1). A second prominent belt extends through the Mediterranean region (Mt. Etna, Mt. Vesuvius), and a third important zone extends the length of the Mid-Atlantic Ridge, nearly 16,000 km long from Iceland to Mt. Erebus, Antarctica. There are also more isolated areas of activity such as the Hawaiian Islands and the African rift system.

### Volcanic Miscellany

During the past 10,000 years, 5564 eruptions have been catalogued. Of these, 96 were eruptions that formed new islands, and 217 were dome-forming eruptions on the 86 lava dome volcanoes (such as Mount St. Helens) (Fig. 5.2). Bezymianny volcano on Kamchatka Island has had 15 dome-forming eruptions since 1955, and the one in 1956 shattered the cone. The Chilean Andes had the highest historic eruption at Mt. Llullaillaco, which has

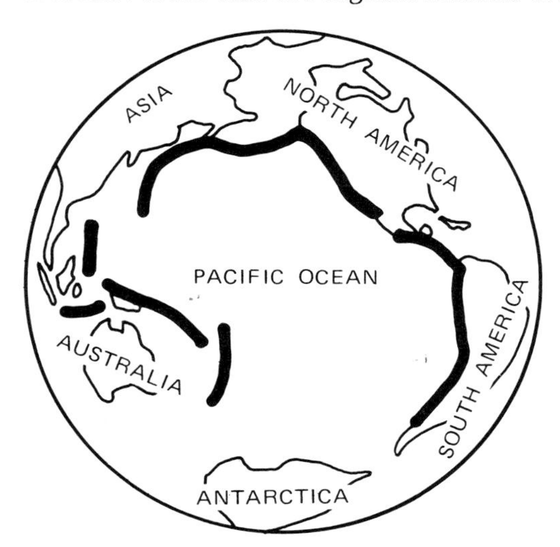


FIGURE 5.1 The heavy lines indicate a belt of volcanic activity that essentially circumscribes the Pacific Ocean, which is known as the "ring of fire." This belt is also the location for a major part of earthquake activity of the earth.

FIGURE 5.2 South side of Mount St. Helens volcano. This photograph was taken five years before the catastrophic eruption of May 18, 1980. The summit consisted of a volcanic plug that formed between 1600 and 1700 A.D. Lava flows on the lower flank in the center of the photograph are marked by sharp, curved ridges that formed along margins of flowing lava. Courtesy of U.S. Geological Survey.

an elevation of 6724 m. The largest historic lava extrusion occurred in Iceland in 1783 with 12.3 $km^3$ of mostly basalt.

Kilauea in the Hawaiian Islands is currently the largest active volcanic mass. However, historic volcanoes, even those such as Tambora, which erupted about 100 $km^3$ in 1815, are small compared with earlier eruptions. About 700,000 years ago, Mono Lake in California exploded 600 $km^3$ throughout a 1200 $km^2$ area; and about 600,000 years ago, the Yellowstone volcanoes erupted 1000 $km^3$ of material, which was followed by massive collapse of structures forming an elliptical caldera. Since 1923, the area has experienced 700 mm of uplift, which may be heralding a new influx of molten rock (magma) into the overlying materials.

Volcanoes can grow suddenly at sites not previously occupied by the principal structure. On February 20, 1943, a new volcano appeared almost overnight at Paricutin, Mexico, in a farmer's field at the base of the old volcano, Tanciaro. While plowing his field, the farmer first noticed a fissure opening with material oozing from it. By the next morning, the cone was 10 m high and after a week it was 150 m and finally in a few months had grown to a height of 370 m with a total production of about 80,000,000 $m^3$ of volcanic materials of all types. Another event occurred on November 14, 1963, 32 km off the coast of Iceland. Here, a volcanic cone pierced the ocean surface, and in five days, this new volcanic island was 600 m long and 65 m high. It was named Surtsey, after an Icelandic god, and became a mecca for study by volcanologists throughout the world.

One of the most recent danger zones for possible volcanism in the United States is the Mammoth Lakes area in eastern California (Fig. 5.3). This site is on the southern edge of a 32 by 15 km caldera, a broad caldera known as Long Valley where a giant eruption 150 times larger than Mount St. Helens exploded 700,000 years ago. Since 1978, the area has been rocked by a series of earthquakes, which reached 6.3 Richter magnitude in May 1980. These spasmodic tremors, the progressive movement of the earthquakes toward the surface, and the formation of a new group of fumaroles caused this area to undergo intensive study by the U.S. Geological Survey. The in-

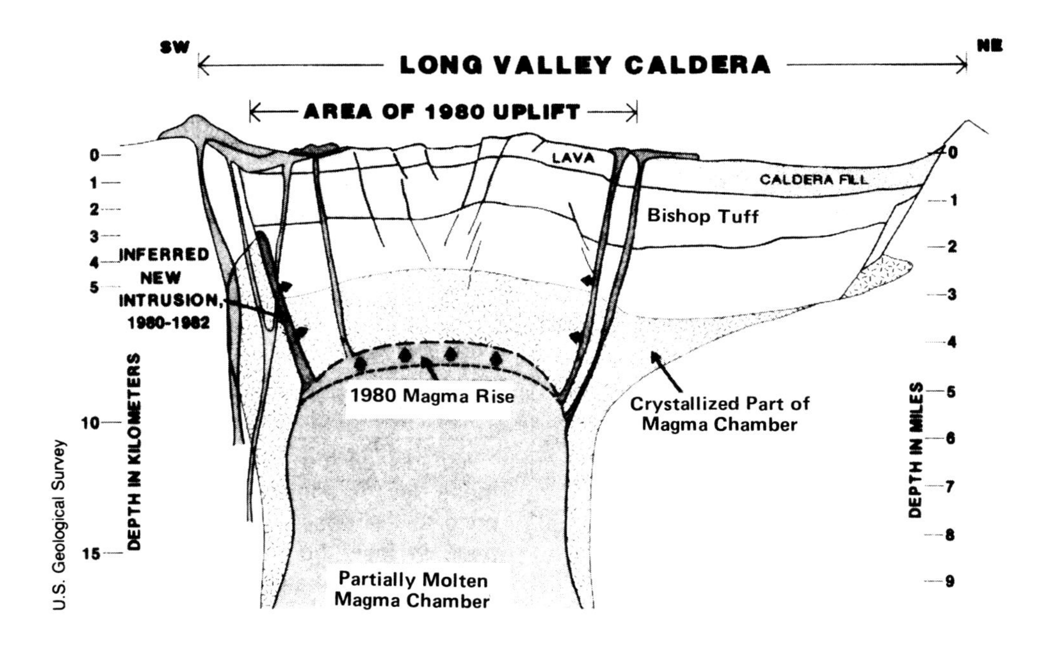


FIGURE 5.3 Geological cross section of Long Valley Caldera, Mammoth Lake, Calif. Geophysical data indicate a vertical intrusion of magma is pushing toward the surface toward the southwest. Three inactive dikes reached the surface during the last 500,000 years. Courtesy of U.S. Geological Survey.

vestigation showed that by mid-1982 there was a magma chamber about 8 km deep that had expanded so that the center of the site had been uplifted 25 cm during the 1975-80 period and rose an additional 7 cm between 1980 and 1982. This evidence prompted the U.S. Geological Survey to issue on May 25, 1982, a bulletin, "Notice of Potential Volcanic Hazard..." for the Mammoth Lakes area.

## Origin

Volcanoes occur wherever magma (molten rock) can penetrate through the crust and reach the earth's surface. To accomplish this, the overlying materials must be weakened, thinned, or subjected to such pressure that the rocks cannot withstand the upwelling of the melted rocks. The currently approved theory to explain a large majority of volcanic and earthquake activity is called "global tectonics" or "plate tectonics." Early in the 1900s, a somewhat similar idea was known as "continental drift" as propounded by the German climatologist Alfred Wegener and the American geologist Frank Taylor. These ideas have in common the gradual shifting and movement of ocean basins and continental masses with respect to each other. The various segments of the earth's crust are divided into a series of plates that move somewhat independently of each other, and the major ones are the Eurasian Plate, Indian Plate, Pacific Plate, North American Plate, South American Plate, African Plate, and Antarctic Plate. Other important ones are the Philippine Plate, Cocos Plate, Nazca Plate, Caribbean Plate, and Arabian Plate. During their travels, the plates may have different types of motion, such as splitting apart, colliding, or grinding against each other.

A favorite location for many volcanoes is an area known as a "subduction zone." This is a site where one lithospheric plate is being "subducted" or is foundering and being displaced under another plate. A typical occurrence is the area along the west coast of the western hemisphere where the Pacific Plate is riding under the continental plates (Fig. 5.4). The lower density of the continental crust overrides the heavier oceanic crust, which is thereby pushed down into the mantle and melted. The molten material, being lighter than the surroundings, rises and melts its way upward into fractures to form such volcanoes as those in the Cascade Mountains (and Mount St. Helens). This same process forms the numerous volcanoes of Japan, Indonesia, South America, the Mediterranean, and others.

A second cause for location of volcanoes in specific zones is lithospheric plate divergence. These zones are somewhat linear areas or ridges in the middle of the oceanic plates where hot magma from the mantle rises and accretes onto the oceanic plate. This process, acting as a gigantic conveyor

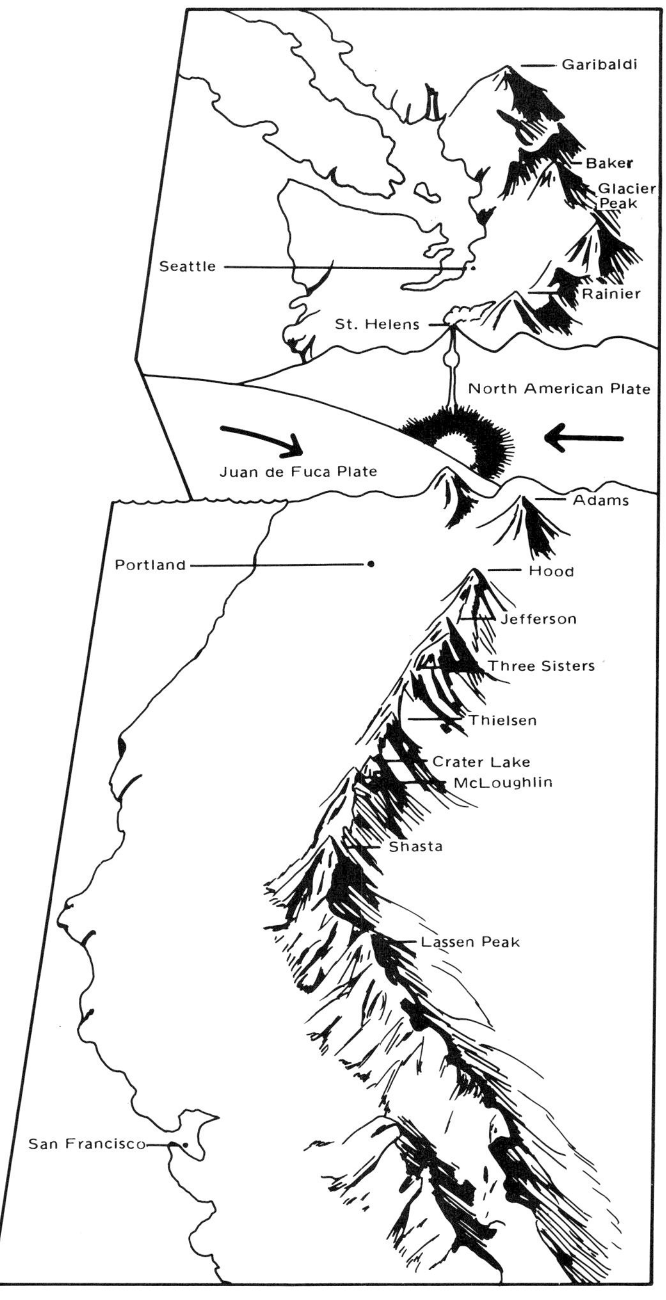


**FIGURE 5.4** Artist's view of Cascade Mountain volcanic chain in the Pacific Northwest. A cross section through Mount St. Helens shows subduction of the Juan de Fuca oceanic plate by the North American continental plate. This on-going process has created magmatic activity, which forms the spectacular series of majestic volcanic mountains.

belt system, produces a spreading in the sea floor in opposite directions. The submarine volcanism, and even some of the surface expressions, as in Iceland, of the Mid-Atlantic Ridge are examples of this type of activity (Fig. 5.5).

It is also possible that a plate boundary will sag at the junction with another, or grind along as with the San Andreas Fault Zone. Such activities when allied with other plate motions are responsible for creating not only most volcanoes, but also earthquakes and mineral and metal deposits. For example, the abrasion of rocks along these giant fracture systems releases energy in the form of seismic waves thereby becoming the focus of earthquake phenomena.

The new theory of "hot spots" had to be developed to explain those volcanoes that did not occur near plate margins. Under this theory, it is proposed that there are immobile sources of magma that rise from the mantle in plumes, melting overlying crust. Because the crust is in motion as an oceanic plate, the hot spot becomes an area of new magma, which forms new volcanoes. Such a process can form long strings of volcanoes, as in the

**FIGURE 5.5** Fissure eruption near the Krafla Caldera, northern Iceland. Courtesy of Axel Bjornsson, Icelandic National Energy Authority.

Hawaiian Islands where the oldest and most dormant ones are at one end and the youngest at the other end. Thus, cut off from their source of supply the older volcanoes become extinct.

## PROCESSES AND CLASSIFICATION

Although all volcanism is related because of magma reaching the surface, there are many types of activity, and fundamental differences exist in the character of the material and the landform created. At one time, it was fashionable to classify volcanoes on their recency of action, calling them active, extinct, or dormant. However, this provides a false sense of security because some of the active ones pose no serious danger, whereas some labelled extinct, such as Pelée and Krakatau, have unleashed great havoc and destruction when they did explode.

Another classification groups volcanoes into the type of landform that is created and the materials in the structure. It is the degree of explosiveness and the composition of the material that determines this. For example, cinder cones are small peaks rarely more than a few hundred meters high and composed of volcanic fragments and ash blown out from a central vent, such as Paricutin and Sunset Crater, Ariz. (Fig. 5.6). Composite or stratocones are perhaps the most spectacular because of their height, beauty, and symmetry (Fig. 5.2). Lava flows alternate with layers of ash and cinders that may form mountains thousands of meters high such as Fujiyama, Mayon, and many in the Cascades. Shield volcanoes such as the Hawaiian Islands are massive, gently sloping mountains composed of numerous basaltic lava flows. Mauna Loa is the world's largest single mountain mass – nearly 10,000 m high from the floor of the ocean where it starts to its summit of about 4000 m above sea level.

When an area contains numerous fissures where lava may flow and cover a large region, such places are called plateau basalts, as in the Columbia and Deccan plateaus (Fig. 5.5). Other landform variations are volcanic domes, where the highly viscous lava may congeal and form isolated conical edifices.

### Products and Effects

Probably the most common conception of volcanic activity, and also the most dramatic, is that of a mountain that explodes with such explosive violence that it scatters its debris throughout the entire countryside. The chief products from such eruptions are ash falls. As magma rises in the conduit, the old hardened material is blasted into the air in the form of pumice

FIGURE 5.6 Sunset Crater area, near Flagstaff, Ariz. The hills in the background are composed of volcanic ejecta and are cinder cones. The darker rocks in the foreground are lava flows.

and fine ash. Such materials can bury nearby areas under a heavy blanket of debris. Probably the most famous case is that of the eruption of Mt. Vesuvius in 79 A.D. and the burying of Pompeii. Not only are there immediate effects to life, but long-range effects also occur when animals eat ash-laden vegetation. Their digestive tracts become clogged and their teeth lose their grinding ability because of the abrasive silica ash.

Glowing clouds, also called *nuees ardentes*, are some of the most terrifying products of volcanism. During the initial eruptive phases of some volcanoes, a large mixture of hot gases and ash can form a dense cloud that races like an avalanche down the mountainside at incredible speed, more than 120 km/hr. These materials can destroy and suffocate all in the path. The prime example of this type of catastrophe occurred in 1902 to the 30,000 citizens of St. Pierre on the island of Martinique. Here in a matter of minutes, the eruption of Mt. Pelée killed all except two of the inhabitants.

Although lava flows can destroy everything over which they move, their slowness generally allows human escape. The lavas associated with the volcanism at Paricutin buried the villages of Paricutin and San Juan Parangaricutiro in 1944, but no lives were lost by this action.

Eruptions can also create secondary effects that result from the volcanic activity. Mudflows, or "lahars," can be triggered. These are characterized by movement of fragmented volcanic materials and debris that have become water-saturated. In 1919, the Kelut volcano on Java erupted and caused lahars that killed more than 5000 people. In this instance, the water source was a crater lake, which subsequently was artificially drained to prevent a repeat of the disaster.

Tsunamis, giant surges of ocean water (improperly called "tidal waves"), can result from volcanic explosions as well as from earthquakes. Although the eruption of Krakatau in 1883 in what was then the Dutch East Indies blew away most of the island, it was the resulting 10–30 m high tsunami that killed most of the 36,000 people, as water swept through low coastal zones throughout the region.

In addition to the direct effects by volcanoes in producing loss of life and property damages, there are the more insidious and secondary effects that may result from weather and climate changes. The year of 1816 was known as "the year without a summer." Vermont and New Hampshire had snow in June. The corn crop failed and sheep froze to death. In western Europe, crop failures caused widespread famine, and food riots broke out in France. The summer of 1816 was the coldest on record in western Europe. Most specialists in weather phenomena attribute these bizzare events to after-effects produced by the explosion of Mt. Tambora in the Dutch East Indies on April 5, 1815. The blast produced 170 $km^3$ of debris that was thrown into the stratosphere and continued to circle the earth for several years. The eruption even had demographic effects because many farmers left New England and headed West. Some even attributed the worldwide outbreak of cholera to this event. Alaska's Katmai Volcano blew its top in 1912, created the Valley of the Ten Thousand Smokes, and ejected huge clouds into the atmosphere. It also caused some abnormal weather effects.

Some scientists believe there may be another side to the relationship of climate and volcanic activity. Rampino and co-workers (1979) have suggested the possibility that climatic change, instead, can contribute to volcanic eruptions. They believe that variations in climate can lead to stress changes in the earth's crust, such as in loading and unloading of ice and water masses and by axial and spin-rate changes that might augment volcanic and seismic potential.

## VOLCANIC DISASTERS

Volcanoes pose hazards not only to nearby inhabitants, but also to people far removed from the eruption (such as the cases of Tambora in 1815, which killed 12,000 by the tsunami and the Krakatau explosion of 1883). People are attracted to volcanic areas because of the richness of the fertile soil. They may be lulled into a false sense of security believing the adjacent mountain is safe. The so-called "dead" or dormant volcanoes may be more dangerous than those that are periodically active, because instead of letting off energy, they store energy until it is suddenly released by a giant upheaval.

A certain romantic mystique surrounds the eruption of Santorini in 1500 B.C., leaving as a remnant the island group of Thera in the Aegean Sea. Because of the possible magnitude of the destruction, some scholars contend it is the lost Atlantis of Plato's writing. This event is also cited as the reason for the demise of Minoan civilization in the region. Where the volcano had once been is a giant coastal bay 80 $km^2$ in size and 400 m deep. The resulting ash buried nearby sites to thicknesses of 60 m, and the tsunami had a runup of 50 m on islands 25 km away, and 6 m even at Tel Aviv, a distance of 800 km!

What is emerging as an equally engrossing mystery is the degree of disruption to Mayan civilization caused by the Ilopango eruption in Central America. In the third century A.D., a massive explosion struck the southeast Maya highlands. Ash deposits still 1 m thick occur at distances of 77 km and *nuee ardentes* type materials moved as far as 43 km. Population of this immediate region would have exceeded 300,000, and those not killed would have had to emigrate from the land because of the immediate loss of the agrarian base to their livelihood.

By the first century A.D., the Vesuvius area in the Bay of Naples had become a favorite resort and playground for the wealthy and aristocratic of Rome. For centuries, the mountain had slumbered, and little thought was given to any possible danger it might possess. Even after a severe earthquake occurred in 63 A.D. with intermittent activity for 16 years, the populace was still complacent and unafraid. August 24, 79 A.D., changed that when severe earthquakes rocked the area followed by the catastrophic eruption on August 25th. Pompeii was buried by 4.5–7.6 m of pumice and ash, and nearby Herculaneum was overwhelmed by a lahar up to 20 m thick. Other towns were also damaged, such as Strabiae where Pliny the Elder died. Two letters written by Pliny the Younger provide the only eyewitness account of the eruption. Of the more than 20,000 inhabitants, most were able to escape, but the disaster became the tomb for about 2000 residents. Many of the skeletons have been found 60 cm above the base of the initial air-born debris, indicating that they didn't immediately succumb to the event (Fig. 5.7). In addition, many bodies were found near the walls that surrounded Pompeii. They had been carrying possessions and were attempting to flee before being overcome by the fine dust and gases.

The area near Java has the dubious distinction of producing two of the greatest volcanic explosions of historic times. The 1815 Tambora explosion not only blew away most of the mountain with ejecta that changed short-term weather conditions, but also created a devastating tsunami that killed 12,000. Even worse, it caused the nearby areas to lose their food-growing potential, which led to death by starvation and disease of 92,000.

Then in 1883, the cataclysm of Krakatau occurred. It created such a powerful blast that windows were shattered to distances of 150 km, and the sound was heard 4800 km away! About 16 $km^3$ were blown away to heights of 80 km. Fortunately, the island was not inhabited, but the ensuing tsunami more than made up for this by drowning thousands on distant shores.

Appendix B shows that during the last several hundred years the god Vulcan or the goddess Pele has not lost touch with how to wreak havoc on the human race. During these times, many hundreds of people have been killed at single events in such widely distant places as the Philippines, the East Indies, the West Indies, the Mediterranean, and the United States.

## Mount St. Helens

For many years scientists had touted the volcanic chain of composite cones in the Cascade Mountains of the Pacific Northwest as being the most likely site for future eruptions in the United States (Fig. 5.4). In a 1975 article in *Science* magazine, geologists Dwight R. Crandell and Donal R. Mullineaux state, "The high probability, based on past behavior, that Mount St. Helens will erupt again indicates that potential volcanic hazards should be considered in planning for future uses of the land that could be affected by an eruption."

On May 18, 1980, without any immediate warning Mount St. Helens made the forecast true. For 123 years, Mount St. Helens had slumbered, but March 20, 1980, heralded an awakening when swarms of micro-earthquakes began. The quakes gradually increased their activity until a small eruption occurred on March 27, sending clouds of smoke and ash more than 6000 m into the sky. A crater formed, and additional activity that extended into March 28 formed another crater until the two finally merged. On April 3, seismic activity developed into harmonic tremors, which signalled the migration of molten rock beneath the volcano. By April 9, the crater had reached a size of 300 m wide, 500 m long, and 250 m deep below the south rim. Total volume of ash ejected had been 70 million $m^3$.

**FIGURE 5.7** Plaster casts of victims of the eruption of 79 A.D. at Pompeii. These casts were made *in situ* outside the Nucerian Gate on the southern side of the city and represent forms preserved by the bodies in the solidified tephra of Vesuvius.

Although there was intermittent activity prior to the fateful day, and indeed, the mountainside bulged more than 100 m, the experts were unprepared for the suddenness and ferocity of the May 18 paroxysm. A 5.0 M earthquake triggered landslides in the bulge area. This was immediately followed by an eruption of slowly rising dark-gray, ash-laden column from the summit crater and a subhorizontal blast of light-gray clouds of incandescent gas and ash (Fig. 5.8). Within minutes, the blast dissipated, and major activity became concentrated in the vertical explosive plume (Fig. 5.9). The compressive explosion was felt up to distances of 425 km. The blast devastated a rugged mountain area of nearly 500 km$^2$. Mudslides rushed into Spirit Lake and then 21 km down the North Fork of the Toutle River, filling it to 60 m depths. The eruptive column reached a height of 23 km with the energy of 10–15 megatons. (The atomic bomb that fell on Hiroshima was only 20 kilotons.) The total ash blasted from the site amounted to 2.7 km$^3$, the same volume as ejected in the Vesuvius eruption (Fig. 5.10). The mountain was lowered 380 m by the crater (Fig. 5.11).

**FIGURE 5.8** The May 18, 1980, eruption of Mount St. Helens, Washington. The initial blast was directed nearly horizontally toward the north. Mt. Hood, another one of the Cascade Mountain volcanoes, is in the far background. Courtesy of U.S. Geological Survey.

**FIGURE 5.9** Mount St. Helens eruption of May 18, 1980, About 2.7 km$^3$ of pulverized rock and ash was blown from the mountain, reaching heights of 23 km. Courtesy of U.S. Forest Service (USDA).

FIGURE 5.10 Dotted line indicates former profile of Mount St. Helens, as viewed from the north side after the eruption. The mountain's peak was reduced by the eruption from a height of 2950 m to 2438 m. Courtesy of U.S. Forest Service (USDA).

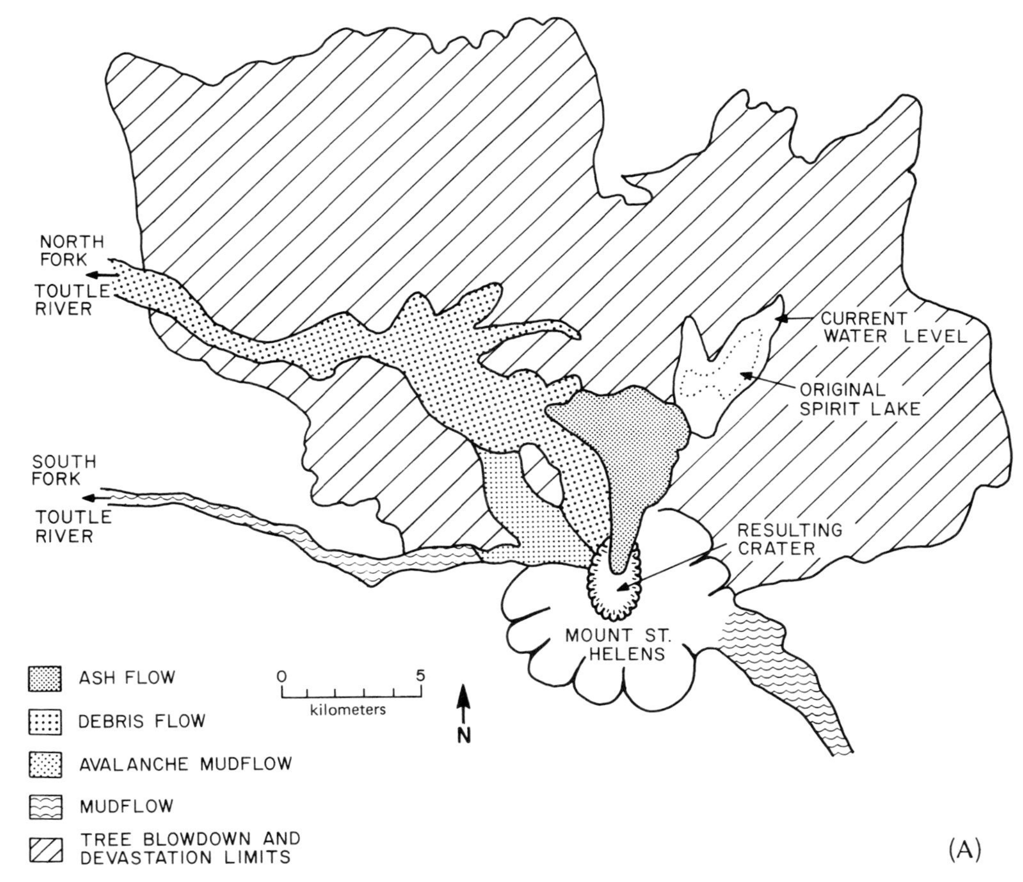


FIGURE 5.11 Artist's sketch of Mount St. Helens eruption and effects it produced. Redrawn from Association of Engineering Geologists Newsletter, U.S. Geological Survey and U.S. Forest Service (USDA) information releases. (A) Map view, (B) Cross view.

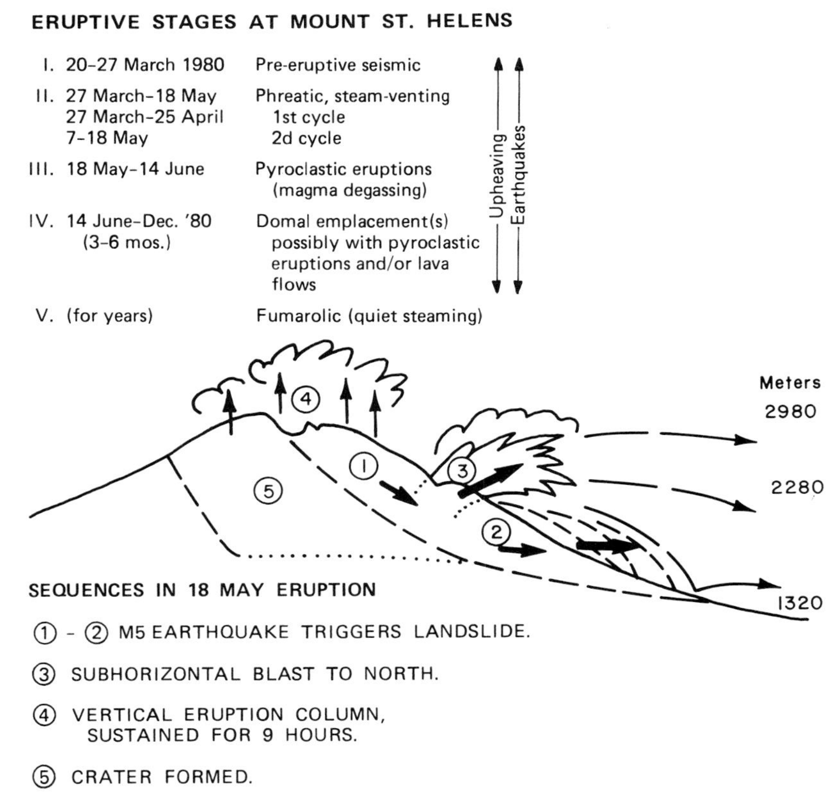


(B)

**FIGURE 5.11 (Continued)**

Additional secondary effects were the silting of main channels of all nearby rivers and tributaries (Fig. 5.12). The muds reduced the 180 m wide shipping channel of the Columbia River from a normal 12 m depth to 4.3 m. Timber loss amounted to 800 million board feet. When all the damages were summed up – costs for dredging, loss of crops, timber, fish, highway and bridge repairs, equipment and machinery, homes, businesses, and many others – the total bill was nearly $3 billion. Surprisingly, loss of life was low, about 50, considering the magnitude of the catastrophe. The government agencies had cleared most people from what was considered to be the area of greatest risk, but the most knowledgeable person of all, David Johnson, was one of the victims.

**FIGURE 5.12** North Fork Toutle River, June 30, 1980. Volcanic mud flow breccia and debris from the Mount St. Helens eruption clog the valley to depths of more than 100 m. Photography by Austin Post. Courtesy of U.S. Geological Survey.

### El Chichon, Mexico

The most recent devastating volcanic eruptions occurred on El Chichon, a late Pliocene stratovolcano located in the southeastern state of Chiapas, Mexico. Historically, this 1300 m mountain had been inactive, displaying only solfataric activity. Chiapas marks the juncture of the American, Cocos, and Caribbean plates, and volcanicity is attributed to subduction of the Cocos Plate beneath southeastern Mexico.

A series of tephra eruptions began on March 28, 1982, and culminated in the most massive eruption on April 4th. Throughout the eruptive phases, there were three types of activity. The early eruptions produced light-gray pumice and ash of andesitic composition of juvenile (or magma) origin. This was followed by pyroclastic flows and surges in addition to ash fallout composed of previously formed rock. There was no lava production. The total number of people killed by this activity was more than 187. In addition, 60,000 people were evacuated from the region, three nearby villages were buried by the ash, and many were left homeless.

Although the total amount of material ejected by El Chichon was less than for Mount St. Helens, the blast was vertical into the sky, rather than horizontal, resulting in much more entrainment of dust, ash, and sulfur dioxide into the atmosphere. The vast clouds that were produced were mapped and traced by the Nimbus 7 satellite. The cloud started out between 22 and 26 km in height and then unexpectedly spread upward to 32 km with the densest portion at 26 km. It circled the globe about every 10 days and gradually moved into more northern latitudes from its original zone of 0°-30°N. The cloud was 20 times larger than that from Mount St. Helens and the largest in the northern hemisphere since the 1912 eruption of Mt. Katmai, Alaska. Nearly all specialists predicted some cooling effects from the cloud but disagreed on the total amount or the lasting effects. Another controversial matter concerns the relationship of the eruption with the unusual coincidence of El Nino, the warming of equatorial Pacific surface waters that decimates the anchovy population of the east Pacific and fosters abnormal weather extremes on adjacent land. El Nino began developing in May instead of October as usual. The last out-of-phase El Nino appeared in 1963, shortly after the eruption of Agung volcano, another massive eruption, which temporarily caused climatic cooling.

## PREDICTION

The question has been asked, "If geologists are so smart, why can't they predict volcanic eruptions?" Indeed, it seems that if man can land on the moon, he should also be capable of understanding the timetable of volcanic activity. However, as with all earth processes, there are numerous factors that contribute to the final production of a volcanic event. Great strides have especially been made during the past two decades, and every new action provides increased information toward the ultimate goal of exact precision in volcanic prediction. What are the symptoms that herald the onslaught of a new catastrophe?

### Seismic Activity

Nearly all great eruptions are preceded by some type of seismic event. The cause of increased earthquakes stems from the movement of magma within the volcano and the rock adjustments that take place in response to such motion. The patterns, frequencies, and magnitudes of the quakes provide special signals that may determine the position of the magma. In 1978, scientists correctly determined the events at La Soufriere on St. Vincent in the Caribbean and predicted the ensuing volcanic activity on the basis of the associated seismicity (Fig. 5.13). An increase in earth tremors was first noted by mid-1978. On April 12, 1979, the first abnormally strong earthquake was recorded, and by nighttime, the two seismographs were saturated with continuous tremors from the mountain. The scientists gave warning to the government, which quickly ordered evacuation of 20,000 people. This was none too soon, because at 5:00 A.M. on April 13th, La Soufriere erupted and produced ash falls and a glowing avalanche. No lives were lost because of the warning and the effective evacuation.

### Hillside Deformation

Changes in the shape of the ground surface can be caused by the rising mass of magma as it swells the land surface. This change and increase in size can be determined by the use of tiltmeters placed on the flanks of the volcano. They record the new distortions, and monitoring of many volcanoes has shown that each has its own particular rhythm and timetable for an explosive or lava-forming event. The longest records have been made at Kilauea on Hawaii, where the tiltmeters have correctly provided data so that forecasts could be made about impending events. Although the Mount St. Helens tiltmeters indicated dramatic inflation of the northern slope, and the bulge was increasing at a rate of 0.5 m per day, the actual timing of the eruption still came as a surprise to the observers.

### Other Changes and Precursors

There are other potential techniques and observations that may yield clues to the impending time for an eruption. At Kilauea, gases emanating from fumaroles changed in composition several weeks prior to eruptive activity. The severe drop in the helium/carbon dioxide ratios was the most sensitive barometer of the new period of volcanic activity. Other changes that can possibly be monitored, which may indicate the increasing movement of magma, include increased geothermal heat as measured by infrared detectors, changes in the magnetic field, and variations in electrical currents. Even erratic animal behavior, as with earthquake sensing, may at times be a positive indicator of renewed activity.

FIGURE 5.13 Crater Lake of La Soufriere volcano, St. Vincent. Photograph taken by David Schwartz in July 1971 before volcanic activity later in the year and in 1979. In 1902, eruption of this volcano killed more than 2000 people.

### Historical Record

Although on a different time scale from other possible precursors, analysis of the volcanic history of the site may reveal significant information about the potential for future activity. Volcanoes that have a recent record of eruptions have always been classed as "active." However, long periods of dormancy do not necessarily indicate a "dead" volcano. Indeed, the volcano Arenal in Costa Rica had been considered dormant because of its 450 year period of inactivity. Yet in 1968, it violently erupted and killed 78 people. Similarly, Mt. Agung had been entirely quiet for more than 100 years, but erupted on February 19, 1963, with additional eruptions in May, killing a total of 161 and leaving 78,000 people homeless. Taal volcano in the Philippines violently erupted in 1965 killing more than 180 (Fig. 5.14). Again, its previous major eruption had been more than 200 years ago. The moral seems to be that volcanoes can be extremely dangerous, regardless of age or previous history.

## HAZARD MITIGATION

Volcanic eruptions cannot be prevented, but in some cases, certain measures can be taken to minimize losses. Of course, the best defense would be restriction of development in high-risk areas. This is virtually impossible in many cases because the richness of volcanic soils provides a food source for millions of people who inhabit such areas. The next best solution to save lives is an effective warning system so that the populace can be evacuated prior to explosive activity. Certain possessions that are transportable may be saved, but there is little help for permanent installations or fields.

FIGURE 5.14 Taal Volcano, Batangas Province, Philippines. This is a resurgent volcano in a collapsed caldera. An explosive phreatic eruption in 1965 blew away all of a former cone, which was rebuilt in 1966-67. Strombolean-type eruptions in 1968-69 formed lava flows, and the eruption of September 1976 blew out lava and cinder cones. These events killed more than 190 people. Photograph taken October 1976 by John Wolfe.

Whenever possible, mankind has attempted to control natural processes that threaten him. The first attempt to control damage from vulcanicity was during the 1669 eruption of Mt. Etna. Although an earthquake had just killed 20,000 people, a lava flow was streaming toward adjacent villages. It destroyed 14 towns in its relentless journey and was moving toward the village of Catania by the Mediterranean Sea. In an attempt to protect their homes, the citizens increased the height of the city wall. For a while, this was successful, but when the lava reached 20 m thick, the wall was overwhelmed. The molten rock invaded the city.

The city of Hilo on the island of Hawaii resides at a vulnerable site on the base of the giant volcanoes of that island. In 1935, a lava flow jeopardized the community, and noted volcanologist Dr. T. A. Jagger convinced the U.S. Army to bomb the flow in an attempt to divert it. The bombing was partly successful, and the flow moved to a more remote area. Also in 1942, another lava stream was bombed. The natural levees that build at the side of such flowing lava provide good targets for possible relief. So the levee was bombed, and the lava spilled through the breach to another site.

The most successful and massive experiment to stem lava flows was done on the Icelandic island of Heimay in 1973 (Fig. 5.15). Here, the town of Vestmannaeyjar with a population of nearly 6000 seemed doomed. Although the citizens had been given sufficient warning and had been safely evacuated, the seemingly relentless lava flows were moving in a direction that threatened to engulf the entire town. The fight against the volcano took two forms — a water-spraying operation and the construction of bulldozed barriers to divert the flows. The purpose of the water spray was to cool the lava so that it would lose its mobility and harden in place. The program was remarkably successful. Whereas before spraying, the flow had greater mobility, after spraying, it became more jagged because the cooling caused the more fluid interior to break upward and disrupt the surface materials, making them more jagged and difficult to move. Although some homes were destroyed, the village was essentially saved. More than 5.6 million $m^3$ of seawater had been sprayed by the end of operations, and the total control effort had cost $1.5 million. However, the port and harbor had also been saved along with the livelihood of the inhabitants who occupy the most important fishing village of the country.

The most recent attempt to reduce damage from volcanic activity occurred in the spring of 1983 at Mt. Etna, Europe's most active volcano. Although eruptions on the mountain in August 1979 were the strongest in 20 years, those in 1983, were even more severe, and moving lava threatened to engulf the villages of Ragalna, Niolosi, and Belpasso. During the early stages of the activity, more than 1.5 million $m^3$ of lava were surfacing per day on the southern face of Mt. Etna. Soon, it had destroyed more than

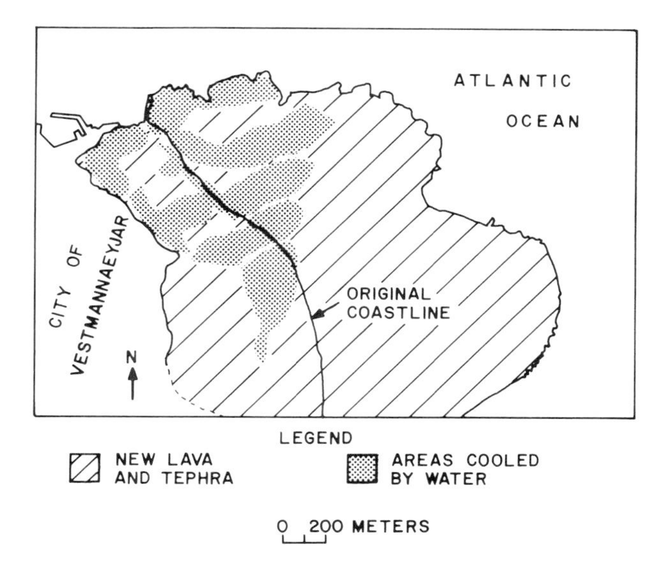


**FIGURE 5.15** Map of Heimaey in the Vestmann Islands, Iceland. This is the site where volcanic activity from Helgafell and Eldfell destroyed much property and threatened the very existence of Vestmannaeyjar, a city of 5000. An extensive project was launched to cool the lava by a giant pumping system. Apparently this effort was successful because most of the city and the harbor were saved from destruction. Courtesy of U.S. Geological Survey.

100 buildings, ruined 150 ha of orchards and farmland, burned forests, and caused millions of dollars of damage. Volcanology specialists from Italy devised a plan to force the lava to flow into a natural depression about 1.5 km above the base of the mountain. About 150 m above the depression, the lava was flowing in an old trough that prevented it from entering the depression. Workers using power shovels and bulldozers dug a new trench from the outer wall of lava in an attempt to divert it. When this failed, it was decided to dynamite the outer lava wall to force the molten material in the new direction. Fifty holes were drilled and emplaced with 8 cm diameter steel tubes that contained 400 kg of dynamite. Unfortunately, before the blasts could be detonated, not all tubes could be used because of the intense heat. The explosives were timed to produce three blasts. For a time, the main flow slowed, but only 20–30 percent of the lava was diverted. Most continued down the old path. However, within 10 days after the wall had been breached, the lava flow had completely stopped. Whether the efforts

to control the flow were instrumental in the halt is still a matter of conjecture.

## BENEFICIAL ASPECTS

To this point, only the bad features of volcanic activity have been mentioned. Unlike other geologic hazards, however, volcanic areas can become exceptionally beneficial to humans. Some of the most highly productive soils in the world form from volcanic materials in warm moist climates. In such terrains, rapid weathering helps provide mineral nutrients necessary for abundant plant growth. Thus, some of the highest agrarian population densities in the world occur where volcanic soils are farmed, and typical of such areas are the islands of Sumatra and Java. For example, Java, half the size of California, supports a population of 100 million people on its rich soil.

Volcanic areas produce a wide range of economic resources in addition to the legacy of rich soils. They can be the site of important rock and mineral products. Sulfur, alum, and boric acid can be exploited for such widely ranging uses as cement and pharmaceutical supplies. Metals such as arsenic, bismuth, and antimony occur in some volcanic regions. Bentonite, the weathered clay-like material composed of volcanic ash, finds wide usage in the drilling of oil wells. Hardened lava can be quarried as building stone or used as aggregate for highways. Pumice is prized as a source for abrasives.

When magma is near the surface, it may produce heat and be a source of geothermal energy. Under particular conditions, steam and gases may rise to the surface and form geyers, as in Yellowstone National Park, Iceland, and New Zealand. In other places, fumaroles occur, and such steam discharge can also be harnessed as a heat source.

Volcanic terranes and mountains are also prized from an aesthetic sense. Such mountains a Fujiyama, Mt. Mayon, and the peaks of the Cascades (Mt. Baker, Mt. Hood, Mt. Rainier, etc.) form the basis for recreation and the resort industry.

From the geological historical viewpoint, volcanic activity was crucial in providing to the earth a unique habitat for life. The degassing of molten materials provided water for the oceans and gases for the atmosphere — indeed, the very ingredients for life itself and its sustenance.

## POSTLUDE

Although study of geologic hazards has become an increasingly popular field for social science research, there is a dearth of data about hazard awareness and perception in dealing with volcanic activity. The Mount St. Helens disaster, however, afforded an excellent opportunity to obtain a strong data base on such topics as the range of steps taken to alert the public to the potential danger, and how such warnings were accommodated (Perry and Greene, 1983).

Moderate earthquakes were first detected under Mount St. Helens on March 20, 1980, and several ash and steam eruptions occurred starting on March 27. Intermittent activity occurred until the cataclysmic eruption on May 18. The first government response was on March 25 when residents who owned cabins near Spirit Lake were advised to evacuate. The U.S. Forest Service also closed the mountain above the timberline, and roadblocks were formed on several state highways and Forest Service roads. The Federal Aviation Administration banned planes within an 8 km radius of the mountain. On March 27, 80 residents on the Lewis River valley were asked to leave because of possible flooding, the Weyerhaeuser Company evacuated 300 loggers from its timbering operation, and 120 residents and workers were evacuated along the north and south forks of the Toutle River. When mudflows occurred on the south flank of the volcano on April 1, the Cowlitz County sheriff asked the governor of Washington to call the National Guard to help maintain the roadblocks. However, the 300 loggers were allowed to return to their jobs. By April 4, federal, state, and local agencies began preparations for plans to enforce if a major eruption occurred. The governor declared a state of emergency, roadblocks were strengthened, and sightseers were urged to stay away. The U.S. Geological Survey distributed pamphlets to the media and public officials that described what to do in an emergency. Although several additional steps were taken to alert residents to the seriousness of the threat, up until the time of the disaster, the actual timing of the blast took both scientists and officials by surprise. Unfortunately, 50 people died in the cataclysm, but much larger numbers would have lost their lives without the various types of governmental actions that were taken.

The assessment of planning during the Mount St. Helens activity and disaster reinforced the ideas of what is necessary for those officials who must manage such problems. For example, perceived personal risk determines a person's behavior. Thus, proper warnings should be in terms of (1) naming the disaster agent that will inflict damage, (2) determining the time of the disaster impact, (3) estimating the severity of the event, and (4) locating the specific area that will be most affected. Emergency managers must provide a variety of services that include organizing community meetings, serving as speakers for the media, submitting articles to local newspapers and magazines, and developing special pamphlets that describe the character of the hazard and the steps to be taken when danger becomes imminent.

CHAPTER 6

# EARTHQUAKES

In 1970, unusual seismic activity convinced scientists that a destructive earthquake might be in the making, so additional recording instruments were brought into the area to monitor earth changes. By June 1974, the pattern of seismicity started to show important changes. In addition, the land surface was undergoing some tilting, changes in water level in wells were occurring, electrical currents in the ground were being modified, and abnormal animal behavior was being detected. Such bizarre events reached a crescendo by winter so that in December 1974 scientists convinced the local government to warn people that a large earthquake could be expected. In January 1975, the populace was placed on alert for evacuation because the earthquake was imminent. There was a gradual increase in earth tremors in early February, but an ominous lull occurred on February 4th. This "calm before the storm" convinced scientists that the situation was now threatening, and a general evacuation of the area was ordered.

The above scenario might sound like a movie script, but it represents the actual chronology of events that occurred in Liaoning Province, China, and describes what happened until the fateful day of February 4, 1975. On that date, indeed, a major earthquake did occur, and because of the sufficient warning and evacuation, many thousands of lives were saved. The 7.3 M (Richter scale magnitude) earthquake caused severe damage throughout a 1000 $km^2$ area, and centered near highly populated Haicheng (90,000) where 90 percent of structures were destroyed. Thus this area, where one-half million people lived, underwent great destruction, but because of the well-planned evacuation procedures, there was little loss of life. Unfortunately, warning signals by both man and nature were inadequate when more than 655,000 people lost their lives in the destruction of the Chinese city of Tangshan. The art of earthquake prediction has come a long way in the past 10 years, but the level of precision for determining in advance all major earthquakes still needs improvement.

Throughout history, mankind has been fascinated by and fearful of earthquakes. Countless early theories were advanced to explain the powerful forces unleashed during these events. Natives on Bali and Borneo believed the buffalo supported Earth and that quakes resulted when the animal shifted its weight from one foot to the other. The lamas of Mongolia thought a frog held Earth, and whenever the frog moved its feet or head, an earthquake occurred at that spot. Other ancients called upon different animals to create the shaking, but by Grecian times, Aristotle (384–322 B.C.) was espousing more naturalistic theories. These ideas of earthquakes being produced by the escaping of unruly winds in subterranean caverns were still the prevailing theories into the 1700s. Since then, a variety of theories have been included, such as those that God was being wrathful because of evil-doers and those that related earthquakes to movement of rock masses within the earth.

## TERMINOLOGY

An **earthquake** is a vividly well-named phenomenon because it is a sudden shaking of the earth. Thus, earthquakes are catastrophic events, which provide evidence of the dynamic character of the earth's interior. The quick release of energy creates shock waves and elastic vibrations (seismic waves) that rebound through the earth in all directions (Fig. 6.1). Measure-

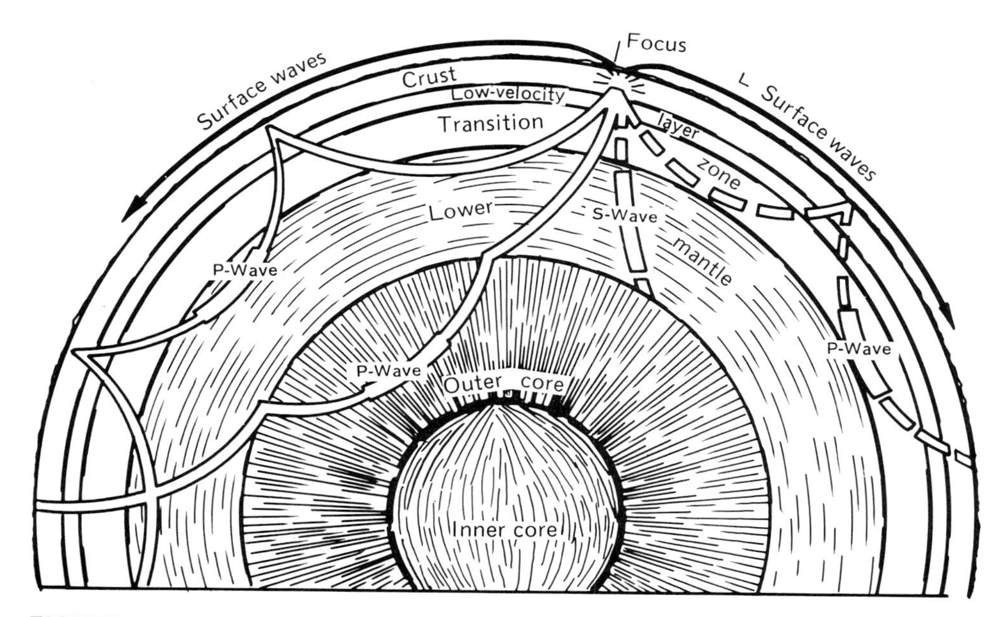


**FIGURE 6.1** Diagram of Earth's interior. The crust is composed of sedimentary, igneous, and metamorphic rocks. A low velocity layer, sometimes called the "asthenosphere" separates the crust from higher density materials in the transition and lower mantle zones. The core consists of exceptionally dense material in a molten condition, which prevents penetration by S waves. Courtesy of U.S. Geological Survey.

ment of seismic waves is accomplished by a **seismograph** (Fig. 6.2), and these are recorded on a **seismogram**. From study of such records, it has been determined that there are three types of seismic waves: (1) the "P" or primary waves are compressional type waves and reach the recording instruments first. (2) The "S" or secondary waves have a shear motion and are the next waves to arrive. (3) the "L" or surface waves travel along the earth's surface and arrive last, but because they have the strongest vibrations, they cause most of the damage.

Earthquakes are usually described in terms of their location, severity, and loss of lives and damages. The **focal depth** of an earthquake is the distance from the earth's surface to the focus, which is the position where an earthquake's energy originates. The epicenter is that point on the earth's surface directly above the focus. Its position is stated in terms of longitude and latitude. There are two scales in common use that provide measures for

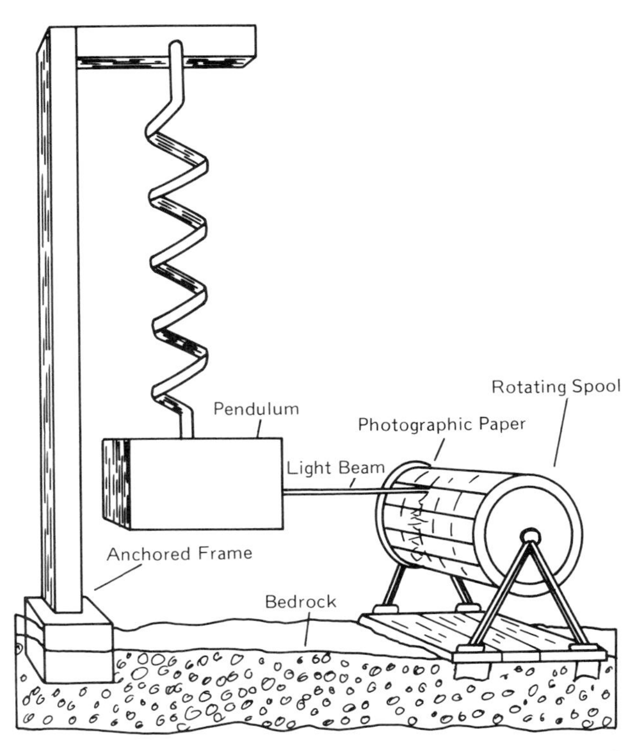


**FIGURE 6.2** Diagram that illustrates the principle of the seismograph. A rotating spool vibrates during an earthquake. A light beam from the motionless pendulum makes an impression on the recording spool. Courtesy of U.S. Geological Survey.

the severity of the earthquake. The Modified Mercalli Scale (1931) is a subjective determination of the intensity of the shock at a specific location. It is stated in Roman numerals (Table 6.1), the higher the number, the greater the "felt" event. The Richter Scale (1935) is a series of numbers that are logarithmic for measuring the magntiude of the earthquake. For example, a quake with a magnitude of 6 is 10 times more powerful than one with a magnitude of 5 (5 M). A recording of 6 M is considered a major earthquake, but it does not necessarily follow that high magnitude quakes always do the greatest damage. The amount of destruction is a function of the local geology and environmental setting of human occupation and the distance from the quake. An area underlain by soft sediments will undergo much more damage than if underlain by granite bedrock. Of course, the final denominator for measuring earthquakes is the loss of life, human suffering, and property damages and destruction.

## DISTRIBUTION

Earthquakes are not uniformly located throughout the world, but instead are concentrated in specific belts (Fig. 6.3). The majority occur in the same circum-Pacific belt as volcanoes, "the ring of fire." A second belt extends eastward through the Mediterranean region. Turkey, Iran, and such Asian mountain ranges as the Himalayas. Thus, most earthquakes are associated with the major topographic features of the earth — mountain ranges, ocean ridges, ocean trenches, and island arcs. Within the last 10 years, many scientists have shown the close relationship of earthquakes near the boundaries of lithospheric plates. Zones that are especially vulnerable are those where one plate, such as the oceanic plate is subducting under a continental plate (Fig. 5.4). However, even inland areas are not necessarily safe from major earthquakes. The largest earthquakes to hit the United States in historic times were positioned within the continental lithospheric plate. Centered near New Madrid, Mo., earthquakes in 1811-12 were felt throughout a 5 million $km^2$ area, and significant topographic changes were caused at distant places along the Mississippi River and the Reelfoot Lake area of Tennessee.

There is also an uneven occurrence of earthquakes within the earth. Foci are distributed in three general depth zones. Shallow quakes originate within 60 km of the surface, intermediate focus earthquakes occur at 60-300 km depths, and deep quakes are below 300 km. It is even possible for earthquakes to develop at depths of 700 km, but the great majority are of the shallow type. More than 1 million quakes occur each year, and although their frequency varies, they do not seem to happen at any particular season or with any type of weather.

**TABLE 6.1** Modified Mercalli Scale of Earthquake Intensity

| | |
|---|---|
| I. | Not felt except by a very few under especially favorable circumstances. |
| II. | Felt only by a few persons at rest, especially on upper floors of buildings. Delicately suspended objects may swing. |
| III. | Felt quite noticeably indoors, especially on upper floors, but many people do not recognize it as an earthquake. Standing motor cars may rock slightly. Vibration like passing truck. Duration estimated. |
| IV. | During the day felt indoors by many, outdoors by few. At night some awakened. Dishes, windows, doors disturbed; walls make creaking sound. Sensation like heavy truck striking building. Standing motor cars rock noticeably. |
| V. | Felt by nearly everyone; many awakened. Some dishes, windows, etc., broken; a few instances of cracked plaster; unstable objects overturned. Disturbances of trees, poles, and other tall objects sometimes noticed. Pendulum clocks may stop. |
| VI. | Felt by all; many frightened and run outdoors. Some heavy furniture moved; a few instances of fallen plaster or damaged chimneys. Damage slight. |
| VII. | Everybody runs outdoors. Damage negligible in buildings of good design and construction; slight to moderate in well-built ordinary structures; considerable in poorly built or badly designed structures; some chimneys broken. Noticed by persons driving motor cars. |
| VIII. | Damage slight in specially designed structures; considerable in ordinary substantial buildings, with partial collapse; great in poorly built structures. Panel walls thrown out of frame structures. Fall of chimneys, factory stacks, columns, movements, walls. Heavy furniture overturned. Sand and mud ejected in small amounts. Changes in well-water levels. Disturbs persons driving motor cars. |
| IX. | Damage considerble in specially designed structures; well-designed frame structures thrown out of plumb; great in substantial buildings, with partial collapse. Buildings shifted off foundations. Ground cracked conspicuously. Underground pipes broken. |
| X. | Some well-built wooden structures destroyed; most masonry and frame structures destroyed with foundations; ground badly cracked. Rails bent. Landslides considerable from river banks and steep slopes. Shifted sand and mud. Water splashed over banks. |
| XI. | Few, if any, masonry structures remain standing. Bridges destroyed. Broad fissures in ground. Underground pipelines completely out of service. Earth slumps and land slips in soft ground. Rails bent greatly. |
| XII. | Damage total. Waves seen on ground surfaces. Lines of sight and level distorted. Objects thrown upward into the air. |

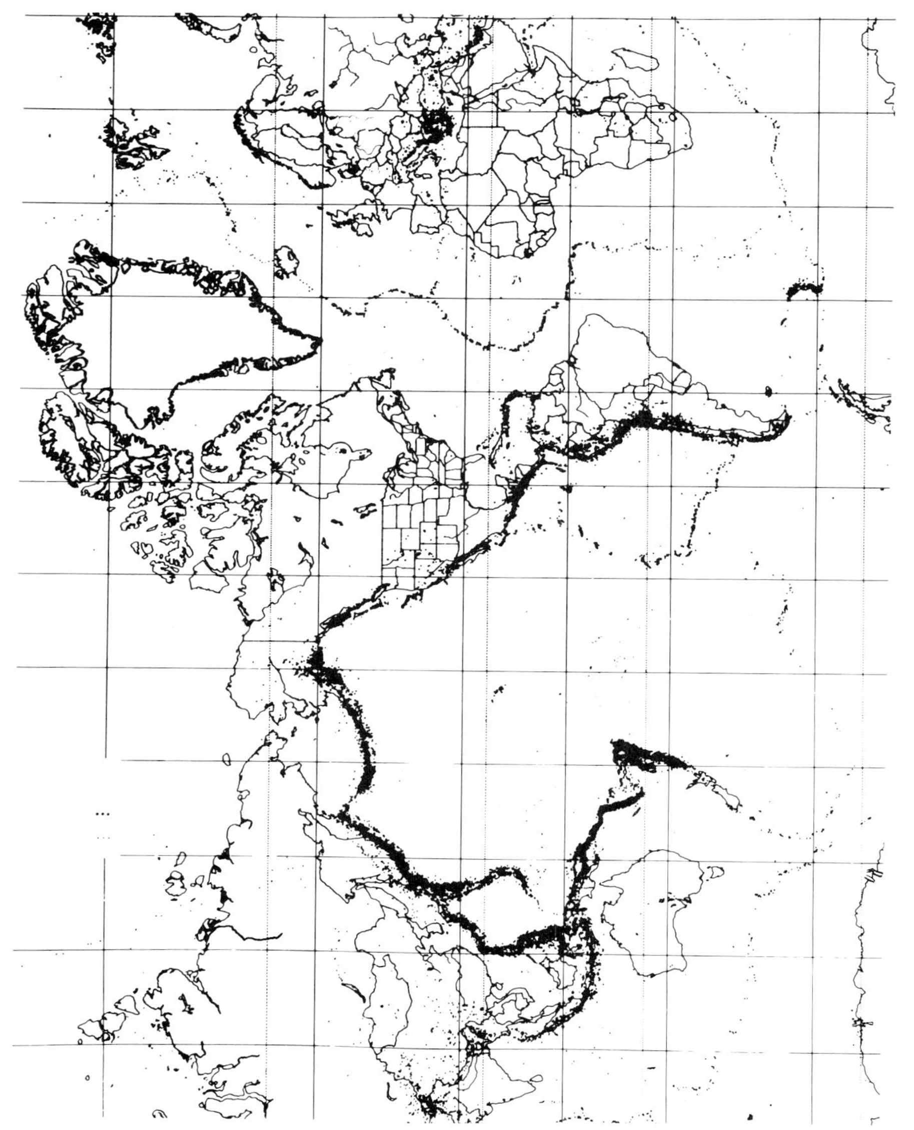

**FIGURE 6.3** World seismicity map prepared by the U.S. Geological Survey. This shows the location of 23,000 earthquakes. The black dots represent the earthquakes and demonstrates the principal occurrence in such zones as the "ring of fire," the Mid-Atlantic Ridge, and the Mediterranean region.

## CAUSES

Shifting of crustal segments of the earth along faults (fractures where one side is displaced relative to the other side) is responsible for most earthquakes. Other quakes originate during volcanic activity, major landslides, and even by human means. The **elastic rebound theory** proposed by H. F. Reid in 1908 has been a very popular and generally accepted model for earthquake development. He postulated that over long periods of time, strain energy increased along a zone of weakness in the earth's crust until the resisting strength of the rock was finally exceeded. A rupture developed, and along this fault, there was rock displacement. The snapping action of this movement produces a type of rebound reaction, which causes the seismic vibrations, thus ringing the rocks like a bell.

Earthquakes can also result as by-products from human endeavors — either inadvertent or deliberate. The engineer's ability to create huge artificial lakes can upset the balance of the underlying bedrock and generate moderate-size earthquakes. This was first discovered when the Marathon Dam in Greece was completed. By 1931, the reservoir had nearly filled and earthquakes were noted. In similar fashion, Hoover Dam's Lake Mead started producing earthquakes in 1936 in an area that previously had been aseismic. Perhaps the most notable destruction attributed to dams occurred from waters in the Koyna Reservoir of India. Here on December 10, 1967, a 6.3 M event killed nearly 200 people and left thousands homeless. More than 40 cases have now been substantiated where reservoirs have triggered earthquakes.

Another major way that mankind causes earthquakes is by injection of fluids into bedrock. To recover additional oil in petroleum wells, flooding techniques are now being used. For example, at the Rangely field, Colo., the Chevron Company started secondary oil recovery by water injection in 1957. In 1962, a new seismological station 65 km away immediately recorded small earthquakes from the Rangely locality. Experiments conducted at the site by the U.S. Geological Survey, with alternate injection and pumping of water, confirmed the relation that seismic activity was caused by lubrication of faults during injection periods.

A potentially more dangerous condition resulted from injection of toxic wastes into a 3671 m deep well near Denver, Colo. Starting in March 1962, the wastes were disposed of in a weathered and fractured group of granite and gneiss rocks. By April 1962, Denver, a seismic quiescent area, felt the first earthquakes. With documentation of linkage of the quakes with the injection program, this disposal method was halted in September 1965. However, there was a lag time of several years before the quakes stopped. During the 1962-67 period, more than 1500 tremors were produced, ranging from 0.7 to 4.3 M. Water injection has also caused earthquakes at Matsushiro, Japan, at the Snipe oil field in Canada, and from hydraulic salt mining at Dale, N.Y.

Seismic activity also occurs during blasting operations as in mining and road construction, and even heavy traffic is capable of yielding microseisms. Some of the largest seismic waves are produced by the detonation of nuclear devices underground, where magnitudes as high as 7.0 have been obtained. Minor earthquakes are deliberately created by geophysical prospecting methods that use the information obtained to determine rock properties and their potential for yielding natural resources. In this technique, a series of dynamite explosions are produced, which create seismic waves that travel through adjacent rock strata. A series of recorders measure the wave characteristics, and interpretation of these seismograms is then made to determine the mineral and oil resource possibilities.

## EFFECTS

There is a large range of earthquake intensity from those barely detected on the most sensitive seismographs to those that destroy life and property over large areas. Many of the results attributed to earthquakes are not caused by the vibrations, but instead, to their side effects. During an earthquake, the earth does not open and swallow people, houses, and buildings, and then close again. Most openings and ground fissures occur on a restricted scale, and it is only when landsliding is produced that there may be gaping holes in the disturbed terrain (Fig. 6.4).

The ground motion from the seismic waves may cause shaking and lurching, tilting of segments of the land, uplift or subsidence, offset along faults (Fig. 6.5), and development of new fracture and fold systems in rock masses. These phenomena can trigger subaerial and submarine landslides, mudflows, soil liquefaction (Fig. 6.6), and tsunamis. Strange lights in the sky sometimes appear at some earthquake sites (Fig. 6.7). These lights may either fill the sky or appear as distinct flashes and streaks. Although scientists do not agree on a mechanistic theory, the mysterious lights are in some way produced by earthquake activity. Sounds may also be associated with impending quakes and also during the major event. These noises are variable, but range from low rumbles or roars, to thunder-like jolts and gunshots, or even the tearing of cloth.

Strange, abnormal behavior of both wild and domesticated animals has now been verified from several localities, which subsequently had earthquakes. Prior to the Haicheng earthquake, hibernating snakes emerged from their holes and froze on roads. Other aberrant behavior included chickens

**FIGURE 6.4** Ground effects from the October 28, 1983, Idaho earthquake. These fissures are along the fault trace and resulted from displacement and a type of "lateral spread" gravity movement east of Highway 93 near Mackay, Idaho. Courtesy of Earl Olson (see also Figs. 1.2, 1.3, 6.5, and Olson, 1983).

**FIGURE 6.5** Effects of the October 28, 1983, Idaho earthquake (Olson, 1983). This is a double fault scarp north of Mackay, Idaho, at the west base of the Lost River Range. Courtesy of Earl Olson.

refusing to enter their coops, pigs rooting at their fences, cows breaking halters, extreme restlessness in goats, dogs, and cats, and rats wandering about as if in a drunken stupor. Unfavorable human response can result during or after the earthquake. People may become dizzy and nauseated and vomit during the event. A more lasting impact may occur with some who undergo psychological distress as happened with uninjured children, and even some adults, who experienced the San Fernando, Calif., quake of February 9, 1971.

## DAMAGES

Unlike other geologic hazards, earthquakes would rarely be lethal if it were not for the side effects that produce a chain reaction of destruction to buildings and installations, landsliding, tsunamis, and the like. Earthquakes

**FIGURE 6.6** Liquefaction of the Bootlegger Cove Clay near Anchorage, Alaska, produced by the disastrous Alaskan earthquake of March 27, 1964. Courtesy of Wallace Hansen and U.S. Geological Survey.

result in physical and psychological injury, loss of life, destruction of property, economic disruption and indirect losses, and ecological changes and damage. Most deaths and injuries are caused by collapsing structures, falling debris such as glass and bricks, inundation of communities by tsunamis; engulfing of structures by landslides, avalanches, and mudflows; and flooding from collapsed dams or levees. Much loss of life and damage in the early twentieth century, as in the San Francisco earthquake of 1906 and Tokyo earthquake of 1921, was caused by fires – many of which were started when wires broke and electrical circuits shorted out.

Earthquake losses depend on such factors as the magnitude of the quake, time of occurrence, stability of underlying earth materials, durability of manmade structures and services, and population density. For example, the Alaskan earthquake occurred on a date and at a time that greatly reduced loss of life. The quake of March 27, 1964, hit at 5:36 P.M. The date was a

**FIGURE 6.7** Photographs of earthquake lights taken by T. Kuribayashi, during the Matsushiro, Japan, earthquake swarm, which lasted from 1965 to 1967. Courtesy of U.S. Geological Survey.

religious holiday so schools were closed (Fig. 6.8). Furthermore, it was at a time of day that most workers were not in office buildings, and the fisheries were closed so there were few people in the waterfront areas. Such cities as Anchorage and Valdez that were especially hard-hit would have suffered many more casualties than the nearly 100 that were killed.

Damage generally occurs because earthquakes exert horizontal stresses against structural elements that are usually designed to only withstand vertical stress. In addition, most structures contain a variety of materials, and where they join, weakness in integrity occurs. Rigid elements are fractured and torn from their moorings. Buildings made of unreinforced masonry are especially vulnerable to damage.

The character of the geologic foundation is a crucial determinant as to how much damage will result from earthquakes. Unconsolidated materials such as alluvium, fill, and sediments transmit a greater severity of motion than bedrock. Thus, most of the structural damage in the Alaskan earthquake occurred in the Turnagain Heights area, which is underlain by the Bootlegger Cove Clay, a glaciomarine sediment. The San Francisco quake deformed particularly that part of the city where artificial landfill had been placed as foundation material.

Environmental and terrain changes can be widespread from earthquakes of high magnitude. The New Madrid earthquake changed land position as much as 6 m over thousands of kilometers. It drained swamps, altered segments of the course of the Mississippi River, and created new lakes. The Alaskan earthquake raised intertidal areas far above the reach of highest tides, causing immediate destruction of aquatic life. Surges of water scoured clam beds and deposited mud and debris on others, suffocating life. Estuaries were uplifted or submerged so they no longer were available as nesting grounds for waterfowl. Intertidal water changes also ruined spawning

**FIGURE 6.8** Destruction of Government Hill Elementary School, Anchorage, Alaska, by the earthquake of March 27, 1964. It was extremely fortunate school was not in session at the time of the quake. Courtesy of Wallace Hansen and U.S. Geological Survey.

grounds for salmon, and other aquatic life was killed by pressure changes within the water.

## DISASTERS

Throughout history, earthquakes have been a scourge to mankind, and their relentless fury has continued unabated in the twentieth century (Appendix C). Indeed, one of the greatest disasters in all recorded time occurred at Tangshan, China, on July 28, 1976, when 655,000 were killed, 779,000 injured, and that industrial city of 1.6 million people largely destroyed. The most completely documented early earthquake occurred near Lisbon, Portugal, November 1, 1755. Shocks from the quake were felt in many parts of the world, chandeliers rattled in the eastern United States, and buildings trembled throughout Europe. Although some of the 70,000 people who lost their lives in Lisbon were crushed by toppled buildings, most were drowned by the tsunami that washed inland in giant waves 5 m above normal high tide. So universal was the event that canals and lakes oscillated at such distant sites as Holland, Scotland, Sweden, and Switzerland.

Although the New Madrid earthquakes were the most widely felt ones in North America, earthquakes in the coastal areas of California and Alaska have caused the greatest destruction to lives and property. The San Andreas fault zone is a particularly hazardous corridor, which has produced the two most harmful earthquakes in the 48 contiguous states (Fig. 6.9). The San Francisco earthquake of April 18, 1906, killed 315 in that city and 385 people in other vicinities along the fault. Much of the city was left in ruins from a combination of fire and toppling of structures built on loose earth materials. Another devastating earthquake was produced by a subsidiary fault of the San Andreas system at San Fernando, Calif., on February 9, 1971 (Figs. 6.10, 6.11). Although all structures were broken in the immediate vicinity of the fault, the severe shaking that was produced caused 99 percent of the damage at sites some distance from the fault. The heaviest loss of life occurred at the Veterans Hospital, 47 out of the total 65 deaths, when patients were unable to move sufficiently rapidly to protect themselves (Fig. 6.11). The early hour of the quake, 6:01 A.M. was responsible for a low death rate, because few people were going to or were at work, and schools were not yet occupied. Damages far exceeded $500 million, but the result was enactment of still more stringent building codes throughout the state.

The Alaskan earthquake on March 27, 1964, was so powerful that it released twice the energy of the 1906 San Francisco one. The ground motion was so severe that tops of trees were snapped off. Because the epicenter was in ocean waters of Prince William Sound and rock rupture extended for

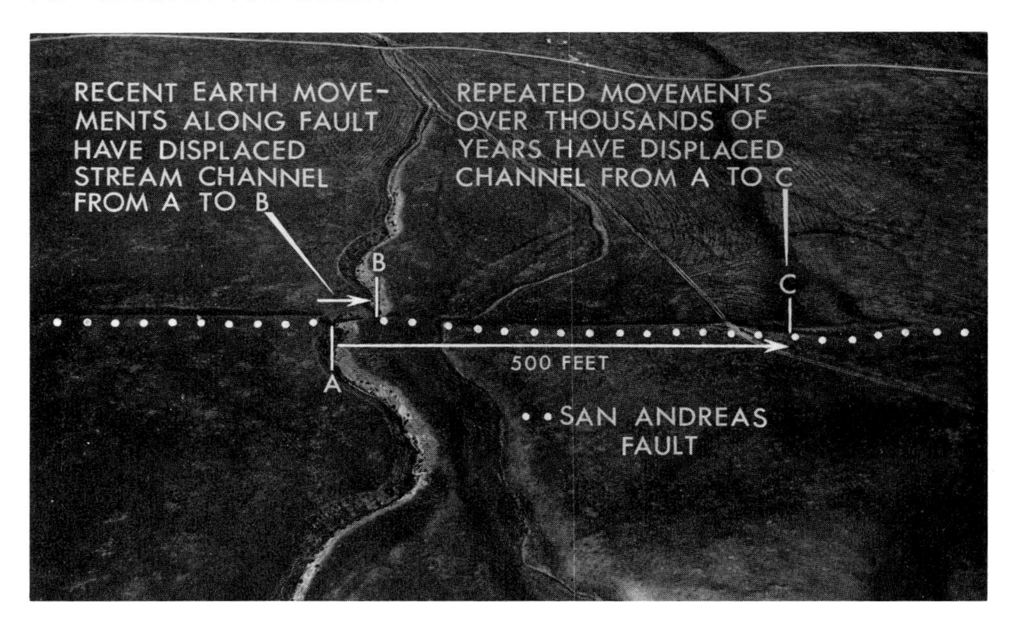


FIGURE 6.9 Aerial view of the Carrizo Plain area, Calif., showing displacements of stream channels caused by earth movements along the San Andreas Fault. Courtesy of U.S. Geological Survey.

800 km roughly paralleling the Aleutian trench, extraordinary changes in sea floor terrane produced a train of seismic sea waves that battered coastal communities. At Kodiak harbor, the first wave did little damage, but one hour later and again two hours later waves 10 m high devastated the harbor area. The resulting tsunami killed 16 people in Oregon and Washington and did damage as far south as San Diego. In Alaska, land was uplifted 15 m in some places, lowered 3 m in other places, and at other sites there were horizontal changes of several meters. Water levels in wells in parts of South Africa were even caused to fluctuate.

The two most recent major earthquakes in the United States occurred in California and Idaho. Coalinga, Calif., with a population of 7000 in the fertile San Joaquin Valley was abruptly hit May 2, 1983, by a 6.5 M earthquake that lasted 15 seconds. It demolished the central part of the town and business district, destroying 560 of the town's 2670 homes and causing damages of $31 million. Although 47 townspeople were injured, miraculously, none were killed. The town was hit with another quake on June 10, 1983, that registered 5.2 M. Although not much new damage was done, several roads were ruptured, rocks tumbled from hillsides, and a power transformer was toppled, which started a grass fire. The Coalinga earth-

**FIGURE 6.10** Road damage caused by the San Fernando, Calif., earthquake of February 9, 1971. Courtesy of Robert Wallace and U.S. Geological Survey.

quakes had no foreshocks, so no warnings were given. However, in retrospect, geophysicists discovered that for eight years, there had been clusters of small earthquakes that marked the edges of the eventual rupture. This first recognition of the classic doughnut pattern of precursory seismicity in California may aid future forecasting in some earthquake-prone areas.

On October 28, 1983, a 6.9 M earthquake hit Idaho, its epicenter was near Challis. This was the strongest quake to hit Idaho since it was settled, and it had the highest Richter magnitude in the West since 1959. It cut a swath of destruction through the region and caused the death of two children (Figs. 1.2, 1.3). Buildings were toppled, boulders were sent crashing into homes and roads, and facilities were greatly damaged. A 3.3 m high fault scarp was formed that could be traced for 37 km from near Borah Peak through piedmont slopes and alluvial fans (Fig. 6.5).

During the past 100 years, 13 of Italy's major earthquakes have occurred on the west coast between Naples and Cosenza. The one in 1980 was the

FIGURE 6.11 Intense structural damage of Olive View Hospital caused by the San Fernando, Calif., earthquake of February 9, 1971. Courtesy of Robert Wallace and U.S. Geological Survey.

most devastating in 65 years, measuring 6.8 M with epicenter at Eboli near Salerno. The reign of destruction raged throughout the regions of Campania and Basilicata, a rugged belt between the Apennines and the Tyrrhenian Sea on the ankle of the Italian boot. Scores of communities were leveled, with a total of 179 villages suffering damage (Figs. 6.12, 6.13). In Naples, 15 percent of the buildings suffered some damage, and the total region affected was more than 26,000 km$^2$. Conza della Campania lost 80 percent of its inhabitants. Throughout the region, more than 3000 people were killed, 310,000 were left homeless, and total damage was in the billions of dollars. The Italian government, somewhat belatedly, approved a $1.5 billion package of aid for the striken communities and people, but the long-range costs will total more than $8 billion.

Other major earthquake calamities in the early 1980s include:

**FIGURE 6.12** Salvitelle, Campania Region, southern Italy. Intensive damage caused by the November 23, 1980, earthquake (Richter magnitude 6.8, Intensity IX). Eleven people died at this site from collapse of low-strength masonry buildings. Courtesy of David Alexander.

1. Al Asnam, Algeria. On October 10, 1980, a 7.5 M event destroyed 80 percent of the city, killed 3000 people, and made destitute 325,000 in the affected region. This was the site of an earlier earthquake in 1954 that also destroyed the city, then known as Orleansville. That quake killed 1600 and laid waste other villages in a 40 km radius.

2. Turkey. An earthquake of 7.1 M along the East Anatolian Fault on October 30, 1983, killed 1339 people and 30,000 livestock, and left 33,000 villagers homeless. The high death rate in this mountainous area occurred because peasant homes are made of rough stones piled together with soil roofs 1.3 m thick, which easily topple and bury inhabitants during earth tremors. In 1939, more than 20,000 perished in the same region.

3. Iran. In the province of Kerman, two earthquakes on June 11 and July 28, 1981, killed 4000 people. The largest quake measured 7.0 M.

4. North Yemen. The earthquake of December 13, 1982, although only of 6.0 M, killed 1340 people, injured 3000, left homeless 400,000, and destroyed 136 villages.

**FIGURE 6.13** Potenza City, Italy, after the November 23, 1980, earthquake. Buildings are temporarily shored to prevent further damage, until entire site can be cleared and restored. Courtesy of David Alexander.

## PREDICTION AND AMELIORATION OF EARTHQUAKES

Earthquake prediction can only be accomplished when most phases of earthquake activity are understood. Obviously no progress could be made as long as mankind believed earthquakes originated from the whimsical movement of some imaginary animal or resulted from a vengeful and angered deity. Although the Chinese and Japanese have a long history of attempting to foretell earthquakes, only within the past two decades have geophysicists finally been able to derive some promising clues that may aid in partially solving such a long and elusive riddle as the prediction of earthquakes.

Prior to the 1960s, the established methods for identifying areas of high seismic risk depended largely on studies of the past performance and occurrence of earthquakes and the mapping of faulted terrane (Fig. 6.14). Such seismic-risk maps, which show the relative vulnerability of zones to earthquakes and the types of damages to be expected, do not provide significant information as to earthquake timetables. For example, it had been realized that sites where past earthquakes had occurred, such as along the San Andreas Fault, could likely also be the site for new seismic activity. Movement or tilting of the land surface, as at Niigata, Japan, had also been recognized as an ominous harbinger of quakes. In the United States, the 1964 Alaskan earthquake brought attention to the importance of devising new methods for early earthquake detection, but only after the San Fernando earthquake of 1971 did the U.S. Geological Survey finally mount a major Earthquake Hazards Program with the express purpose of developing earthquake prediction capability. This has become a search to determine the reliability of precursory events for the prediction of the location, time, and severity of major seismic disturbances.

A significant breakthrough in recognizing earthquake behavior was finally reached in the late 1960s when scientists of the Soviet Union reported differences in seismic-wave velocities prior to some earthquakes, contrary to prevailing opinion that such motion was constant. This discovery was reached in the Garm area, a region of high seismicity, and one which the Soviets had been monitoring for 25 years. Further documentation of precursory events prior to earthquakes had also been collected, which showed (1) electrical resistivity changes in earth materials, (2) changes in seismic activity and migration and reorientation of rock stress, and (3) radon gas increase in well water. Scientists now believe that all of these observations can be explained by a phenomenon called **dilatancy**. This occurs when stress builds up, and rock near the fracture zone develops small cracks that produce an expansion in rock volume. Triggering of all these phenomena is therefore preparatory to the final major fracture along the fault, which then releases the principal seismic shocks that produce the earthquake. Verifica-

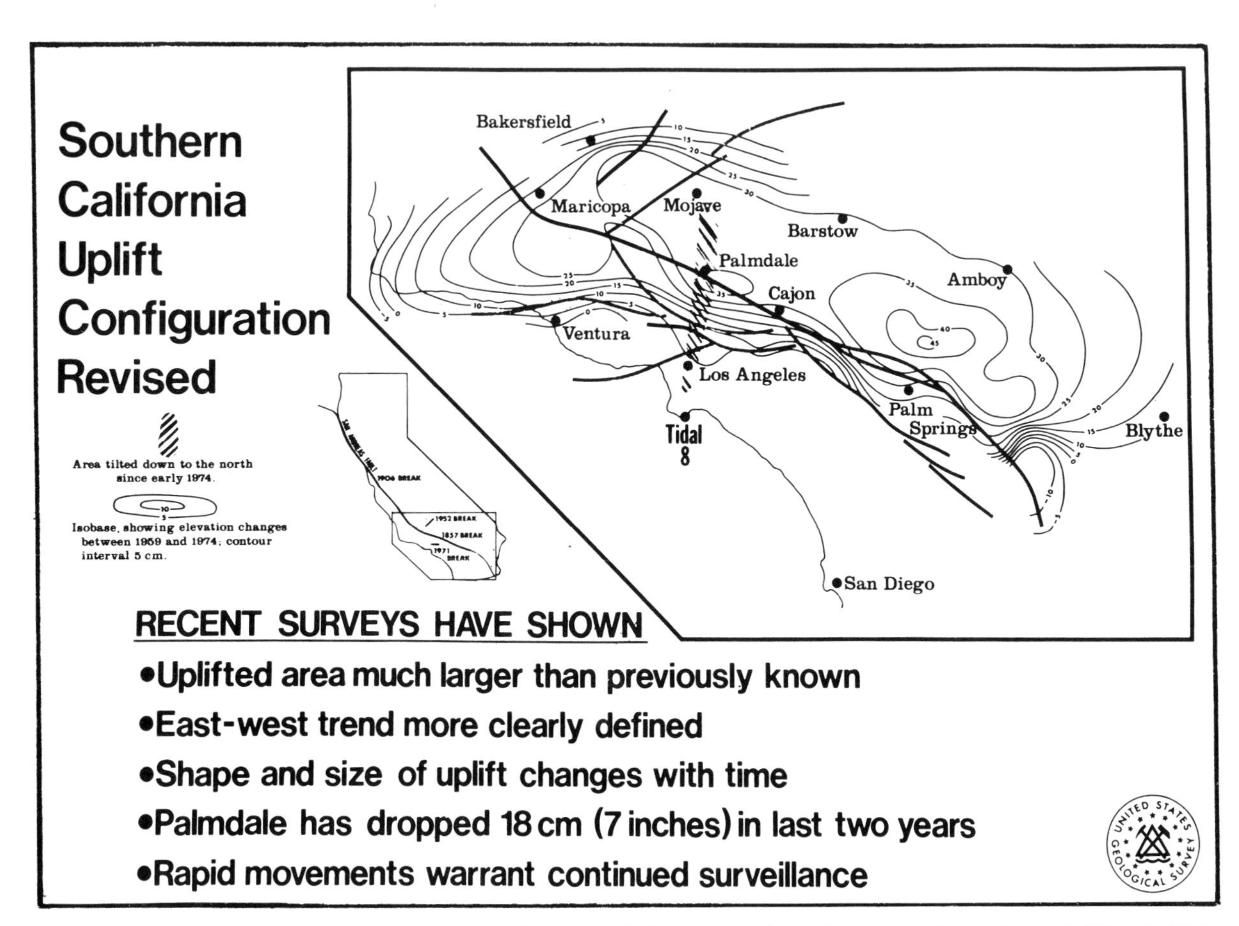


FIGURE 6.14 Map of faults and associated features of the San Andreas fault zone in southern California. U.S. Geological Survey.

tion of these findings was forthcoming, in hindsight, when it was later shown that the San Fernando earthquake, indeed, had velocity anomalies in seismic waves prior to the major earthquake. In addition, a small earthquake was accurately predicted at Blue Mountain Lake in the Adirondack Mountains in 1973, when velocity anomalies occurred. Also, a resistivity anomaly was observed prior to a 3.9 M earthquake in central California in 1973, and a magnetic anomaly was created before the Hollister, Calif., 5.2 M earthquake in 1974. Both quakes were roughly predicted because of such precursory events. Of course, as indicated by the opening comments of the chapter, a crowning achievement was reached with the prediction of the Haicheng earthquake in 1975. This event provided the most thorough case history of aberrant animal behavior.

At about the same time as the development of ideas that culminated in the theory of dilatancy and the range of precursors caused by this phenomenon, another idea was uniting the observations of earthquake frequency and plate tectonics – the **seismic gap theory**. According to the dynamics of plate tectonics, the earth's outer shell, including the crust and the upper mantle, is composed of large semi-rigid plates that are in constant slow motion. Most of the world's earthquakes occur where the plates slide against each other, or where one slides over another along a dipping subduction zone. The term seismic gap refers to regions along these active plate boundaries where (1) large earthquakes have occurred in the past but not within 30 years, and (2) nearby parts of plate boundaries have either released strain energy within the past 30 years or have no history of great quakes. Thus, a seismic gap is a zone where there is likelihood of stress buildup for many years at a site that is known to have large earthquakes (Fig. 6.15). More than 10 gaps have been identified where the last great earthquakes occurred more than 100 years ago, and other gaps are known where the quakes happened more than 30 years ago.

Perhaps the earliest use of the gap theory was in 1965 when S. Fodotov of the Soviet Union predicted earthquake activity for some of the seismic gaps along the Kuril Island chain. Within eight years of these forecasts, large earthquakes had ruptured five of the seismic gaps. The most spectacular success of the seismic gap theory was achieved when it was used to predict Oaxaca, Mexico, earthquakes. Recognition that this was a seismic gap area was first made in 1973. This awareness prompted the installation in 1978 of a local seismic network to record the patterns of low-level seismic activity that might presage a large quake. Although a 7.5 M event had been forecast for the area, the use of sensitive seismographs was necessary to narrow the timing of a possible quake. The instruments showed that even micro-earthquake activity was greatly subdued within a 100 km zone centered around the gap. This low level activity remained until about a day before the prin-

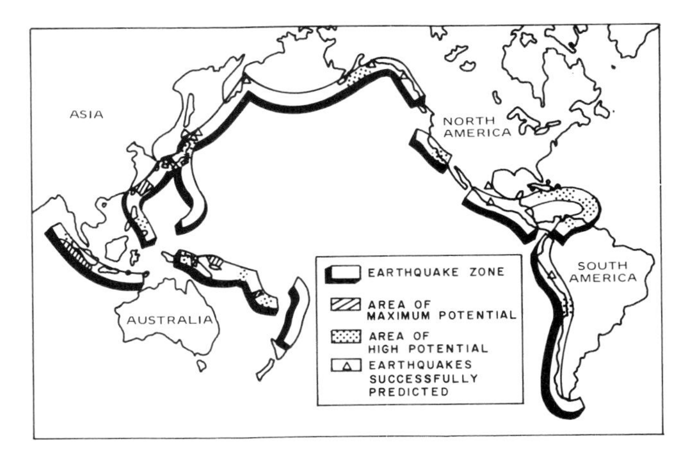


**FIGURE 6.15** Map of Pacific area showing principal belts of earthquake activity and location of breaks in such continuity, which are called "seismic gaps."

cipal earthquake. At that time, a center of activity with 3.6 M developed at one edge of the gap and slowly migrated during the next 12 hours to the final location of the large event. When the quake struck 18 hours later, the magnitude was indeed 7.5–7.8 as forecast. In this case, the precursory behavior of seismicity could only be detected by the very sensitive seismic network that had been installed. Such networks are costly and can generally only be installed at those localities where there is a good chance of prediction at sites where man has important developments and population centers. However, when all of the types of possible precursors are used in a systematic way, it brings closer to reality that final goal, of scientists, which is the prior detection and prediction of severe earthquakes to allow time for appropriate safety measures to be implemented.

The realization that humans create earthquakes by injecting fluids into rock, filling reservoirs, and exploding nuclear devices provides a possible means for control of some shallow seismic activity. The aim of earthquake reduction is to release energy by physical means and in small increments so as to initiate only minor quakes instead of allowing buildup of stress that would produce a major event. Some argue this is possible under the proper conditions whereby controlled injection of fluids will sufficiently alter pore pressure to allow minor slippage along known faults. However, such a technique is fraught with fearsome consequences, both legal and physical. For example, if the geotechnologist miscalculated and thereby caused a larger damaging event than predicted, who would be liable, how would damages be recovered, and what would be the long-range consequences in loss of confidence by the general public to the earth science fraternity? However, if progress in knowledge of earthquakes continues at the same pace as in the last 20 years, it might not be too long before mankind has a much better chance to save lives than we did in the past thousand years when more than 3 million people were killed by earthquakes (Appendix C).

Of course, one of the biggest questions facing environmental managers in California is, when and where is the next major earthquake going to occur along the San Andreas Fault. In an attempt to provide an answer, Raleigh and co-workers (1982) have approached the problem in terms of the seismic gap model. They concluded that a mature seismic gap now occurs along the fault in southern California. From San Bernardino north to the Carrizo Plain, the long-term displacement is 25–45 mm per year. They state, "...great earthquakes yielding 4 m or more in displacement are now possible given the 125 years since the last great earthquake (here) in 1857." The prime candidate for a great earthquake is the region from Cajon Pass near San Bernardino south to the Salton Sea. The event has an annual probability of about 13 percent per year.

Within the last few years, a range of other types of prediction criteria for earthquake activity have been developed.

1. Kiyoo Mogi (1979) of Tokyo University first drew attention to the geometry of 5 M earthquakes that served as his basis for the 7.2 M Songpan, Japan, earthquake of 1976. He predicted this eight months in advance, and this method is referred to as the "Mogi doughnut," whereby the earthquakes ring the area most susceptible for quake activity.

2. Wakita and his co-workers (1980) have studied the Yamasaki Fault zone in Japan, paying special attention to volatile gases of the region. This zone has produced numerous earthquakes with magnitudes over 4.0. They found excessive concentrations of hydrogen along the fault zone, which they interpret as a degassing phenomenon of earth materials caught in the shear motions along the fault.

3. Sugisaki (1981) analyzed variation of the helium/argon ratio of gas bubbles in a mineral spring along a fault zone near Mizunami, Japan. He found that variations in the ratio coincided with fluctuations of areal dilation induced by earth tides. Higher ratios occur when they are squeezed out by stress preceding an earthquake, so he concludes, "Continuous observation of gas quality at a location geochemically sensitive to stresses at depth could therefore be meaningful for earthquake prediction."

## POSTLUDE

Earthquakes and volcanoes have several elements in common: they provide the most awesome verification of nature's power; they are unevenly distributed throughout the world; they commonly can occur in the same region; and they constitute the most feared events on the planet. Because of the continuing tragedies they cause, scientists are spending more time and money in attempts to solve their riddles and are especially engaged in research aimed at providing greater precision in predicting their occurrence. In some regions, however, we may be running out of time. For example, the San Andreas Fault, which has been referred to as "the smoking gun," could create massive quakes anywhere along its 1000 km trajectory in the United States. However, the degree of preparedness by most communities is slight to nonexistent. The art and technology of earthquake engineering has come a long way in the past several decades, but the great majority of structures will still fail with only minor shaking. A NOAA (National Oceanographic and Atmospheric Agency) study of Los Angeles and Orange counties showed thousands would be killed by a moderate earthquake. Disruption of public services would be widespread, and pipelines, rail lines, and aqueducts would be severed. The shaking would break water mains at 1200 places, and gas lines at 1500 places. The quake would knock out 50 percent of electrical transmission lines and 50 percent of sewage and pumping systems. Landslides would block mountain roads, and many city streets would be impassable due to fallen debris. The Federal Emergency Management Agency placed a $17 billion price tag on such a disaster, which does not include damage to transportation or communication systems, dams, and military bases. Indeed, such a scenario should prove sobering to even the most faithful optimist.

The science of earthquakes is so complex that it requires data from seismology, geology, soil mechanics, geophysics, hydrology, and engineering. If hazards are to be reduced, earth-science data must be translated from technical language to information readily understandable by environmental managers. One of the best guidelines for this type of work has been that performed by the U.S. Geological Survey in the San Francisco Bay area and its adjacent nine counties. Such data are now available that should be translated into action by the appropriate governmental and societal agencies. Only through such cooperation can communities learn to deal with hazard problems and their mitigation.

CHAPTER 7

# LANDSLIDES

Some of the worst disasters in human history have been caused by landslides. Entire cities have been obliterated as happened in 1970 in Peru where 21,000 people were buried in the ruins of the Huascaran area. Even greater loss of life resulted from landslides at Kansu, China, in 1920 where 200,000 were killed. Both these events were initiated by earthquakes that jarred unstable slopes and released deadly cascading earth materials. Thus, landslides pose severe hazards in mountainous and hilly terrain, but they can even occur in more gently sloping terrain as we shall soon learn.

The incidence of landslides has been increasing throughout the twentieth century and has become part of a double jeopardy syndrome. With expanding urbanization, more construction enterprises and highways, and because of restrictions on use of floodplains, many building sites are now forced to occupy sloping terrain. Thus, properties that may have been subjected to flooding, if built on low ground, can face landsliding dangers when steeper slopes are used. It is a "no win" situation. Much of the most suitable land has already been developed in choice areas, so building costs are greater when occupancy becomes necessary on inferior ground. Damages in the United States caused by landslides, when coupled with the cost of measures to control them, exceed $1.5 billion annually.

The term **landslide** is used to describe both the process of rapid downslope movement of earth materials, as well as the topographic landform that results from such action. Landslides constitute the fast-motion end of gravity movements whereas creep processes comprise very slow displacement of surface materials. The terms **mass movement** and **mass wasting** used by some people are synonymous with **gravity movements**. Furthermore, the term landslide is really more inclusive than the name implies because it has traditionally been used to cover falling and flowing types of earth materials, in addition to those moving by sliding. However, landslides differ from another family of gravity movements, those referred to as subsidence, collapse, and settlement. Such terms denote vertical change in the ground surface, whereas landslides are a down and out motion.

## EFFECTS

Landslides produce both primary and secondary impacts. Those changes and damages caused by the earth materials are primary, and secondary effects may occur when the materials charge into water. For example, a 1964 landslide at Lituya Bay, Alaska, sent a massive wave across the bay that reached heights of 510 m on opposite hillslopes! The landslide that plummeted into the Vaiont reservoir, Italy, in 1963 was responsible for creating a 100 m high wave that overtopped the dam and killed 2200 inhabitants in the valley below (Figs. 7.1, 7.2).

Landslides not only destroy property and lives, but landslide-prone terrane can also be a financial burden to communities and people in other

**FIGURE 7.1** Shear plane of the Vaiont Reservoir rock avalanche located on slopes of Monte Toc in the Piare Valley, Italian Alps. Lithology is limestone with clayey and marly interbedding. It slid *en bloc* as a flexible sheet, up to 30 m/sec. Courtesy of David Alexander.

FIGURE 7.2 At the time of the rock avalanche, the Vaiont Dam was the world's largest double-arched reinforced concrete dam. It remained intact in spite of the 260 million $m^3$ of rock that avalanched into the reservoir and the creation of a 100 m high wave. Courtesy of David Alexander.

ways (Appendix D). Such lands increase the tax base for other lands, cause higher insurance rates and utility rates, and lower property values. From a geomorphic viewpoint, landslides and associated phenomena may produce major landform changes in a region. In 1935, New Guinea landslides denuded 130 $km^2$ of tropical rain forest. Studies show that about 12 percent of lands in the adjacent regions are denuded by landslides every century, compared with weathering and stream-wash processes that denude only 3 percent per century. A 54 $km^2$ region in Panama was greatly changed by landslide processes in 1976, and there are many other areas in Central America where landsliding is the dominant erosive force on hillslopes.

Although landslides are usually conceived as detrimental to society, occasionally the landform may become useful. For example, Bonneville Dam on the Columbia River, Farmers Union Dam on the Rio Grande, and the Cheakamus Dam, British Columbia, are all situated on ancient landslide deposits.

## LANDSLIDES AND RELATED PROCESSES

In this chapter, we will include a wide range of gravity movement phenomena, regardless of their character or constituents, with the exception of running water. However, some of the topics will be further addressed in other chapters. If gravity movements are considered to include all those earth movements that involve dislodgement or migration of materials to lower elevations, then both slow and fast processes are involved, vertical motion is included, as are various types of material from soil and regolith to rock and snow. Streams and glaciers are the exception in this listing.

The process of creep is nearly a universal force operating on hillslopes to move materials from higher to lower elevations, so quantitatively, it is very important in the total erosion–denudation process (Fig. 7.3). Solifluction, or soil movement due to very high moisture content, is especially large in arctic regions, whereas mudflows can become a major method for transportation of materials in arid regions.

### Loss of Vertical Support

When there is a loss of support for surface materials and rocks, they readjust by gravity and cause lowering of the land surface. If the rate of lowering is rather slow and affects large areas, the process is usually called **subsidence** (Appendix E). However, for more rapid to instantaneous motion and for smaller features, the term **collapse** is used. **Settlement** is produced by buildings whose weight causes them to gradually sink into the ground (Fig. 11.20). The Leaning Tower of Pisa is a good example of differential settlement. Many homes exhibit evidence of settlement such as wall-floor separation, cracked walls and broken mortar, and windows and doors out of plumb.

FIGURE 7.3 Displacement of concrete walls by earth creep south of Hollister, Calif. Courtesy of Robert Wallace and U.S. Geological Survey.

Natural subsidence and collapse is most commonly developed where soluble rock units underlie shallow surface materials. Circulating acidic groundwater is very effective in dissolving great volumes of carbonate (limestone and dolomite) and rock salt. Collapse failure, which produces most sinkholes, creates roughly circular depressions and is common in parts of Florida, southern Indiana, central Kentucky, eastern Tennessee, central Missouri, southwest Georgia, central Alabama, and south-central Kansas. Construction of the Upstate Medical Center in Syracuse, N.Y., in 1962 encountered problems of collapse and ground displacement that resulted from solution of underlying salt beds. Expensive cement-caisson construction was required to insure a safe foundation.

## CLASSIFICATION

Because landslides cover a broad range of earth movements that include falling, flowing, and sliding, schemes that attempt to provide an integrative classification may seem complex. The task is compounded when considerations involving the type of earth material, its water content, and the rate of motion become other components of the categorization. Figure 7.4 provides one approach for such groups. **Falls** are abrupt free-fall displacement of earth materials from very steep slopes and cliffs. **Slides** constitute those gravity movements where there is a well-developed shear surface under the ruptured mass, and the material moves downslope with some degree of coherence. Two types of slide motion can be singled out. **Slump** involves a rotational motion whereas a **planar** or **translational** slide involves displaced materials that move roughly parallel to the sheared surface (Figs. 7.5, 7.6). **Flows** involve movements in which the displaced materials behave like a viscous mass. The mechanics of flow is a controversial topic, especially for rock avalanches — the largest of all landslides (Fig. 7.7). Their speed poses a problem because they move more than 100 km/hr. This has caused some specialists to believe they move on a cushion of trapped air, whereas others argue their extraordinary mobility stems from buoyancy produced by some type of fluidization phenomenon within the displaced mass.

| | Type of Materials | | |
|---|---|---|---|
| Type of Landslide Movement | Rock and Bedrock | Coarse Sediment | Fine Material |
| Fall | Rockfall | Debris fall | Earth fall |
| Slide: Rotational / Translational | Rockslide<br>Block slide | Debris slide | Earth slide |
| Flow | Rock avalanche | Debris avalanche<br>Debris flow | Earthflow<br>Mudflow |

(increasing moisture ↓)

**FIGURE 7.4** Classification of landslide, largely based on type of movement and materials.

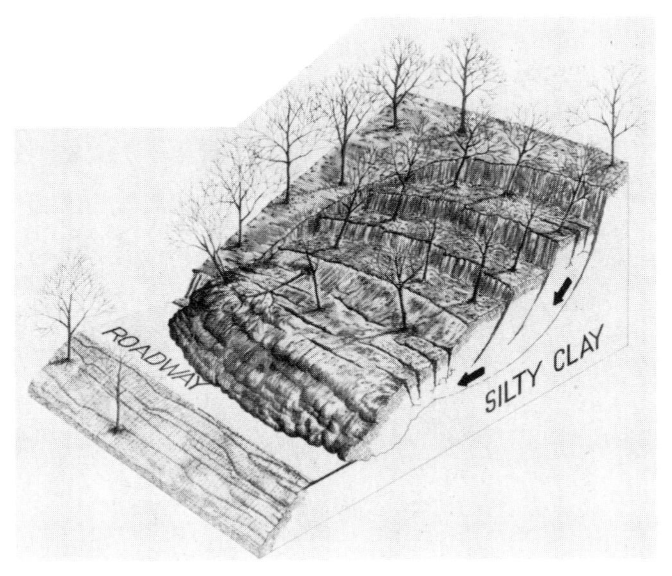


**FIGURE 7.5** Rotational type of landslide. Drawing by David Royster of landslide in Moore County, Tenn. (Royster, 1979).

Another difficulty in attempting a landslide classification is that in many instances the gravity movement may change character so that there is one type of motion in the upper part, which changes to a different type of motion in downslope parts. For example, some sediment may begin as a **sediment slump** but end as an **earth flow**. Or bedrock displacement may start as a **rockslide** but may turn into a **rock avalanche**. **Snow avalanches** may also incorporate rock material and possess compound motions.

## NATURAL CAUSES

For landsliding to occur, a threshold must be reached in the earth materials to initiate their sudden failure. This happens when the forces of earth and slope stability are exceeded by the driving forces seeking to produce change. The character of materials are crucial in determining the possibility for a landslide event. Broken rock is more easily displaced than solid rock. Thus, the number and type of fractures and rock discontinuities, and rock composition and degree of lithification help determine their stability. Other factors that are important are the steepness of hillslopes, type and amount of vegetation, and moisture content of materials. All of these qualities are controlling factors for susceptibility to failure.

**FIGURE 7.6** Spanish Fork landslide near Thistle, Utah. Brabb and Fleming (1983) called this the single most costly landslide in U.S. history. The following description has been provided by Earl Olson, U.S. Forest Service, Ogden, Utah:

Located on west side of Spanish Fork Canyon, T.9S., R.4E., Section 29.

*Ancient format:*

| | | |
|---|---|---|
| Head elevation about | 6200 ft | 1890 m |
| Toe elevation about | 5040 ft | 1536 m |
| Elevation difference | 1160 ft | 354 m |
| Slope pre-1983 movement | 11.7° | |
| Area | 135 acres | 54.6 h |
| Original terminal depth | 120-130 ft | 36-39 m |
| Estimated volume | 12-15 million $yd^3$ | |
| | 9.2-11.5 million $m^3$ | |

Multiple flowslides in existing drainageway called accumulate flowslide. Late Pleistocene in age. Mobilized North Horn formation (Upper Cretaceous and Paleocene). There have been several dozen Holocene flow format landslides on the same track. There are several dozen landslides in the Wasatch Plateau that have mobilized materials from the North Horn formation.

*Superlatives:* Secondary creep the greater portion of this century. Fore-toe deformation deformed Highways 89, 50, and 6, and railroad. Fifth largest landslide in coterminous United States this century. Probably second most costly because of blockage of highway

**FIGURE 7.6 (Continued)**

and railroad (estimated million or more dollars per day were lost by coal industry). May have had effect on international balance of payments? No one hurt. Town of Thistle, Utah, destroyed by lake water that developed from the blockage of canyon streams.

*Present format:* Translational slide. A swarm of flow format landslides at head. Landslide movement noticed at railroad track Tuesday, April 12, 1983. Wednesday, 8 ft (2.4 m) of lift at tracks. Thursday night (April 14), road closed. Terminal lift of mass Monday, April 18, 6:15 P.M., 2 ft (.61 m) per hour. Notice, over 85 percent of driving force is above the terminal buttress.

FIGURE 7.7 Turtle Mountain landslide, Frank, Alberta, Canada. This rock avalanche on April 29, 1930, displaced 36.5 million $m^3$ of rock from the mountain face and slid 3.2 km into the valley in 100 seconds, killing 66 residents. Courtesy of Canadian Geological Survey.

Although a hillslope may be stable under usual conditions, a failure mode may be produced by the three most common trigger events – earthquakes, excessive precipitation, and man's activities (Fig. 7.8).

**FIGURE 7.8** Craco, Basilicata region, Italy. This ancient village was totally destroyed by a landslide in December 1983. The houses were built on jointed Pleistocene conglomerates that failed along clay beds. Courtesy of David Alexander.

There are many ways that natural processes can produce landslides on what formerly were stable hillslopes. Excessive precipitation and earthquakes cause the largest and most numerous landslides. Undermining of hillsides by rivers eroding the base has created spectacular landslides in the Fraser Canyon, British Columbia. Other normal weathering and erosion processes that operate on the hillslope may eventually produce that extra amount of stress whereby the threshold barrier is surpassed, setting in motion the landslide.

## Excessive Precipitation

Water can act to destabilize earth materials in at least three ways. The extra moisture can increase the weight of the total mass thereby creating a surcharge, which is abnormal for the environmental setting. Penetration of water to zones below the surface can provide a type of fluid lubrication along which slippage can more easily occur. The filling of soil and rock openings with water increases pore-water pressure, which in turn reduces grain-to-grain strength and lowers resistance to movement of the earth materials (Fig. 7.9).

**FIGURE 7.9** Landslides in the Philippines produced by excessive precipitation on hillslopes largely denuded by deforestation practices. Courtesy of John Wolfe.

In southern California, unusual rainstorms in 1951 and 1952 caused landslides and mudslides in the Los Angeles area with damages in the millions of dollars. These events had a beneficial result because there was sufficient public outcry that the first Grading Ordinances were passed in 1952. This led the way for involvement of geologists in the planning and policy stages for construction of developments on hillsides. In the Rio de Janeiro region of Brazil, exceptionally heavy rains in 1966 and 1967 devastated a 170 $km^2$ area with earthflows, debris slides and avalanches, mudflows, and rockfalls and rockslides. The storms killed more than 2700 people and caused staggering property losses, hundreds of millions of dollars. The greatest disaster by landslides in the United States resulted from precipitation spawned by Hurricane Camille on August 10–11, 1969. In Nelson County, Va., alone, debris avalanches and flooding killed 150 people and caused property losses of $116 million (Figs. 7.10, 7.11). Asia has been the site for numerous disasters produced by rainfall from typhoons. For example, at Kobe, Japan, in 1939, rockslides and mudflows killed 461 people and destroyed 100,000 houses. In 1945, Kure had similar landslides from a typhoon that killed 1154 people, and in 1958, the Tokyo area had a typhoon that created 1029 landslides, which killed 61 people. Taiwan, the Philippines, and Hong Kong have also suffered from landslides created by excessive precipitation.

**FIGURE 7.10** Torrential rainfall from the remnants of Hurricane Camille on August 19–20, 1969, produced extensive debris avalanches such as this in central Virginia. Courtesy of G. Williams and H. Guy (1971).

**FIGURE 7.11** Torrential rainfall from the remnants of Hurricane Camille on August 19–20, 1969, produced extensive debris avalanches such as this in central Virginia. Courtesy of G. Williams and H. Guy (1971).

## Earthquakes

In the Gros Ventre Valley of Wyoming, the largest historical rock avalanche in the United States occurred on June 23, 1925 (Fig. 7.12). Within a three-minute period about 40 million $m^3$ of weathered sandstone, limestone, and water-saturated sediments roared across the valley at a speed of 160 km/hr to heights of more than 80 m on the opposite side of the valley. The Gros Ventre River was dammed, and its water formed a large lake. Early studies of the landslide attributed the failure to heavy rains and melting snow, which had saturated many of the earth materials. However, recent work by Voight and others attribute the real trigger to earthquake activity, which shocked the materials into a liquefied condition, thus causing progressive strain that finally triggered the event.

The Madison rock avalanche that occurred on August 17, 1959, had many similar elements in common with the Gros Ventre event. It was also produced by an earthquake that shocked weathered schist, gneiss, and dolomite on a mountain. The dislodged material sped across the valley and dammed the Madison River with a lake, 10 km long and 60 m deep. The

**FIGURE 7.12** Gros Ventre rock avalanche, Wyoming. This major landslide occurred on June 23, 1925, caused by a combination of heavy snow-melt and rain that saturated the ground. A light earthquake was sufficient to trigger displacement of 38.2 million $m^3$ of valley-dipping shales and sandstones. The slide speed was 160 km/hr and rode to a height of 80 m above the valley on the opposite side of the river. Courtesy of Wyoming Travel Commission.

main landslide involved nearly 30 million $m^3$ of material and killed 28 campers.

The Alaskan earthquake of March 27, 1964, produced a large array of terrain changes and earth disruptions throughout a 129,000 $km^2$ area. None were as catastrophic, however, as damages sustained in the Anchorage area. Here, nearly 100 people were killed in the Turnagain Heights section when the earth materials underlying the terrace were shocked into a state of liquefaction creating flows and slab slides (Fig. 6.6). This resulted in large surface ruptures and displacement of materials, crushing property and people. The earthquake also produced other landslides throughout the region, such as the Sherman slide, and immense subaqueous slides in the Gulf of Alaska, which are the largest such features on Earth with 590 billion $m^3$ of displaced material.

## HUMAN INTERFERENCE

There are many past and present examples where human activities have sufficiently changed the stability of earth materials so that landslides have been produced. Because of increased pressures from expanding population for construction on hillslopes, the pace of human-induced landslides will continue to escalate. For example, estimates in California for landslide damages from 1970–2000 are nearly $10 billion unless proper measures are taken to lessen man's developmental tendencies on hillslopes.

### Hillslope Disturbance

The likelihood of landsliding is increased whenever hillslopes are steepened such as occurs with highway construction, emplacement of buildings on sloping terrain, and mining (Fig. 7.13). The landslides that may be produced cover a wide range of types and sizes — from those that involve only tens of meters of cubic rock to those that aggregate thousands or even millions. The inability to handle landslide problems even has historical significance. For example, the French firm headed by Ferdinand de Lesseps went bankrupt trying to dig the Panama Canal because of their inability to control landslides caused by the digging process.

Hillside cuts required for highway construction often become sites for landslides. Each year more than $100 million is expended in attempts to remedy and control these landslide problems. A study of 104 landslides that occurred from 1966 to 1968 in the Province of Cosenza, Italy, showed that all had occurred because of road building. Highway cuts, and indeed all manmade incisions into earth materials, produce instabilities because the lateral gravitational component is increased. In addition, subsurface drainage is affected and provides lowered viscosity and increased mobility of the upslope landmass.

Mining along hillsides also poses threats to the balance of earth materials. The famous disaster in Switzerland produced by the Elm rock avalanche in 1881 was reported to have been caused by quarrying activities. The landslide started with the development of small rockslides at each side of the quarry. A few minutes later the entire rock mass above the quarry cascaded down

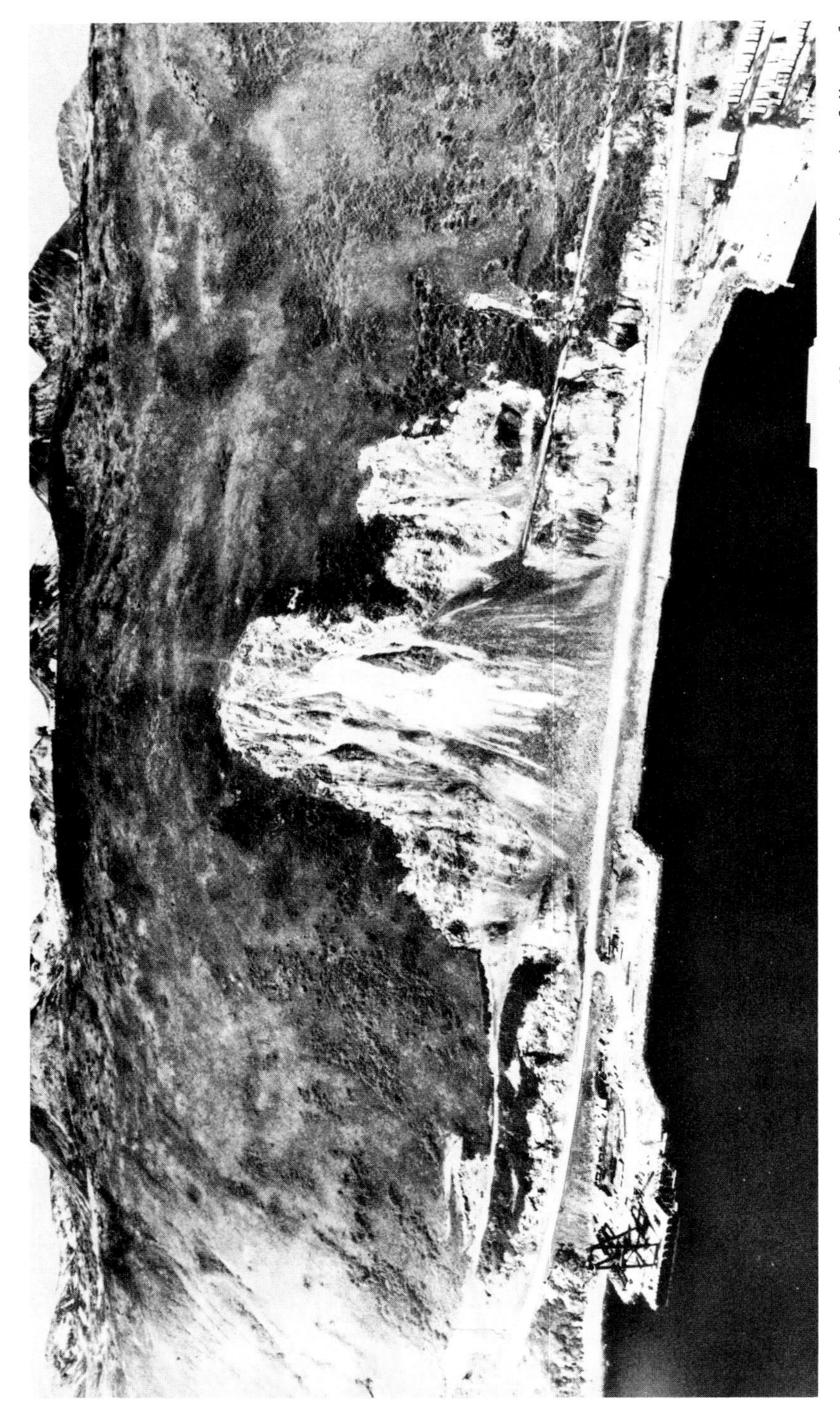

**FIGURE 7.13** Wright Mountain landslide, Calif. An original landslide at this site occurred about 500 years ago with massive failure of the bedrock ridge. That landslide was 400,000 $m^2$ in extent and displaced 14 million $m^3$ of rock. This view shows the scar that resulted from remobilization of the slide caused by exceptionally heavy rains and also undermining of the toe by road construction. Courtesy of U.S. Geological Survey.

the mountain and shot and flowed across the valley. During one 40-second burst, what has been described as a **sturzstrom** travelled a distance of 2 km. The motion had been likened to a torrential flood. However, the interstitial "fluid" between the colliding rocks and material was not wet, but instead was a dry dust (Hsu, 1975). Clay mining at Haverstraw, N.Y., led to landsliding on January 8, 1906, which killed 20 people and did extensive property damage.

### Vegetative Changes

The type of timber harvesting methods that are used can lead to accelerated erosion of hillslopes. Forest removal by clearcutting reduces the rooting strength of trees and alters the hydrologic regime. Swanston and Swanson (1976) describe a large range of gravity movements that result from such practices in the Northwest Pacific area of the United States (Figs. 7.14, 7.15). These effects include creep, slump and earthflows, and debris flows and avalanches.

Another type of vegetation modification occurs when land managers change the type of vegetation to improve watershed, range, and wildlife and environmental conditions on wildlands. J. V. DeGraff (pers. comm. 1982) has shown by studies in the Sheep Creek watershed of Utah that the conversion from a tree and brush cover to grasslands increased soil moisture, decreased rooted strength of the vegetative mat, and led to a 300 percent increase in landslide activity. Forest fires and brush fires, as in southern California, have claimed tens of thousands of hectares with a large increase in hillside erosion by both wash and gravity processes.

### Pore-Water Changes

There are still additional ways that human activities increase water content within earth materials, thereby causing changes in pore-water conditions. The added water lowers rock and regolith strength and can become critical in determining the threshold of stability. Increased moisture in earth materials in the Palos Verde Hills, Calif., caused by lawn water and effluent from septic tanks was the most probable cause for reactivation of the Portuguese Bend landslide in the late 1950s. Although property losses amounted to millions of dollars, the courts ruled that highway construction had initiated the landslide and held Los Angeles County accountable to pay property owner damages of $5.3 million.

**FIGURE 7.14** Landslide, Drift Creek, Ore. This is a rotational slump caused by clearcutting timber operations (Swanston and Swanson, 1976).

Water-level changes in reservoirs can also create modifications of pore-water pressure and also of the water table. The disastrous Vaiont Dam landslide, in which 2200 people perished from the 70 m deep wall of water that overtopped the dam, probably stemmed from multiple causes. Earth tremors had been noticed prior to the landslide. Also the period before the event had been one of higher precipitation than normal. In addition, water level of the reservoir had been frequently changed, which produced modification in the adjacent water table and accompanying pore-water pressure. It is no wonder that the rocks underlying the hillslopes could not adjust to such a battery of stresses. Thus, 260 million $m^3$ was dislodged from the hillside and plummeted into the reservoir creating a 100 m wave that surged over the dam (which was not destroyed) (Figs. 7.1, 7.2). Jones and co-workers (1961) described the case history of landslides produced by the waters impounded by Grand Coulee Dam. They studied more than 300 landslides, which were caused by saturation of earth materials adjacent to the reservoir. Although damages ran into the millions of dollars, no lives were lost from the events.

## LANDSLIDE DESTRUCTION

The average annual property loss attributed to landsliding in the United States has now passed the $1.5 billion mark. Throughout the world, the figure is about 10 times this figure. Furthermore, annual losses are increasing

FIGURE 7.15 Chute and debris avalanche channel in clearcut area of Oregon. Photograph taken in 1975. Courtesy of Fred Swanson.

because of the expanded use of hillsides for developmental purposes. The following disasters made the headlines during the 1980–1983 period:

1. 1980
   a. June 25, Himalaya, India, landslide swept away a village killing 50.
   b. August 5, Mount Fujiyama, Japan, rockslide killed 12 mountain climbers.
   c. September 7, Himalayan foothills, India, landslides caused by monsoons killed 33.
   d. October 5, Sichuan Province, China, landslides and flooding killed 240 and left 100,000 homeless.

2. 1981
   a. March 22, Belgrade, Yugoslavia, a landslide toppled two railroad cars into Morava River killing 12.
   b. June 21, Mount Rainier, Wash., ice and boulders smothered 11 hikers.
3. 1982
   a. January 31, Salzburg, Austria, avalanche crushed a school group of 10 students and 3 teachers.
   b. February 1, northern California, landslides and mudslides killed 31 people and caused $280 million damages in 230 km long area.
   c. February 2, Kashmir State, India, landslide killed 23 following heavy rains.
   d. March 14, Grenoble, France, avalanche killed 11 skiers.
   e. March 21, Nagano, Japan, 13 died in an avalanche.
   f. Mt. Ararat, Turkey, a snowslide killed 15.
   g. July 24, Nagasaki, Japan, torrential rains and landslides killed 42.
   h. October 6, Nye Nye, Liberia, 45 miners smothered by tons of debris when iron ore waste pile failed.
4. 1983
   a. March 25, Tornamesa, Peru, avalanche of mud and rocks killed 90 and swept two buses into Rimac River.
   b. April 29, Quito, Ecuador, landslides buried more than 100 in the Andes.
   c. July 30, Bogota, Colombia, rock- and mudslides roared down a rain-ravaged Andian mountainside and killed 160 workers at the Guavo dam.

## LAND MANAGEMENT

Education of the general public must be part of any program that seeks to address problems associated with landslide-prone terrane. There is an increased awareness by policy makers as evidenced by larger numbers of laws and ordinances treating the topic of unstable ground. These judgments are little by little becoming better known by people who are making investments on potentially dangerous lands. For example, the 5-year joint program of the U.S. Geological Survey and the Department of Housing and Urban Affairs for evaluating physical factors that are environmentally important in the San Francisco Bay region has had a big impact on future planning for the region. Their studies paved new ground in assigning the importance of landslide potential to the region. In similar manner, evaluation of landslide risk for the urban areas of West Virginia by the State's Geological Survey has done much to alert the populace to the significance of such features.

## Recognition

The initial phase for any planning on hillsides must be identification of potentially troublesome sites. Such work must involve geologic, hydrologic, geomorphic, and when possible, soil-rock mechanics analysis. Each environmental setting has its own distinguishing characteristics that mark it for possible future ground failure. Clues for discovery of potential landslide danger include: (1) abnormal configurations in the terrain, (2) unusual occurrence of wetted surfaces, (3) break in hillside at upper part of the slope, (4) lateral tears at right angles to terrain contours, (5) downslope bulge and pitted surface, and (6) disruption in vegetation.

## Legal Affairs

An important antidote to minimize potential landslide damages is passage of laws and ordinances and institution of regulations that prevent or restrict usage of those lands recognized as being hazardous areas. The California style of grading ordinances first passed in 1952, with later amendments, as in 1963, was the first attempt at regulation of development on hillsides. These require site investigations and reports that cover the character of earth materials and procedures that must be followed if construction is allowed at the location. Geologists and engineers are some of the types of professionals that are mandated to be involved in such investigations. Slosson (1969) reported on the remarkable success of the program by showing a phenomenal decrease in damages on those properties covered under the ordinances, as contrasted to those built prior to the regulations. More recently, the federal government, with passage of amendments to the Flood Disaster Protection Act, for the first time included mudslides (which in the language of the Act means landslides) as applicable under the stipulations provided in the Act. So although insurance can be obtained, the same as flood insurance, in order to qualify, the community must make certain safeguards for protection of properties and to disallow new construction at unfavorable sites.

Brazil was the first foreign government to pass laws relating to landslides. The Forest Law of 1959 specified which slope levels were considered safe, and construction was not permitted above such elevations. The Law of License of Construction in Uneven Terrain of 1967 required that contractors must supply proof of hillslope stability for all sites suspected of being potentially hazardous. These laws were amplifications of government decrees dating from 1955 aimed at reducing damages for slope instability in developing areas.

## Remedies and Control

Before any engineering operation is mounted, the characteristics of a potential landslide must be assessed. Such factors as the safety ratio need to be known. This is the relation of the resisting earth forces to the processes likely to create movement. Of course, the larger the ratio is above 1.0 the safer the site. In addition, some type of benefit/cost ratio must also be evaluated to determine whether the total cost of providing for hillslope stability is worth the investment in terms of properties and installations to be protected. If it is then determined that some action must be taken, there are a variety of strategies that can be adopted, which depend on the size of the area, type of landslide that might occur, and the frequency with which motion might be initiated. The following list includes some of the plans for action:

1. Avoidance methods. A possible alternative for new construction is simply to avoid using the site if mitigative measures are impossible or the size of the development does not warrant protection.

2. Water control methods. A dominant cause for the majority of landslides is high moisture content within the earth materials. Thus, it is important to decrease the amount of water that can enter the slide area, or to release water that is already incorporated within the earth materials. To do this, diversion ditches above the slide can help prevent runoff from entering. The surface of the slide in some instances can be regraded to allow for surface water to drain away from the danger area. Subsurface waters may be pumped, siphoned, or drained by collection systems that remove the water.

3. Excavation methods. For small slides and where slopes may already have failed, it may be possible to remove all or part of the slide. If partial excavation can be accomplished, this is most effective at the top of the slide to lessen the stress on downslope materials. Hillside benching offers another technique wherein it may be possible to cut a series of benches (Figs. 7.16, 7.17). This helps terrace the mass so that total failure cannot occur. The benches may also serve as drainageways so that water can be diverted from piling up as a surcharge of mass.

4. Restraining structures. These consist of a wide range of devices to inhibit movement of the earth materials. They are most commonly used as either a last resort when there are no other options for protection, or as an integral part of the initial construction. Retaining walls are probably the most common technique whereby some type of wall, bulkhead, dike, or dam is emplaced as a barrier at the foot of the slide. To be effective, such structures should be complemented with an adequate drainage system, so water cannot be imponded behind the barrier (Fig. 7.18). The structures should also be linked to stable ground at the ends. Such walls may consist of timber, concrete, grout, stone, metal ribbing, gabions, or other combinations of material. A variation on such protection is installation of a buttress whose

**FIGURE 7.16** This failure, a translational-slump flow, developed along relict bedding that dipped directly toward the roadway in Tennessee. Courtesy of David Royster (1979).

**FIGURE 7.17** This aerial view shows the engineering repair of the landslide shown in Figure 7.16. Much of the displaced earth material was removed, regraded on a 2:1 slope, and a bench with toe buttress added for additional restraint against further mobilization. Courtesy of David Royster (1979).

**FIGURE 7.18** Gabion-type rock walls were one of the types of structures used in the repair of several landslides that had occurred along Interstate 40 near Rockwood, Tenn. Benches also help prevent long motion of sliding. Courtesy of David Royster (1979).

purpose is to hold back the downhill force of earth materials by a massive embankment of earth and rock. The sheer weight of the material acts as a counterforce to continuing motion of the landslide. At times, rock bolts can be used if the bedrock has planes of discontinuity, and by binding them together, their movement will be inhibited. Rock bolts more than 14 m in length have been successfully used at many sites in the Alps (Fig. 7.19).

5. Miscellaneous methods. At times, other procedures have been used with success to stop potential landsliding. However, most of these have been employed under rather unique circumstances and at sites where other methods had little likelihood for success. Grouting is the introduction of cement or chemicals into earth materials for the purpose of hardening the materials or accelerating the release of pore water (Fig. 7.20). Portland cement may be effective in granular sediment whereas such chemicals as sodium silicate may react to form a stiff silicate gel in some siliceous minerals. Lime has also been used in some highly plastic clays to reduce landslide hazards in such places as Panama, Brazil, and Oklahoma. Other rather exotic methods used in unusual cases include freezing of soils, such as at Grand Coulee Dam to allow for construction, and heating of soil, as in Romania where loess was baked near boreholes to prevent movements.

FIGURE 7.19 Landslide preventative measures. Rock bolts and concrete emplacement are used on this rock cut in Austria. A rock wall, another restraining structure is visible in upper left of the photograph. Courtesy of Allen Hatheway.

FIGURE 7.20 To reduce potential rock falls, the mesh is grouted with Shotcrete applied to rock outcrops at this Marion County, Tenn., site. The purpose here is to intercept excesses of downward and laterally migrating water before it enters destabilizing zones. Courtesy David Royster (1979).

## SNOW AVALANCHES

Snow and ice avalanches are becoming a hazard of increasing scope as more people become involved in winter sports and recreation (Figs. 7.21, 7.22). Snowmobiles transport people into more remote areas, and skiing and travel and occupancy of sites in rugged terrain is becoming more commonplace. Of course, snow and ice have always posed problems in mountainous areas in the Alps and the Andes. On January 10, 1962, a massive ice avalanche occurred in Peru when the front of a hanging glacier on Mt. Huascaran fell. The 3 million m$^3$ mass of ice also dislodged bedrock of the steep valley walls, which cascaded 4000 m, demolishing everything in its path. Nine small towns were destroyed, 4000 lives were lost, and cultivated fields were devastated in the valley formerly famous for its beauty and fer-

**FIGURE 7.21** Snow avalanche that buried this road near Alta, Utah. Photograph taken May 17, 1983. Courtesy of Earl Olson.

**FIGURE 7.22** One of the aftermaths from the snow avalanche shown in Figure 7.21. Fortunately, no one was killed in this event. Courtesy of Earl Olson.

tility. Although the neighboring city of Yungay, protected by a hill, was spared in 1962, it was not so fortunate in 1970 when it was completely demolished by a rock avalanche set in motion by an earthquake.

Destruction by snow and ice avalanches in North America has not yet produced major disasters. Although the United States has about 10,000 avalanches per year, property damage is less than $1 million per year and an average loss of life of seven people. The worst disaster occurred March 1, 1910, in Washington when 96 people died when their snowbound train was swept off the track in a steep canyon. During that same month, 62 workmen perished in Rogers Pass, British Columbia, during an attempt to rescue a train stalled by another avalanche. Other notable snow avalanches occurred on March 22, 1915, when 57 miners in British Columbia were buried by an avalanche, and on February 17, 1926, when the mining community of Bingham Canyon, Utah, was hit by an avalanche, killing 40.

There are two types of snow avalanches — (1) Loose snow avalanches are those caused by a small amount of cohesionless snow that becomes dislodged and helps move additional snow downslope. (2) Cohesive snow avalanches are those where failure begins with brittle fracturing of highly compacted and slab-like masses that may range up to thousands of square meters in area and 10 m in thickness. These are the most dangerous and, worldwide, have produced the greatest losses.

The prevention of damage by snow and ice avalanches depends on prior recognition of hazardous sites and dangerous conditions. Land and air reconnaissance are necessary to locate the crucial risk areas. Potential losses can be reduced by structural or conservational measures. They may consist of walls or baffles to protect property and divert the flow of snow. Snow fences and wind baffles may be emplaced to prevent buildup of snow. On some slopes, reforestation may aid in prevention. Explosives can be used in some instances to initiate small avalanches at opportune times, thus preventing buildup of major avalanches. Of course, avalanche planning is the most effective tool so that roads and structures will be sited at non-hazardous locations, and when coupled with land-use legislation to control development, these methods can forestall great tragedy.

## GOVERNMENT LANDSLIDE PROGRAMS

As previously mentioned, the Los Angeles, Calif., program of 1952 was the first governmental attempt to address the landslide problem by instituting a series of grading ordinances. These codes were modified in 1963 whereby improved technological guidelines were incorporated. These have been instrumental in reducing landslide damages to one-tenth of those developments constructed with minimal regulations. In the San Francisco Bay region, some local governments have supplemented such programs by developing a set of criteria of permissible land use based on a mapping program that indicates susceptibility to landsliding. San Mateo County, Calif., has adopted a zoning map that controls the permissible density of development on lands that may be landslide-prone. Since the 1975 adoption of this zoning, there have been no landslides on 1055 properties. More than three-fourths of all communities in the San Francisco Bay region have made use of landslide-hazard information provided by the U.S. Geological Survey.

In Fairfax County, Va., a different approach has been used in areas undergoing development. Maps have been prepared that outline varying degrees of hazard for different earth materials. Developers are required to obtain professional engineering information and site investigation reports, and plans are required to be submitted to the County. This program has greatly reduced landslides in the area.

Starting with the 1980 fiscal year, the U.S. Geological Survey has combined existing programs in engineering geology and arctic studies into a formal program entitled "Construction and Ground-Failure Hazards Reduction." Its mission is to provide research and data on a wide range of processes that cause ground disturbance including landslides, subsidence, swelling clay-shales, and construction-induced rock deformation.

Other countries also have national programs. France has ZERMOS (Zones Exposed to Risks of Movements of the Soil and Subsoil) plan, which is responsible for producing maps of 1:25,000 scale or larger. These portray degrees of risk of various types of slope failure and include rate of motion and potential consequences of damage. Japan has a national program for landslide control that was initiated after World War II. Initially, the landslide control activities were linked to other erosion control legislation. However, by 1958, the Landslide Prevention Law was passed and culminated in the 1969 Slope Failure Prevention Law. The legislation provides for government aid when disasters are not caused by human intervention. Such funds not only pay for repair and restoration of property but also for work that will prevent future landslides. The costs are split between the central government and local prefectures, and costs are estimated to be $600 million annually.

New Zealand has a national insurance program that assists homeowners when dwellings are damaged that the owner could not have prevented. The program is an outgrowth of the Earthquake and War Damage Act of 1944.

It is now standard procedure in nearly all landslide prevention programs to undertake mapping programs that assess location and degree of potential damage that may result from the destabilization of the hillslope. A wide range of techniques are used in the mapping and inventory process.

## POSTLUDE

Landslides share a common bond with other geologic hazards because they can produce grievous losses, and the tragedy they create can come with sudden fury. However, they also have the distinction of having important differences when compared with some of the other hazards, such as volcanic activity, earthquakes, and floods. Thus, landslides are:

- More predictable. They occur on slopes, so losses can be minimized by staying away from slopes.
- Cover less area. The effects from other hazards may cover thousands of square kilometers. Unlike the low viscosity and mobility of water in

floods, the cohesion of most landslide materials is more viscous and can travel only short distances.

- Subject to more preventative measures. Volcanic action and earthquakes are still beyond the scope of engineering to mitigate because of the awesome energy of these processes. Although the massive forms of gravity movement such as the rock avalanches cannot be tamed, there are many other types of landslides that can be controlled or stabilized.
- More capable of being caused by human activities. Volcanicity is not caused by societal intervention, and with few exceptions, neither are most earthquakes. However, as indicated in Chapter 8, flooding is another exception and can be aggravated when land-use practices or structural engineering projects are faulty.
- Becoming more costly in terms of property damages, along with other hazards, but loss of life is generally less.

CHAPTER 8

# FLOODS

Webster defines a flood as "a great body of water overflowing the land; an inundation." Thus, flooding can occur because abnormal streamflow spills out of the channel, or when lake or ocean waters are driven onto adjacent lands by unusual storms or tidal action. Floods may be either natural events or produced by human interference. Indeed, floods are the price mankind must pay for occupancy of low-lying topography. Such lands were created in valleys by the rivers that occur there. This very normal relationship becomes hazard-prone terrane to the communities that have settled into such an environment.

Throughout history, mankind has had an awareness of floods, and their destructive tendencies have been chronicled in the folklore and legends of many peoples. Of all disasters, floods are probably the best known and least feared, but yet affect more people than all others. The majority of floods, exceptions being cloudbursts and dam failures, are quite predictable, so with proper warning, people can usually escape with their lives and some belongings. However, flood damages keep escalating year after year because floodplains have always been a preferred location for farms and cities. More than 25 million people in the United States live at sites that are in jeopardy from large floods. Most rivers in the United States flood an average of once every 2.33 years, and total yearly damages are more than $3 billion.

Floods are perceived differently by various groups of people. The city dweller may view floods as an inconvenience or adopt the attitude that his property will not be damaged – the "it can't happen to me" syndrome. The farmer may be concerned about excessive erosion of his lands, or losses because of inability to either plant crops or harvest them. However, these are dues he must pay for the chance to use the richest types of soil, alluvial soils with high nutrient content. Commercial people become distraught because of loss of business, and government officials are concerned about damages to roads, facilities, and services and the financial losses that will be passed on to citizens by higher taxes. These burdens are of worldwide extent and are now especially escalating in some of the developing nations.

## RIVER SYSTEMS

Streams occur in channels that occupy the lowest position of river valleys. Because of vagaries in weather and climate, there are times when the channel cannot handle all the runoff being supplied, so water overflows the banks onto the adjacent topography. If these contiguous lands are relatively flat, the feature is described as a **floodplain**. Flooding is a natural way for rivers to accommodate themselves to high moisture conditions. Thus, *floods are a normal behavior of all streams*, and furthermore, are essential components in maintaining a large variety of plant and animal life and in enriching the floodplain with nutrient-rich sediments. The ancient Egyptians depended upon the yearly flooding of the Nile River to maintain the nutrient balance in their soils and supply plants with life-giving moisture.

Floodplains are of different size depending on the environmental setting. In mountainous terrane, there may be very little floodplain, and the river may occupy the entire valley floor. Such areas are subject to flash floods because there is no safety-valve area to receive high discharge, so rivers quickly occupy even much higher ground. However, where the topography is more subdued, floodplains can be exceptionally large and nearly flat, thus providing a big storage area for even large floods. Of course, the residence time of floodwaters may be long because drainage is impeded by the flatness, so it takes time for the water to drain off or percolate into the soil.

Deposits on floodplains may consist of **overbank silts and clay** that settle out of suspension during quiet water times. They may also contain **lateral accretion materials**, which form along the river banks at meander bends where sediments form on the inside of the curves. The floodplain has a central stream channel, and when the topography is exceptionally flat as in the lower Mississippi River, it may contain a **natural levee** built up along the banks. The main avenue of floods is called the **floodway**, and where this adjoins the hillsides is the **floodfringe** (Fig. 8.1).

When the river is at the threshold before flooding and occupies the entire channel, it is said to be in **bankfull stage**. For American rivers in the eastern United States, this condition is reached on an average of about once every year and a half. For planning purposes, it is necessary to know what the recurrence interval is for floods of different magnitude. Thus, floods are often described as being the 50-year flood or the 100-year flood. This quality is important if structures are to be emplaced, and some laws, such as the Flood Disaster Protection Acts, and those that depend upon National Flood Insurance programs, consider the 100-year flood as their management instrument.

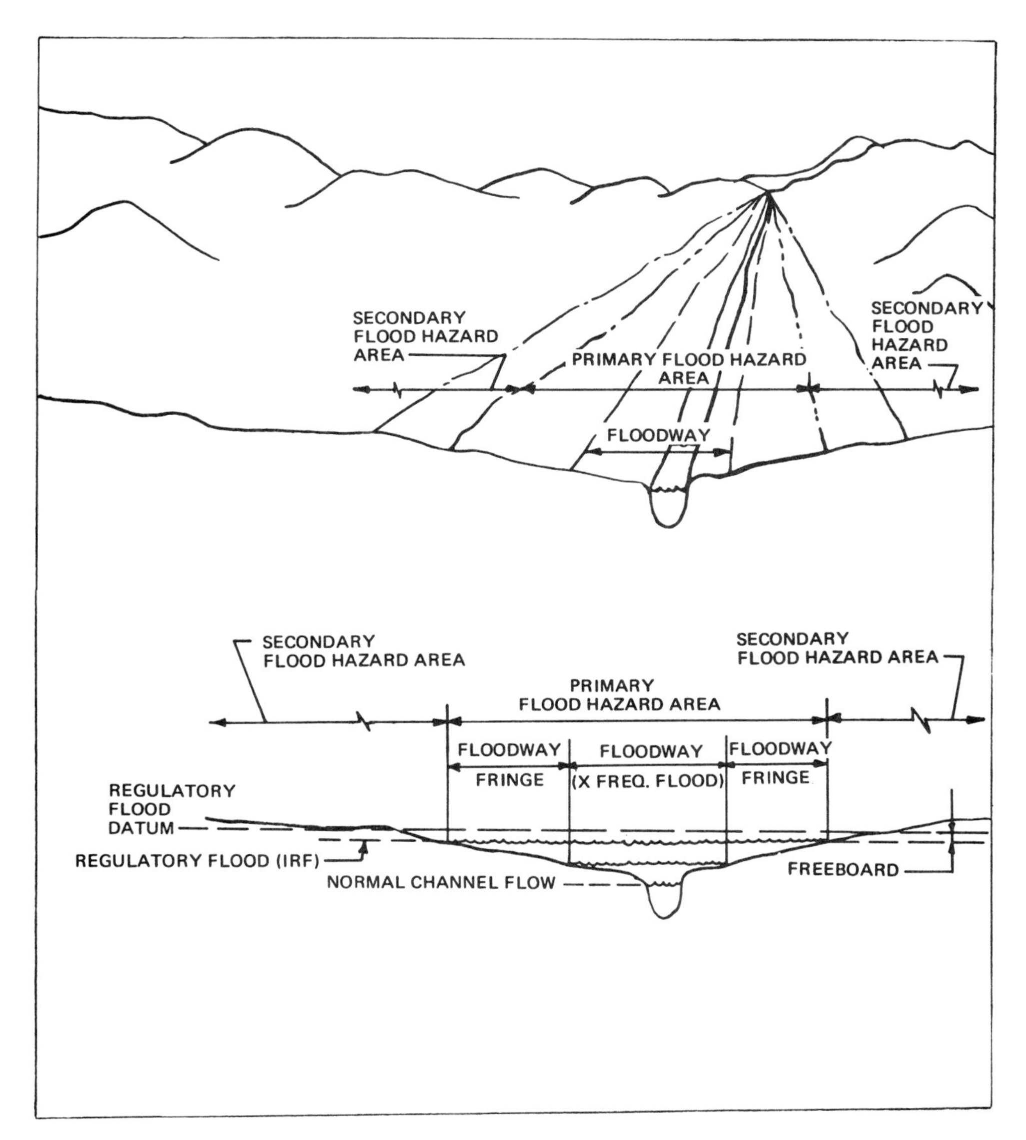


FIGURE 8.1 Typical data and terminology used for flood-hazard management. Courtesy of U.S. Army Corps of Engineers.

## WHY DO WE HAVE FLOODS?

Flooding is caused by both natural processes and human activities. The majority of floods are caused by excessive moisture (See Appendix F). This can occur as abnormally heavy rain in a short period. The effects are especially felt in areas with steep slopes and small floodplains. Big Thompson Canyon, Colo., felt the fury of this type of "flash flood" on July 31, 1976, when 140 persons died and millions of dollars of property were damaged or destroyed (Fig. 8.2). Many floods are also the result of rainfall over a prolonged period. For example, even inland flooding can occur from the aftermath of major hurricanes when their tracks veer onto the land (Fig. 8.3). Hurricane Agnes in June 1972 intruded the United States and lingered several days before leaving in its wake more than $3.5 billion in damages and 118

**FIGURE 8.2** The village of Waltonia on the Big Thompson River, Colo., is shown in these two aerial photographs taken before (A) and after (B) the devastating flood of July 31–August 1, 1976. Here, two large motels and other buildings were washed away, but buildings on high ground remained intact. In this valley, 140 people were killed by the flooding. Courtesy of U.S. Geological Survey.

**FIGURE 8.3** Hurricane Camille did extensive damage to inland eastern United States in 1969 (see also Figs. 7.10, 7.11). A total of 154 lives were lost. In Virginia, damages were $100 million, and along Davis Creek, 50 were killed by the flooding. Courtesy of Virginia Division of Mineral Resources.

casualties – the highest financial flood losses ever suffered in the country (Fig. 8.4).

Snowmelt in the spring can also cause flooding. This is almost an annual event in the Red River valley of North Dakota. When snowmelt combines with spring rains, flooding becomes even more of a problem. Prolonged periods of rain and melting snow throughout the upper Mississippi River basin in the winter months of 1972–73 combined to produce the costly Mississippi River flood of spring 1973. Flooding occurred from March 11 to May 4, causing 68,000 km$^2$ to be inundated and crop and property losses in excess of $1.5 billion (Fig. 8.5).

In northern latitudes, ice dams may produce flooding. Many rivers in Alberta Province, Canada, have such seasonal flooding. An ice dam in the Susquehanna River near Binghamton, N.Y., in 1981 caused extensive flooding in the Conklin area and routed hundreds of families from their submerged homes. Ice dams in Poland in January 1982 led to widespread flooding, and the ice plugs were dynamited in attempts to reduce the buildup of water behind such barriers. Landslides may also dam rivers and lead to flooding when the earth materials are breached and the imponded waters released suddenly.

**FIGURE 8.4** Flooding in Wilkes Barre, Pa., from Hurricane Agnes, June 1972. Damages from this flood were the greatest of any single disaster in the United States. The staggering loses amounted to more than $3.5 billion, and 118 people were killed. Pennsylvania suffered the most, but a total of 20 counties and 27 cities were declared disaster areas by presidential proclamation. Courtesy of Pennsylvania National Guard.

**FIGURE 8.5** Farming losses were exceptionally great during the Mississippi River springtime flood of 1973. In many places, the river was out of its banks for several weeks and prevented spring planting of crops. Courtesy of U.S. Army Corps of Engineers.

Human activities have added a new dimension to the threat of flooding. There are many ways that mankind alters the normal hydrologic regime of river systems. Perhaps the most obvious changes are dams that prohibit normal streamflow. As shown in Chapter 11, however, their construction is not infallible, and dam failures have led to grievous loss of life and property. Another deliberate change in rivers occurs when stream flowpaths are altered as invariably happens in urbanized areas or on those rivers where channelization has been performed. For example, in Houston, Tex., studies have shown that the increase in impervious areas caused by city development increased the magnitude of the 2-year flood peak by nine times, and of the 50-year flood by five times (Fig. 8.6). Channelization may be an effective method for evacuating water more swiftly from the area of construction, but may exacerbate flooding of downstream areas.

Many other human actions within the watershed may also produce profound alteration of hydrologic systems. Gilbert (1917) showed the relation of increased floods in the Sierra Nevada to hydraulic mining in the region. Deforestation and improper soil management on farms increase the incidence of flooding. Removal of trees and vegetation cause increased runoff and erosion. Siltation clogs stream channels thereby reducing their capacity to

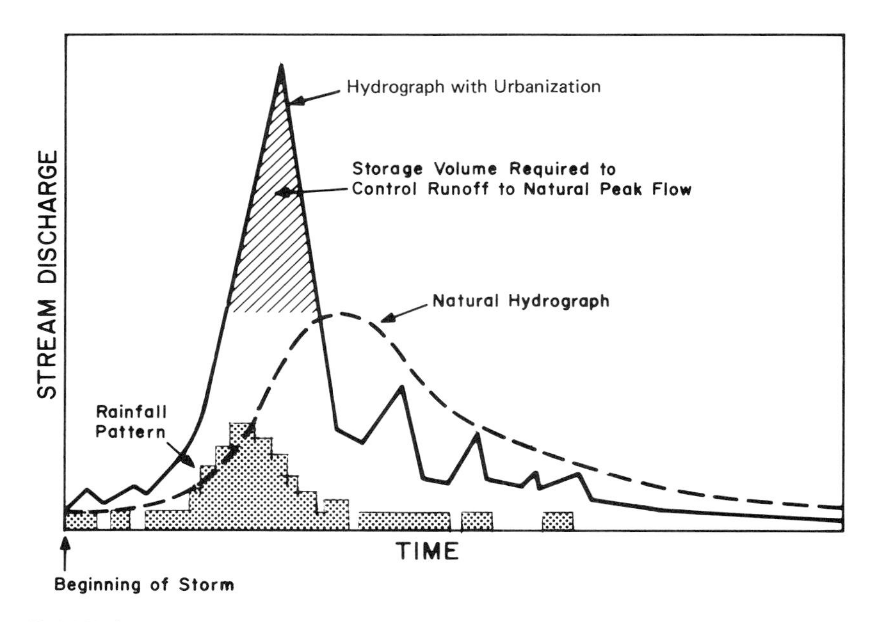


FIGURE 8.6 A typical urban hydrograph. Urbanization changes the natural flow regime of rivers by creating excessively high flood peaks and reducing base-flow discharge. Courtesy of U.S. Geological Survey.

transport water. The rapid pace of deforestation on the subcontinent of India, Pakistan, and Bangladesh has increased flooding incidence by a factor of three during the last 15 years.

Perhaps the most controversial issue arises when trying to relate flooding to man's attempts at weather modification. The case usually cited where there is some correlation is the cloud seeding in South Dakota that preceded the exceptional rainfall and resultant flooding at Rapid City, S. Dak. Here, on June 9–10, 1972, floodwaters killed 245 people and caused damages of more than $200 million. The question of cause–effect is still being debated.

## Environmental Factors that Contribute to Flooding

The extent and duration of flooding is determined not only by weather conditions but also by the physical setting of the watershed and its materials. Antecedent moisture conditions in the ground are an important consideration. Dry soil can absorb new moisture, but if already saturated, the slightest amount of new rainfall may quickly wash away and contribute to flooding potential. Composition of the soil and bedrock determine percolation rates and thus the amount of water than can be absorbed quickly. Fine-grained soils and rocks such as clay and shale are quite impermeable, whereas sandier materials permit greater penetration by water. Thus, flood peaks are reached more rapidly in terrain underlain by shale than by sandstone. The steepness and length of hillslopes also determine runoff character. Even the shape of the drainage basin is important. For example, elongated basins such as occur in folded rock structures (called trellis systems) can produce larger flood peaks than watersheds with dendritic systems. Of course, the type and amount of vegetation and farming practices are important components in the total equation of hydrologic properties.

## EFFECTS OF FLOODS

Flood losses are measured in thousands of lives lost yearly and worldwide damages of many billions of dollars. The annual crop and property losses are approaching the $4 billion mark in the United States alone. The impacts from flooding can be classified into the following categories:

1. Primary effects are those losses that occur due to direct immersion by the floodwaters. The long list of losses includes people and animals killed, and buildings, services, and croplands ruined (Figs. 8.7, 8.8). The swift-moving waters may cause structural damage to roads, bridges, and railroads and destruction of power lines, gas mains, sanitary pipelines, water services, and telephone facilities (Fig. 8.9).

2. Secondary effects include additional damages not directly in contact with water. Malfunctions may occur in service lines, trade and commerce can be interrupted, and business transactions stagnate. Contamination and disease may contribute to severe health problems. Thus, the life-style of nearby communities is completely disrupted.

3. Tertiary effects are those future effects that occur to wildlife, whose habitat may be destroyed, and to the natural character of the water conveyance system. For example, a severe flood may make it even easier for tomorrow's flooding to occur by changing the channel geometry or position, by clogging portions with debris, and by erosion of vegetation so that the protective cushioning of it is lost.

### Damages and Case Histories

The most frequent and devastating floods occur in Asia and the Far East. Taiwan is commonly hit as is the Philippines because of abnormally heavy rainfall on steep and fragile terrain. India and Bangladesh are other lands that suffer grievous flood losses where the annual toll in lives is nearly

**FIGURE 8.7** Debris left by the Sheep Creek Canyon, Utah, flood of 1965, which killed seven people. The rod is 8 ft (2.44 m) high. Courtesy of Earl Olson.

1000. China perhaps has the most costly of all floods because the two great rivers the Yellow (Hwang Ho) and the Yangtse are so huge, and much of the flow is in a channel considerably higher (in some cases 15 m) than the surrounding lowlands. Thus, few countries can feel complacent about floods. Flooding affects more lives than all other geological hazards combined. In the United States, 20 million ha are subject to flooding.

The Mississippi River drains 41 percent of the 48 contiguous states and has flooded repeatedly throughout human history (Fig. 8.10). The greatest floods occurred in 1849, 1850, 1882, 1912, 1913, 1927, 1935, 1936, and 1973 (Fig. 8.11). The disastrous 1927 flood inundated 67,300 km$^2$, killed hundreds, and damages were in excess of \$2 billion (1982 dollars). It prompted political action such as passage of the first major national Flood Control Act of 1928. Large floods in 1935 and 1936 in the upper Mississippi basin tributaries caused further federal action when the Flood Control Act of 1936 was passed, giving the Corps of Engineers the authority to undertake flood-control measures on navigable rivers throughout the country. The 1973 flood was particularly vexing because by that time billions of dollars had been spent in attempts to harness the Mississippi River, and yet a flood with much less river discharge than the 1927 flood immersed even more land, 68,000 km$^2$ (Fig. 8.12).

**FIGURE 8.8** Bridge collapse from failure caused by flooding of November 1979, Province of Matera, Italy. Poor design of the span, coupled with an unstable foundation on Plio-Pleistocene sediment contributed to the damage. Courtesy of David Alexander.

In California, floods and mudflows combine for a one-two punch to damage that state. The devastating results of these processes especially affected southern California in 1951 and 1952, leading to passage of the grading ordinances (see also Chapter 7). Heavy rains in the 1968-69 winter season again caused extensive damage from flooding and landsliding, which amounted to $25.4 million in the San Francisco Bay area, and $6.5 million in the Los Angeles area. In late February 1980, southern California was deluged six times within nine days with 33 cm of rain. More than 24 people

**FIGURE 8.9** Caving of banks in the Atchafalaya River, La. Floods not only cause damages to humans and their property, but also can produce accelerated erosion of the terrain. Courtesy of U.S. Army Corps of Engineers.

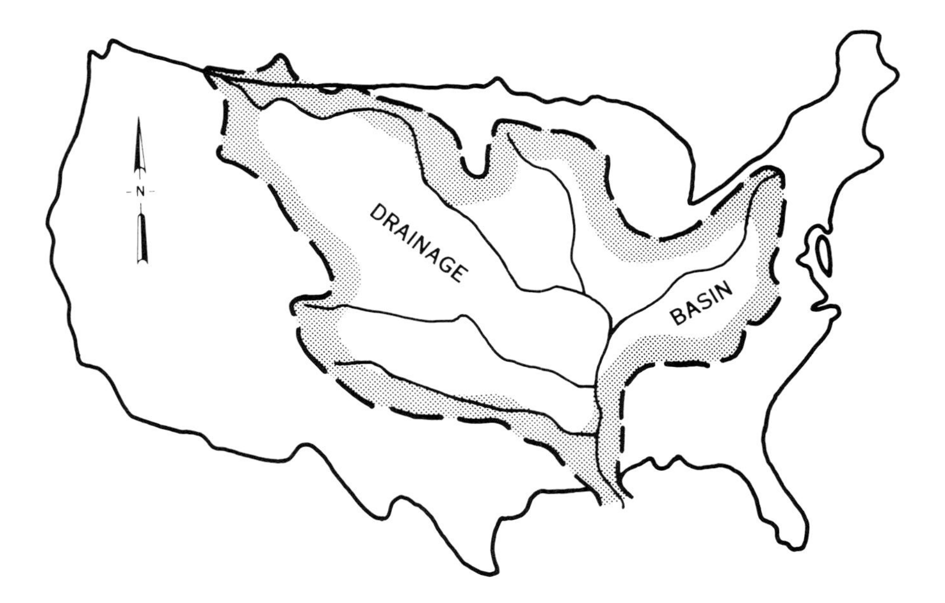


**FIGURE 8.10** The drainage basin of the Mississippi River comprises 41 percent of the lower 48 states of the United States.

**FIGURE 8.11** The Mississippi River flood of 1973 caused more than $1.5 billion in damages. It was unusual that there were no deaths attributed to the flooding. U.S. Army Corps of Engineers photograph near Memphis, Tenn.

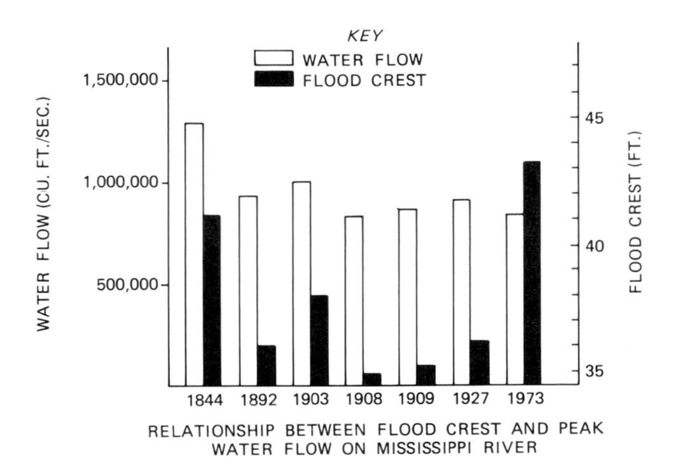


**FIGURE 8.12** Graphic plot of the relationship between flood crest and peak water flow of the Mississippi River at St. Louis, Mo., 1844–1973. Although the cause–effect relationship is heavily debated, many argue that flood-control projects have reduced the carrying capacity of the river, so that there has been an increase in flood crest height with the same amount of discharge in the River. Data from U.S. Geological Survey Water Supply Papers.

were killed, and damages exceeded $400 million. The State was declared a "disaster area" to accelerate relief operations and help the victims. However, two years later, California was once again hit with the awesome powers of nature. This time the damage was focused on the region surrounding San Francisco. Several days of heavy rains early in January 1982 produced devastating results. Mudslides, mudflows, and river flooding all combined to cause more than $500 million in damages and to kill 31 people. Water contamination forced 50,000 people to go on water-rationing, and the governor of the State declared a "state of emergency."

Of course, worldwide flooding constitutes a problem for most countries. For example, the typical record of floods in 1980 reads as follows:

1. March 31, Turkey. 75 people killed in four days of flooding and mudslides.
2. April 21, Peru. 160 killed by floods and mudslides in the central part of country.
3. May 7, Argentina. 26 drowned and 6 million ha of farmland inundated with losses.
4. September 9, Northern India and Bangladesh. Monsoon rains produced floods and mudslides fatal to 1500 people and widespread crop destruction.
5. September 25, Venezuela. 20 people killed from floods and mudslides near Caracas.
6. October 21, Colombia. More than 40 killed during flooding from heavy rains.

Although these were all great tragedies, they still pale in comparison with the magnitude of losses in China. For example, the Yangtse flood of 1931 affected the lives of more than 60 million people and killed many tens of thousands. Disastrous flooding has also hit in the 1980s. Extensive deforestation has been blamed for contributing to the excessively high flood states, and a massive reforestation program is now underway in an attempt to prevent future catastrophes. The once clear Yangtse is presently as muddy as its counterpart, the Yellow River, because of the large amount of silt washed from the denuded uplands.

Floods during the 1981–83 period continued unabated, both in foreign lands and in the United States. In 1981, Szechwan Province, China, was hit with upper Yangtse River floods that left 3000 dead, 100,000 injured, and 400,000 homeless. In Kwangtung Province, China, in October 1981, 70 more were killed and thousands left homeless. In 1982, the Chontayacu River flooded in Peru killing 600, washing away 16 villages, and leaving thousands homeless. Once again, Kwangtung Province, China, was hit by torrential rains and floods that killed 430 and stranded 450,000. The worst floods of the century in south Sumatra, Indonesia, killed 225 and ruined thousands of hectares of crops. Monsoon rains at Nagasaki, Japan, produced the largest floods there in 25 years, killing 307 and causing damages of $1.18 billion, the most expensive ever. The adjacent countries of El Salvador and Guatemala suffered severe rainstorms in September 1982 that killed 1160 and left 30,000 homeless. Even 1983 was no different. Quito, Ecuador, had a gigantic rainstorm that killed 65, left 200,000 homeless, and damaged $350 million in structures and crops. Gujarat State, India, had extensive flooding from heavy rains that marooned 130,000, and at last count, there were 935 dead or missing.

## Colordao River

Disastrous floods in southern California from 1905 to 1907 prompted the United States government to find a way to control the Colorado River. The answer began at Black Canyon, Nev., 50 km southeast of Las Vegas. Here, engineers sculptured the mammoth Hoover Dam in 1935. Since that time, from its headwaters in the Rockies to the estuary in Baja some 2400 km away, the Colorado River has become one of the most carefully controlled and heavily managed rivers in the world. It is now dammed, pumped, diked, and rationed to preserve every precious drop of water for the seven thirsty western states and the farmland of northern Mexico.

With the completion of Glen Canyon Dam and its reservoir, the program seemed finished, until the advent of the Central Arizona Project to siphon off more water. There are now four major dams and reservoirs (Fig. 8.13): Hoover Dam and Lake Mead, Davis Dam and Lake Mohave, Parker Dam and Lake Havasu, and Glen Canyon Dam and Lake Powell. Even all of this

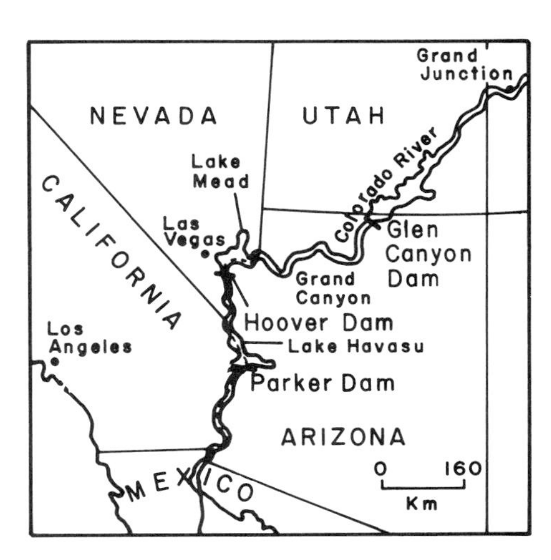


**FIGURE 8.13** Major dams on the Colorado River that control discharge characteristics of the river.

proved insufficient in 1983 when heavy rainfall and snowmelt from the upper basin streamed into the reservoirs, bloating them to capacity. By June 1983, they were full, which forced engineers to release millions of acre-feet of water in a manmade flood that was called a "controlled disaster." The resulting floodwaters killed five in Mexico and caused crop damage of $80 million there. Six counties in California were declared disaster areas with many millions of dollars lost in crop and installation damages.

## MANAGEMENT OF FLOOD HAZARDS

Flooding is not such a mysterious force as volcanoes, earthquakes, or even landslides, so everyone has their own preferred manner for dealing with floods. Is it any wonder that because so many consider themselves experts that methods for flood prevention, reduction, and control are highly controversial? Principal differences arise as to which methods should be adopted to deal with potential problems. These techniques fall into such categories as structural control, legal actions, land-use policies, and loss absorption (Table 8.1). All of these have been used in various degrees in the United States and throughout the world. For example, in the United States, the earliest flood-control measures were always in the realm of engineering structures. In the 1930s, land-use policies started to become an important component in flood-control planning. During the last two decades, legal actions have taken on increasing importance. Regardless of the final decision, nearly everyone agrees to the necessity for developing a strong base in hazard perception education. These are measures whereby those threatened can take appropriate steps to mitigate damage. The long-range objective, furthermore, is to alert people sufficiently in advance so that they do not invest in property development in high-risk zones. The old tale of "it can't happen to me" must be overcome with documented persuasion that indeed, flooding will eventually affect all who occupy floodplain areas.

### Structural Measures

*Local flood protection.* Such structures as levees, dikes, and flood walls are designed to prevent inundation of developed areas by acting as physical barriers (Figs. 8.14, 8.15). These devices are only effective in restricting streamflow when it is less than the design height. When breached, damages can be considerable. Channelization is another technique whose purpose is to increase the capacity of the reach to discharge high flows more quickly. As mentioned in Chapter 11, such changes in flow regime simply transfer the

**TABLE 8.1** Advantages and disadvantages of different floodplain management methods

| Method | Advantages | Disadvantages |
|---|---|---|
| Land use regulations | Low cost<br>Can be used immediately<br>May prevent further damages<br>Can integrate other land uses | Existing flood damages not reduced<br>May inhibit development<br>May lower taxation base<br>Not applicable to all installations |
| Dams, all types of flood walls | Reduce flood losses<br>Protect existing property<br>May be multipurpose<br>Increase development and taxes | High cost and maintenance<br>Proper site may not be available<br>May fail, false sense of security<br>Produce harmful environmental feedback |
| Land treatment | May reduce flood levels<br>Can promote soil and water conservation<br>May be moderately costly | May not be applicable or possible<br>Effectiveness limited to minor storms<br>Need cooperation of most people |
| Public open space acquisition | Can reduce flood losses<br>Achieves broad community involvement for diverse benefits | Acquisition is costly<br>May create land shortage for development and other income purposes |
| Flood insurance | Helps promote community involvement<br>Inexpensive to policy holder<br>Alerts people to dangers, so can act to inhibit rampant growth | Only available to participating communities<br>Indemnification is limited<br>Flood damages are not reduced |
| Warning systems | Economically feasible<br>Prevent loss of life<br>Reduce damages<br>Residents not forced to move property | Community response uncertain<br>Require continuous information<br>Damage potential remains<br>Effectiveness may diminish with time |
| Floodproofing, site engineering | Reduces damages<br>Can remain at site<br>Flood insurance premiums are lowered | Damage potential remains<br>Limited to certain types of structures<br>May be costly with false sense of security |
| Relocation of existing structures | Flood damage eliminated<br>Provides open space<br>Improved hydraulics for floodflow<br>Safety is assured | Costly<br>Advantages of floodplain site are lost<br>Limited to certain structures<br>Resistance by owners |

**FIGURE 8.14** Sloughing conditions and repair of the Nairn Levee during the 1973 Mississippi River flood. Courtesy of U.S. Army Corps of Engineers.

**FIGURE 8.15** Sand boils and seepage water on the mainline Mississippi River levee near the Greenville Bridge, La. The high volume of water during the 1973 flood exerted great pressure on underlying materials and produced a hydraulic head that caused subsurface entrainment of sediment (Kolb, 1976).

problem to downstream areas. When the geography of the area permits, river diversion, as in the lower Mississippi River, may be an alternative whereby the high flows are diverted into huge holding basins, which then become manmade lakes (Figs. 8.16, 8.17).

*Small upstream reservoirs.* In some settings, these can provide relief from small potential floods. This strategem became an important technique for the Soil Conservation Service when they were empowered with building dams for limited areas under the Watershed Protection and Flood Prevention Act of 1954. To have an impact on major flooding, there must be numerous dams in a single watershed, because an upstream reservoir only controls runoff from a small portion of the area. However, such dams can be multifunctional by providing other benefits that can make them cost-effective in some instances. Small reservoirs may silt up faster than large ones because they are higher in the watershed, where incoming streams have higher gradients and can transport large loads in comparison to their size.

*Major reservoirs.* These can be effective for inhibiting many floodwaters simply because of their large storage capacity. However, to finance the huge cost of massive dams the benefit/cost ratio formula generally requires that they be multipurpose — used for water supply, hydropower generation, recreation, etc. The TVA (Tennessee Valley Authority) is the prime example (Fig. 8.18). From a flood-control view, the maximum protection is offered when water levels are low. But for electric power supply and recreation, users want high water levels, so these benefits are at cross purposes. The numerous environmental impacts from dams is further discussed in Chapter 11. An important concern about construction of such dams, and indeed also for building floodwalls and levees, is that people thereafter congregate adjacent to such sites in the belief that they are free from all disasters. For example, with completion of the flood-control project of the Trinity River, Tex., in 1957, the $172 million property value escalated to more than $1 billion with new construction in the floodplain. Thus, a flood if it occurs would do six times the damage as in pre-dam times.

## Land Treatment Measures

These are methods that are used on the land and hillslopes, not in or along the river channel. The objective of these measures is to improve the ability of the land to retain water for a longer residence time and to release it only slowly into stream channels, thereby dampening the effect of high-moisture conditions. The entire range of soil conservation methods are employed for this purpose. Reforestation and reseeding denuded areas can be a powerful runoff deterrant. The hillslopes can be treated by reducing

**FIGURE 8.16** Site of Old River Control Structure where the Atchafalaya River separates from the Mississippi River. This structure helps control the amount of water released during flood stage of the Mississippi. Photograph taken by the U.S. Army Corps of Engineers on April 16, 1975.

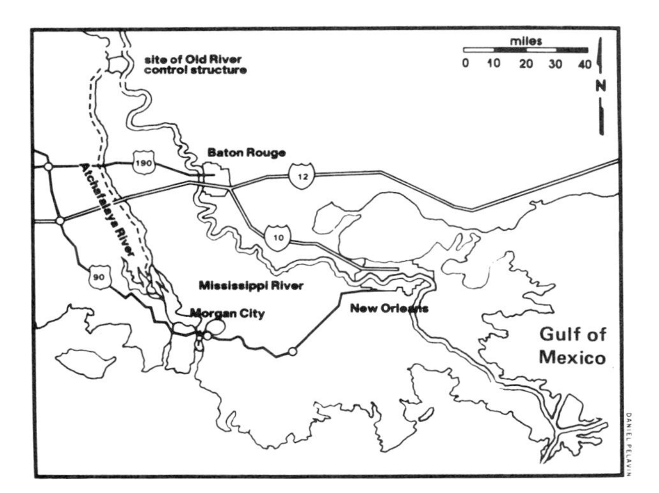


FIGURE 8.17 Map of the Lower Mississippi region in Louisiana. There is increasing apprehension that the Mississippi River is attempting to abandon its present course and discharge down the Atchafalaya River. At present, the sill structure shown in Figure 8.16 helps contain the Mississippi.

gradients, by terracing, and by building contour ditches and small check dams. Contour plowing, strip cropping, rotation of crops, planting of cover crops, and mulch emplacement on barren surfaces can all help prevent erosion, siltation, and excessive runoff.

### Legislative Measures

Governments can enact a variety of rules and regulations that restrict development and occupancy of flood-prone areas. **Zoning ordinances** can be passed that prohibit certain types of development in high-risk areas. **Building codes** can be established and **building permits** given only when construction is completed in accordance with strict guidelines that protect property and people. **Floodproofing** of buildings can be mandated so that any possible damage is minimized. For example, buildings should be reinforced, services that could short-out or cause fires should be placed in upper stories, steel bulkheads emplaced in doorways, and openings sealed with hydraulic cement.

**FIGURE 8.18** TVA's Kentucky Dam (built 1938–44) helps regulate floods on the lower Ohio and Mississippi rivers. The dam creates a 300 km long reservoir with a volume of more than 6 million acre-feet, which can cause as much as 1.2 m difference in downstream flood crests. Courtesy of TVA.

In urban areas, the acquisition of public **open space** can act as a buffer zone to inhibit flood potential. For example, Milwaukee has purchased virtually all floodplain areas for park and recreational use. These sites serve other functions such as control of private development, and they provide locations for nature preservation as well as for recreational use by the public. If the community does not want to buy open space, the property owner can be provided with certain tax incentives to encourage the retention of such sites in an undeveloped state.

In 1968, the federal government took an entirely new direction in its role as trustee of the safety and welfare of the general public. Prior to this time, legislation had been largely aimed at structural control measures, or ways to prohibit erosion on farms and timberland. Furthermore, disaster victims were commonly reimbursed from general revenues by some sort of subsidy or monies from disaster relief funds. For example, rebuilding could be accomplished by lower interest loans than were available for other forms of investment. The National Flood Insurance Act of 1968 made it possible for private insurance companies to write policies for those in flood-prone areas, providing certain conditions were fulfilled. To qualify for these policies, communities had to adopt approved land-use controls that met standards provided by the Department of Housing and Urban Development. Additional legislative amendments and passage of the Flood Disaster Protection Act of 1973 provided coverage to flood related erosion areas and for victims of mudslides. These also substantially increased the ceilings for federal subsidies and made mandatory that inhabitants take out such insurance if they hoped to obtain federal relief funds. If the district was approved as being eligible for insurance and the local government met all guidelines of the program, and homeowners didn't subscribe to the insurance program, they would receive no federal assistance for any flooding damages.

The first step in the program is the delineation by the federal government of the position of the 100-year flood. Those properties within the zone then become eligible for the program after the local government has received approval of plans to institute all measures as required by HUD (Department of Housing and Urban Development). The actuarial rates are established by the government. By September 1975, 609,000 policies with indemnification of $16 billion had been sold. Communities are required to participate in the program one year after the flood hazard area has been identified by mapping. If a community chooses not to participate by the deadline, federally related financial assistance for acquisition or construction purposes within the hazard area will not be available. More than 20,000 communities have subscribed to the program.

Unfortunately, not all homeowners take advantage of the low cost flood insurance. Even though the average annual premium was only $89, in Fort

Wayne, Ind., only 778 of the 10,992 property owners eligible in the floodplains had policies. The $22 million they received from the severe floods in March 1982 was only a small fraction of the total uninsured damage the community suffered.

## GEOLOGICAL DATA

The geologist needs to be involved in many aspects of problems and their solutions associated with the hazard of floods. He can provide the information that determines whether a particular river has serious flood potential. His calculations can tell the recurrence intervals for floods of different size and stage: for example, the characteristics of the 5-year, 10-year, 50-year, or 100-year flood. For planning purposes, geologists delineate these risk zones and show the information on appropriate maps. Geoengineers can determine the safest sites to build dams, and other structural controls if they are deemed necessary by the environmental managers. Upstream properties of the watershed can also be evaluated to determine the rates of erosion and siltation that might affect which planning strategem to adopt. They can also help in designing works that will create the least amount of environmental disruption to the hydrologic system. These data are therefore necessary before a well-planned and integrated management program can be adopted that seeks to minimize flood loss and environmental deterioration.

Societal awareness and preparedness must be paramount considerations for all ranges of flood-prevention management. From 1972 to 1982, the Federal Emergency Management Agency reported that floods in the United States killed 2000, caused property damage of more than $20 billion, and forced 300,000 to leave their homes. The sad part is that most deaths were attributed to people refusing to heed flood evacuation warnings, driving in flooded areas, entering flood-prone places, camping next to watercourses that might flood, and children playing near storm drains and culverts.

## POSTLUDE

Any assessment of water resources must account for floodflows in river systems. Facilities that temporarily impound floodflow may provide storage of water for other use, and thus are directly related to comprehensive plans and control of water and related land resources. Controlled use of floodplains including flood insurance, flood warning services, channel improvement and structural engineering, zoning, and other mandated governmental measures may help alleviate adverse effects of floodflows. Thus, the control of flood damages by preventative measures is closely related to comprehensive management strategies of the environment.

Flood control in the United States began as a matter of local concern, but by 1917, federal support for preventative measures was advanced for the Sacramento and Mississippi River basins. Since 1936, flood-control legislation in the nation has broadened from a single-purpose effort to reduce flood damage to consideration of flood control as a major element in the multipurpose development of the nation's rivers. Furthermore, planning and coordinating management plans has meant increased cooperation among federal, state, and local agencies because all have a stake in reaching the best possible way to coexist with flooding, which is a process totally natural, expectable, and oftentimes predictable.

CHAPTER 9

# COASTAL ENVIRONMENTS

The coastal zone is rapidly becoming the favorite habitat of mankind. By 1990, 75 percent of the population in the United States will live in this corridor where land meets the ocean or the Great Lakes. The invasion of the coastal zone was particularly accelerated after World War II. The reasons were both commercial — business and industry — and recreational — for second home sites, sports, and aesthetics.

Coastlines may seem impregnable, but they actually are some of the most fragile landscapes on Earth. There are vast differences in the 160,000 km of beaches, bays, estuaries, and wetlands that line the shores of the 48 states. And volcano-influenced coasts of Hawaii and glacial terrane shores in Alaska offer still more diversity. Each coastal area has its own particular rhythm and regime, and the complexity of these systems is only now being partly realized by scientists. Pounding surf, changing tides, and raging storms all cause vast, and at times sudden, modification of coastal terrane. However, when man enters the picture with his developmental projects and engineering structures, coastal processes invariably are altered, which results in accelerated degradation of this environment (Fig. 9.1). Therefore, it becomes necessary to understand more completely this very special landscape, where the land meets the sea.

## THE SEASCAPE

Unique landscapes occur and are highly varied in the coastal zone. Here, land and sea environments have molded the shoreline into a nearly endless series of topographic differences. The rockbound coasts of Maine, the barrier islands of the southeast and Gulf Coast states, the cliffed shores of California and Oregon, the fiords of Alaska, and volcano-shrouded slopes of Hawaii all give their indelible stamp on this land-sea interface. Deltaic regions such as the Mississippi River and the highly embayed and estuarine environments of the Chesapeake and Delaware bays also provide their own personality in these marine vistas.

**FIGURE 9.1** Damage to houses on Westhampton Beach, Long Island, N.Y. These houses are downdrift from a field of 11 groins that trapped sand, thereby producing excessive coastal erosion of adjacent downdrift properties.

The materials at the shore can also be highly variable – everything from massive rock cliffs, to sandy beaches, but also coral and organic materials as in the Florida Keys, or mud flats and mangrove wetlands along many southern shores. Both natural and human forces are at work reshaping and changing shoreline configuration, perhaps the most vulnerable terrain on planet Earth.

A prominent feature of many coastlines is the beach, the site where unconsolidated materials occur in the zone between low and high tides. On fully developed beaches (Fig. 9.2), there will be a shallow sloping underwater component that may contain a submerged bar. The foreshore is that part of the beach between low and high tides where the swash and backwash of waves reworks the material. The berm is a benchlike structure separating the foreshore and backshore areas. Its landward terminus may be a sand dune, that position on the beach beyond the normal reach of all but the greatest storm waves where wind processes modify features more than water does. Even sand dunes do not protect inland areas during large storms (Fig. 9.3), when the water surge penetrates the duneline producing a washover.

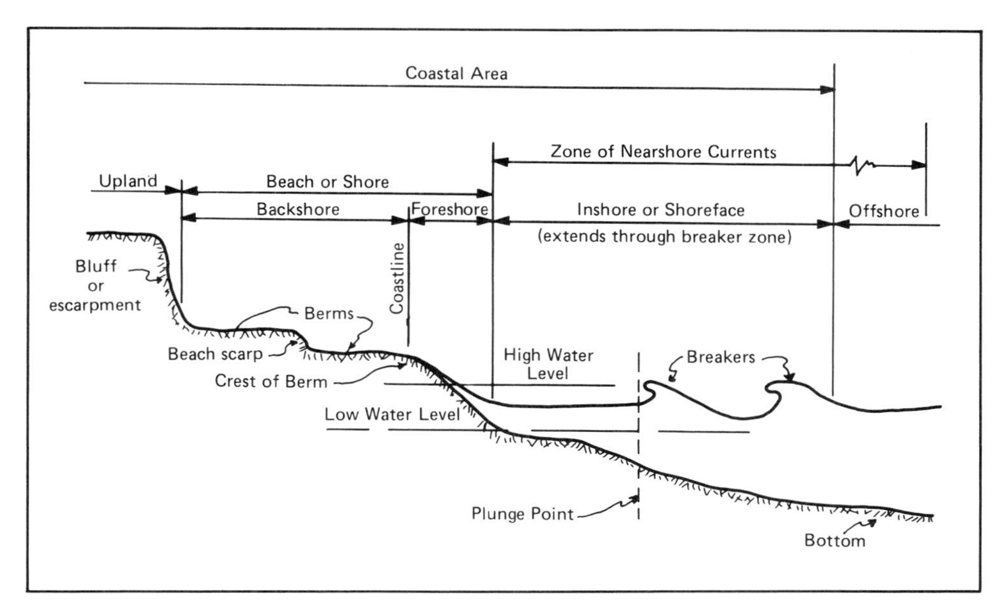


**FIGURE 9.2** Characteristics of the beach environment. Courtesy of U.S. Army Corps of Engineers.

## COASTAL FORCES

The principal processes sculpturing shorelines are waves, tides, and currents. Their effectiveness for creating rapid change is governed by land characteristics and the magnitude and frequency of the water forces. Waves are the single most important geomorphic agent that changes the coast. Waves are an indirect product of solar energy. Differential heating of the water surface produces pressure cells in the atmosphere, which cause wind. Thus, waves are oscillations of the surface water driven by wind, and are the result of two mechanisms. A pushing effect is created on the windward slope of each wave crest, whereas frictional drag is exerted by air moving over the surface. Wave height is a function of wind velocity and duration, and the distance it blows (fetch).

The shape and patterns of waves are changed as they approach near and on shores. With decrease in water depth, the wave crest rolls forward as a breaker (Fig. 9.4). This surge of water moves to the beach and upon encountering the land moves as swash up the beach and entrains material downslope upon return as backwash. The constancy of this motion influences beach shape and acts as a giant mill in movement and erosion of sand and other rock particles.

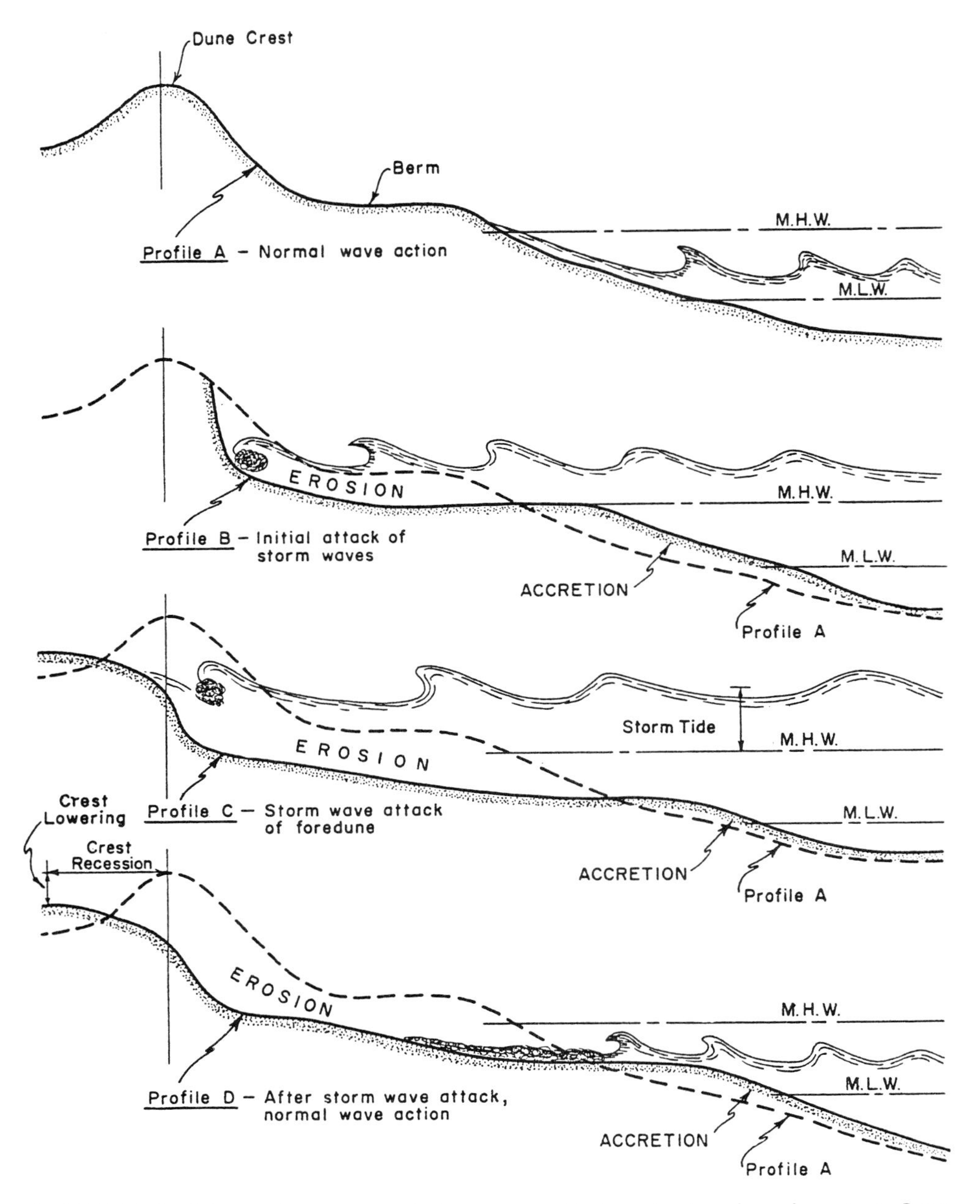


**FIGURE 9.3** Schematic diagram of storm wave attack in the beach environment. Courtesy of U.S. Army Corps of Engineers.

**FIGURE 9.4** Waves breaking on Nauset Beach, Cape Cod, Mass. Extensive damage resulted to beach-front houses and property from the destructive northeaster of February 5, 1978. Courtesy of Dick Kelsey Airviews.

When there are promentories that jut out into the ocean, wave attack becomes focused at such protrusions and develops a pattern of refraction whereby the envelope of greatest energy bends around such obstacles. On straight or gently curving shorelines, the direction of wave approach influences coastal processes. When waves strike the beach at an angle, the backwash becomes concentrated in the nearshore zone as a river of water moving parallel to the land. These longshore currents are held in this position by the mass of waves moving from the open ocean, and such currents become a powerful force transporting huge amounts of sediment as littoral drift. In many East coast areas, such movement amounts to 500,000 $m^3$ per year, as at Fire Island, N.Y., and on the West coast, material in transport along beaches may be more than 1 million $m^3$ per year.

Coastlines experience nearly two low and two high tides daily. This rising and falling of sea level is caused by gravitational attraction of the moon and sun and constitutes the most regular water change in coastal systems. Although the direct tidal force does not create spectacular landforms, two important by-products are created: (1) Rising tides allow waters to reach farther inland, thereby making possible greater erosion of coastal features. The Bay of Fundy tidal range is 15 m, and large tides commonly occur where coastal configuration is dominated by funnel-shaped estuaries. (2) Spring tides occur once yearly and these abnormally high tides, when coupled with storms at sea, have produced some of the most destructive disasters in coastal zones. For example, what has been termed the "Storm of the Five Highs" in March 1962 caused more than $500 million in damages and killed 32 in northeastern U.S. coastal areas. If a similar storm were to hit today, the costs would be 10 times as much.

Tropical storms (hurricanes in the Atlantic Ocean, typhoons in the Pacific Ocean, and cyclones in the Indian Ocean), along with tsunamis, have accounted for the major devastation of coastal communities in the past several centuries (Appendix G). Although warning systems have now been perfected that predict these ruinous storms, they continue to take staggering yearly tolls of human life and property. Tropical storms, so named because they originate in the tropics, are huge vortices of whirling air with speeds more than 120 km/hr. This dynamo sets waves in motion, which pile up on land to depths of 10 m and wreak havoc on all but the sturdiest of structures (Fig. 9.5).

Tsunamis, misnamed "tidal waves," are not generated by tides, but instead are caused by submarine faults, volcanic action, or landslides. The deceptive ocean swells that form near the submarine event measure only a meter in height, but as the motion speeds across the ocean at velocities of 700 km/hr waves on distant lands may be more than 15 m in height. Such unusual water heights were created by the tsunami that destroyed much of Lisbon in 1755 and killed 60,000 people and the Krakatau tsunami that drowned 36,000 in 1883.

Tsunamis are governed by the mechanics of shallow-water waves — those waves that are normally more than four times as long as the depth of water over which they pass. Thus, the wavelength of a tsunami swell is roughly equal to the length of the ocean bottom displaced by the fault (which caused the triggering earthquake). Since this distance is often 100 km or more, it is always far greater than the average ocean depth of about 5 km. The speed of shallow-water waves is proportional to the square root of water depth, which produces the phenomenal speeds of tsunami waves (Fig. 9.6). However, as they approach the shore, the shallowness causes the wave to slow. This inhibition dams a vast quantity of water, creating a gigantic wall that moves inland with a devastating surge. This event is preceded by a dramatic ebb of water from the shore, which exposes the nearby sea bottom. This was the cause for 159 deaths during the Hawaiian tsunami on April 1, 1946. Curious islanders ventured into this zone as the wave rushed from the Aleutian underwater earthquake.

In the eastern Pacific, the ocean can generate an unusual phenomenon, which although subtle in its formation, may lead to coastal and inland

**FIGURE 9.5** The September 1938 hurricane, the first to hit New England in several decades, produced widespread destruction on the barrier beaches of Long Island, N.Y. This aerial view at the west end of Fire Island shows devastation of summer homes on this part of the island. Courtesy of Fairchild Aerial Photographs.

devastation as severe as tropical storms, called the "El Nino effect." This bizarre phenomenon was first studied about 30 years ago, but its ramifications are so complex that it is still not well understood. About every five years, pressure readings in the southeastern Pacific begin to fall, trade winds become gentler, and warm water forms off the coast of Peru. These events generally occur in late fall before Christmas and alter anchovy supply and

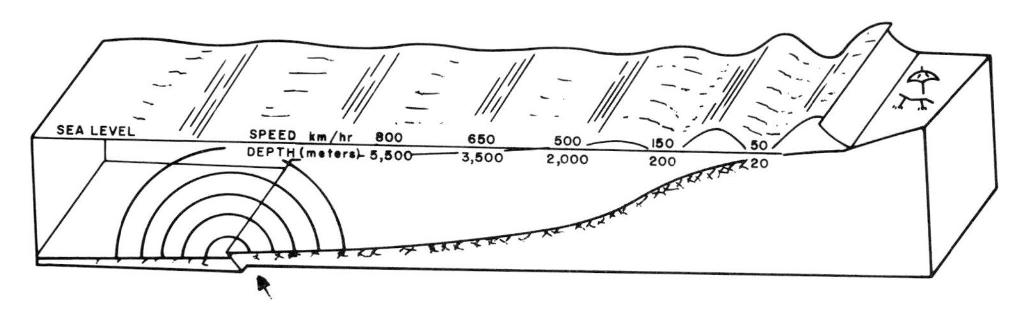


FIGURE 9.6 Schematic diagram showing evolution of a tsunami. A fault at the earthquake site thrusts the seafloor upward creating a surface bulge of water. Gravity helps produce a trough, initiating a tsunami swell. Toward land, the wave slows but wave height increases with shallower water, and it ultimately surges on shore with great power.

change tuna migration. Well documented El Ninos occurred in 1972, 1976, and 1982. They have even been blamed for such past events as the fall of the Peruvian kingdom of Chimu in A.D. 1100. The weather it created caused flooding that destroyed the elaborate canal system that irrigated the arid land. Crops failed, famine followed, and the weakened nation fell to conquerers from the south.

In May 1982, scientists at NOAA (National Oceanic and Atmospheric Administration) noticed ocean temperatures were more than 1°C above normal. By June, the air pressure had continued to fall, and trade winds were subsiding. In September, the conditions had produced a full-blown El Nino, several months earlier than usual, and the heat eventually reached the Gulf of Alaska, the farthest north ever encountered. Parts of the Pacific water became 5°C warmer, which in turn, heated the atmosphere more than usual. Apparently, these events also influenced the jet stream to take a more southerly course, thereby trapping a high-pressure system in Mexico that usually shielded California from the full fury of winter storms. In 1982, however, the northern low-pressure air in the Pacific Northwest pushed to the south, causing furious weather in California. Widespread flooding and landsliding occurred causing loss of life and damages of hundreds of millions of dollars.

Some experts say the principal culprit for the abnormally early El Nino was the explosion of the El Chichon volcano in Mexico in the spring of 1982. They suggest that the fallout into the atmosphere absorbed solar energy, thereby weakening the normal air circulation patterns that feed the trade winds. Others believe the two events are only coincidental. Whatever the cause, Canby (1984) reported the total worldwide destruction produced by the 1982 event amounted to $8.5 billion. Included in these losses were flooding, landslides, fish and crop destruction, and additional land degradation.

Indeed, coastal forces are constantly at work eroding, transporting, and depositing rock fragments and always changing the landscape. Whenever the shoreline zone is developed by man, there is always risk involved in the stability of his structures, and of life itself, if caught in the maelstrom.

## HISTORICAL BACKGROUND

For thousands of years, humans have accidentally or deliberately changed coastal processes and beach configuration. By 2000 B.C., the Mediterranean shores were studded with harbors that influenced coastal waters in Greece, Phoenicia, Egypt, and Syria. By 600 B.C., breakwaters were constructed in 30 m deep water at Eritrea with lengths exceeding 600 m. Duing Caesar's rule of Britain, reclamation of coastal marshes was started and by 1885, 20,000 ha had been reclaimed in the Romney Marsh of southeast England. Extensive coastal changes have been made throughout the Low Countries, and nearly one-third of the Netherlands is the hard-won trophy of the deltas in the North Sea. The sediment yields along shorelines, where two opposing influences occur, have been vastly changed by mankind. Accelerated erosion of inland areas produce extra sediment transport and ultimate deposition along shorelands. Paleolithic hunters in Europe set giant forest fires to drive game, which resulted in accelerated erosion of forestlands and increased sediment yield in rivers emptying into the North Sea. Destruction of forests and improper land cropping in the Fertile Crescent of the Tigris and Euphrates rivers four millennia ago helped produce the inland growth of the delta in the Persian Gulf of about 300 km. The opposite effect has been occurring in many coastal areas during the twentieth century because of extensive damming of rivers. Sediments from upstream erosion are now deposited behind the dams, thus impoverishing coastal beaches. Deprived of these sands, numerous beaches are now undergoing erosion — at times as much as 10 m per year, as in a 19 km stretch in the delta region of the Los Angeles River.

## CURRENT STATUS OF SHORELINES

In a 1971 study of U.S. coasts, the Corps of Engineers calculated that 22 percent were undergoing significant erosion, and that most needed remedial action for their protection. With the continual rise of sea level of 0.3 m per century, the preservation of beaches seems to be a "no win" situation.

Eastern shores are eroding at a rate of 1.5 m per year. With such sobering figures, it seems hard to reconcile the private development and government expenditures in such localities as barrier islands. These beaches are in high demand, not only in the United States but throughout the world, and constitute one of the major landforms of today's coastal zone. The largest string of these narrow islands, separated from the mainland by a bay, occurs along the southeast coast of the United States where many large cities, such as Atlantic City, Daytona Beach, Miami Beach, and Galveston, are built on barrier islands. They constitute 650,000 ha of total land, with 115,000 ha already developed and 300,000 ha unprotected by any type of land-management program. As late as 1976–79, the federal government invested $500 million in these ultra-vulnerable lands, and paid out another $400 million in coastal flooding insurance claims.

## BARRIER ISLANDS

During the past decade, one of the hottest issues in coastal management has been the status of the system of barrier islands (also called barrier beaches and offshore bars) in the United States. They stretch from Maine to Mexico along 1170 km of beaches and provide the first line of defense against Atlantic and Gulf of Mexico coastal storms. They are a succession of narrow low-lying islands, spits, and bay barriers generally located roughly parallel to the mainland coast and comprise about 650,000 ha in 400 units and clusters (Fig. 9.7). They are geologically recent in origin, formed by the interaction of rising sea level, waves, currents, tides, winds, and sediment. On their seaward side, they are generally fronted by a beach, whereas the mainland side may be highly variable, with zones of wetlands, lagoons, and bays.

Since World War II, federal programs have provided increasing support for expansion of residential and commercial development. Most projects have shifted from the private to the public sector to obtain monies for building roads, bridges, causeways, water-supply systems, wastewater treatment facilities, and shore protection, and for disaster relief and flood insurance. During the last 30 years, devlopment has grown at a rate that would completely consume all areas in the next 20 years. In 1950, only 36,000 ha had been developed, but by 1980, the figure had more than tripled. Nearly half of the islands are under public ownership, which includes 12 national seashores and recreational areas, 38 national wildlife refuges, and a number of state wildlife areas and parks.

Unfortunately, these fragile bits of land are extremely vulnerable to inundation, wave action, storm scour, long-term erosion and migration, and

**FIGURE 9.7** Barrier beach, at Fire Island, N.Y., before the September 1938 hurricane. This view is toward the west showing the narrow barrier island separating the Atlantic Ocean from Great South Bay and the mainland of Long Island. Courtesy of Fairchild Aerial Photographs.

high winds (Fig. 9.8). Such conditions are in direct opposition to the human desire for stability and property protection. Therefore, inhabitants have enlisted and received huge public expenditures to help ensure their safety. A variety of techniques have been employed in attempts to decrease the hazards from migration of old inlets, formation of new inlets during storms, preventing storm surge and overwash, and minimizing scour of the beach.

Although barrier islands act as natural sponges to absorb much of the fury from coastal storms, they suffer huge losses in property and real estate

**FIGURE 9.8** Fire Island, N.Y., after the September 1938 hurricane. This house formerly had a lawn and was situated behind a primary sand dune. It was the site for a washover. Courtesy of Fairchild Aerial Photographs.

in the process. Such destruction, however, has often not deterred continued development. Between 1955 and 1975, the amount of urban developed land increased 153 percent. On Westhampton Beach, Long Island, N.Y., where the 1938 hurricane left only 26 of 179 houses standing, there are now 900. On Beach Island near Wilmington, N.C., where Hurricane Hazel in 1954 left only 5 of 359 houses intact, there are now more than 2000. Condominiums continue to be constructed on Padre Island, Tex., at locations where inlets were filled after Hurricane Beulah in 1967, and again filled after Hurricane

Allen in 1980. In 1979, Hurricane Frederic leveled Dauphin Island, destroying a bridge and most of the houses in a resort of 1600. Fortunately, no one died, but the federal government spent $80 million to rebuild the bridge connecting the island to the mainland, and also reimbursed the homeowners. During winter storms of 1982-83, 150 houses were destroyed or badly damaged on Cape Hatteras, N.C. Nauset Inlet, Cape Cod, Mass., has lost three lighthouses during the last several decades, and in January 1980, storms at Fire Island, N.Y., cut a 250 m wide inlet that took $11.2 million and 760,000 $m^3$ of sand to close.

## HUMAN INTERVENTION

There is a long history in the United States for purposeful alteration of the coastal corridor. More than 40 percent of coastal wetlands in the 48 states have already been destroyed and continue to be lost at a rate of 120,000 ha per year. Hackensack, N.J., meadowlands have been filled as a solid landfill site, and 42 percent of the original San Francisco Bay has been reclaimed by landfill to create new bayfront property (Fig. 9.9). Dredging is another method for destruction of wetlands, which causes elimination of vital habitat for wildlife, and deterioration of water quality and hydrologic flow regimes. The Corps of Engineers dredges 300 million $m^3$ of sediment each year, and such a large volume also creates disposal problems. The other aspect to human changes in the coastal corridor concerns engineering structures that are emplaced to halt erosion and stem the flow of natural processes to protect property investments.

## COASTAL ENGINEERING

The emplacement of engineered structures in the shoreline zone is one of the strategies that may be adopted in any coastal management scheme. Two principal arguments are used to justify the building of structures: (1) the need to protect property, and (2) the need for preservation of sand as a resource. The type of program adopted for man's redesign of natural terrain depends on which of the following features need implementation: (1) shoreline stabilization, (2) backshore protection, (3) inlet stabilization, or (4) harbor protection. Before a project is designed, those who plan such structures should make sure the intended structures will accomplish their purpose, the construction will be sited at locations that produce the least environmental harm, and unnecessary feedback will not damage other localities.

**FIGURE 9.9** A wetland and estuarine environment was lost at Ocean Isle Beach, N.C., when it was filled and converted to a housing development with marina-type facilities. Courtesy of John Clark and the Conservation Foundation.

When a decision has been reached that natural conditions are intolerable and that manmade interference is necessary, the protection schemes can be placed into three main categories: (1) those that inhibit direct wave attack, which include seawalls, bulkheads, revetments, and breakwaters; (2) those designed to inhibit the transport of sand by currents, which include jetties built at bays and inlets, and groins; and (3) those that change the beach-zone topography, such as sand dunes and artificial beach nourishment.

### Seawalls, Revetments, and Bulkheads

These structures are built parallel to the shore, separating the land from the wave action. Their purpose is to absorb wave energy, thus preventing direct attack on human installations. Sloping seawalls produce less adverse reactions because basal scour is reduced by the more natural contours. These structures are often the last line of defense when there is little or no beach left, and man's developments such as buildings, homes, roads, and factories are adjacent to ocean waters (Figs. 9.10, 9.11). Materials used in seawalls and revetments, and their design, are highly diverse. For example, they may consist of concrete, riprap, steel, and timber (Fig. 9.12). Attrition is often high and constant maintenance is usually required.

**FIGURE 9.10** Jupiter Island, Fla. Severe beach damage and erosion caused by Hurricane Donna in 1960. Courtesy of Carthage Mills.

**FIGURE 9.11** Jupiter Island, Fla. A ramp-type sea wall revetment has been emplaced after Hurricane Donna in an attempt to prevent the apartments from being undermined by coastal erosion. Courtesy of Carthage Mills.

**FIGURE 9.12** Wooden sea wall at Fire Island, N.Y. This structure was built west of Ocean Beach in an attempt to protect homesites from accelerated erosion caused by two groins built into the Atlantic Ocean opposite the water tower in the background. The sea wall was a failure.

## Breakwaters

Breakwaters are structures that protect a harbor portion of the shoreline. They are constructed offshore, although one part may be shore-connected, and their purpose is to absorb and dampen wave energy. Because they impose artificial conditions on approaching waves, there is refraction and diffraction of the waves, which reduces the energy available for sediment transport. This commonly results in deposition of material on the lee of the breakwater. Periodically, the site needs to be dredged to keep the channel open. Beaches downdrift from the structure may be excessively eroded by the sand-starved currents. Although most breakwaters are constructed for permanency, floating or adjustable breakwaters have been in use at some localities where massive structures are not needed, as at some localities on Lake Ontario and in Rhode Island and Florida (Fig. 9.13).

## Jetties and Groins

These structures have many similarities — they are anchored to land but extend into the offshore zone and invariably are constructed of concrete, steel, or riprap (Fig. 9.14). Jetties are built at inlets to protect navigation channels, whereas groins are placed to extend out from the beach for the protection of the beach or adjacent property. To prevent the longshore material from entering the channel, jetties are often built beyond the breaker

**FIGURE 9.13** A flexible floating-tire breakwater. These have been installed at several marinas and beaches on Lake Ontario and in Rhode Island and Florida. Courtesy of Florida Sea Grant Program.

**FIGURE 9.14** Groins at Ocean Beach, Fire Island, N.Y. These two groins are constructed of concrete tetrapods. They have contributed to downdrift coastal erosion (see also Fig. 9.12).

zone and are sufficiently high to completely obstruct most sediment transport. Groins, which likewise may be considered as dams, can be classified as permeable or impermeable, high or low, long or short. The design depends on funding, sea conditions, and the purpose to be accomplished. For example, permeable groins permit the passage of more sand to downdrift beaches, thus causing less environmental change.

Jetties and groins cause major disruption of coastal processes. They both act as dams for longshore currents, which deposit sand on the updrift side. The loss of such sediment enhances the erosional ability of the currents in a downdrift direction, thereby contributing to beach loss in the nearby area. Because of this, single groins are rarely installed. Instead entire groin fields develop, as along the New Jersey coast, Miami Beach, and Long Island, N.Y. For example, the numerous Miami Beach jetties were installed after the 1926 hurricane, but only succeeded in hastening beach deterioration (Fig. 9.15). The Corps of Engineers has recently pumped 11 million $m^3$ of sand to the beaches at a cost of $80 million and will need to continue to add 150,000 $m^3$ each year to replenish the sand being lost by erosion. The groin field at Westhampton, N.Y., was largely responsible for causing $2 million in property damages to downdrift beaches during storms that occurred in Novem-

**FIGURE 9.15** Groin field at Miami Beach, Fla., prior to multimillion dollar beach nourishment project by the U.S. Army Corps of Engineers. The groins only accelerated the complete loss of beach along this row of hotels.

ber 1968 (Fig. 9.1). Jetties built in 1935 to stabilize the inlet south of Ocean City, Md., trapped the strong longshore drift of sand. Whereas the shoreline advanced on the north side of the jetties, beaches to the south at Assateague Island were eroded 450 m within a 20-year period.

## Beach Nourishment and Sand Bypassing

The process of artificially placing sand on beaches to reconstruct them has become more popular during the past 20 years (Fig. 9.16). Such nourishment can be especially effective as a desirable alternative to structural methods at those localities where humans have disturbed normal beach processes. Advantages of beach nourishment include: (1) a system that does not disrupt natural processes, (2) a beach suitable for recreational purposes, (3) a buffer against inland coastal erosion, (4) a supply of sand to adjacent beaches, and (5) a more aesthetically pleasing environment.

It is important that the sand emplaced on the beach should contain those grain sizes similar or slightly larger than materials previously eroded. Thus, selection of a suitable dredge site is vital to the success of the project. Sand bypassing is a method that is becoming more common because it can solve two problems at once. By dredging materials that may be clogging navigation inlets, such sites can be continually open to commerce, and at the same time, the material can be pumped to those nearby beaches that have been deprived of the sediment.

**FIGURE 9.16** Buxton, N.C. At this Cape Hatteras site, a washover is being rebuilt by an extensive beach-nourishment program. Note the huge pipes that deliver sand and water hydraulically pumped from an offshore barge pumping station.

When dredging sand for use as a resource, it is important that the extraction site is not from a locality where accelerated erosion will result. This occurred in 1894 at Hallsands, England, when 500,000 tons of sand were mined from an offshore position. Because wave energy was no longer dampened by the shallow underwater profile, severe erosion to the village resulted. However, successful beach nourishment programs have been conducted by the Corps of Engineers on the Connecticut shore at Prospect Beach, Seaside Park, and Sherwood Island State Park. Sand bypassing at this date has also been a success with the dredging of Fire Island Inlet and the construction of Gilgo Beach, N.Y.

## Sand Dunes

Sand dunes are generally considered important bulwarks in the coastal environment (Fig. 9.17). An unexpected and unfavorable chain reaction occurred on the Outer Banks of Cape Hatteras, N.C., however, when an extensive manmade system of dunes was constructed starting in 1936. Here on Bodie, Hatteras, and Ocracoke barrier islands, a massive 150 km long new duneline was installed and covered with 2.5 million shrubs and trees. Unfortunately, the high dunes led to a steeper foreshore, increased wave energy, and caused beach narrowing (Fig. 9.18). For example, where dunes had been

FIGURE 9.17 Artificially created sand dunes on Fire Island, N.Y. This dune has built to a height of 5 m by the successive emplacement of the snow fences. When one line of fencing becomes covered, another line is placed on top, or near it.

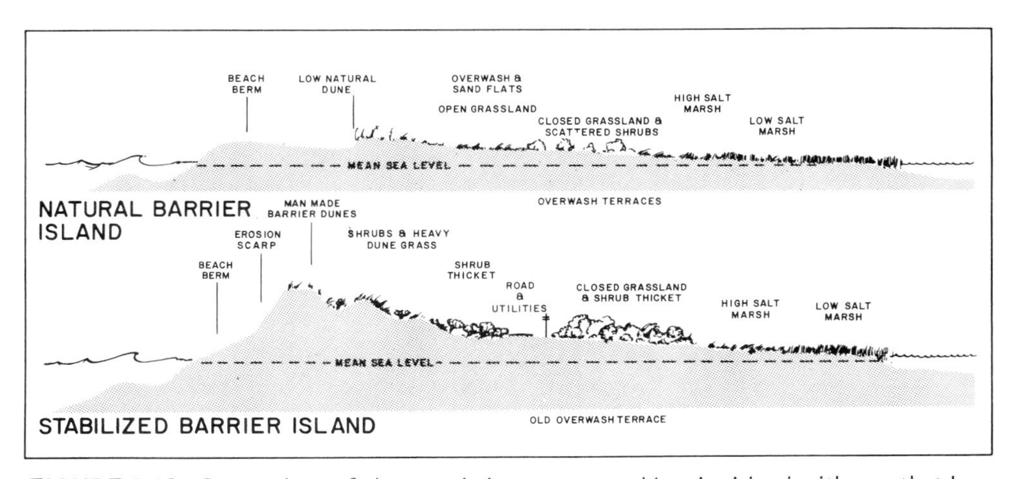


FIGURE 9.18 Comparison of characteristics on a natural barrier island with one that has been artificially "stabilized" by construction of manmade dunes (Godfrey and Godfrey, 1973).

built, the beaches by 1972 were only about 30 m wide compared to beaches at Core Banks (unmolested barrier beaches south of Ocracoke), which were 125 m wide (Godfrey and Godfrey, 1973). Thus, what had been viewed as a conservational measure turned into a nightmare of human-induced accelerated erosion. However, this is now a hotly debated issue because in other areas manmade dunes do not seem to have produced deleterious effects (Leatherman, 1979).

### Off-Road Vehicles

On some beach-dune systems, the use of off-road vehicles (ORV) has been shown to be particularly damaging to normal coastal processes. For example, at Fire Island, N.Y., there were 95 washovers (locales where storm surge waves penetrate through the duneline) that occurred as a result of the storm of the Five Highs in 1962. Detailed study of the washover localities showed that 57 of the washovers were at positions where the duneline had been deliberately cut through by humans, and most of these were attributed to vehicle cut-throughs. Prior to 1972, ORVs helped produce a recession rate of the duneline that was five times greater than along dunes where pilings protected against vehicle intrusion (Fig. 9.9). Therefore, with this great variety of possible ways whereby mankind can degrade the coastal corridor, is it any wonder that legislation has now become a high priority for the salvation of these fragile lands?

## COASTAL LAW

In A.D. 534, Justinian provided the basis for the common law of Rome regarding citizen's rights for the enjoyment of shorelines. This set the precedent for the Public Trust Doctrine of government responsibility under riparian principles. However, this was taken to the extreme degree in a decision rendered by Lord Tentreden of England in 1820. He rejected a landowner's case against a village whose groins had caused erosion of his property. The judge ruled he was not allowed damages and that if an owner wanted to protect his land he should erect his own fortification structures.

In the United States, coastal legislation and court rulings have become detailed and far-reaching, and are a vital component in coastal-zone management. In no other landscape theater has there been such a merging and cooperation of local, state, and federal statutes. The earliest significant legislation was the federal Rivers and Harbors Act of 1899. Under the Act, a permit must be obtained from the Corps of Engineers prior to release of pollutants into navigable waters. Permits are also required for all construction activities such as dredging, filling, and excavation. For more than 50 years, the Act was rarely enforced, but by the 1960s, it had become one of the primary statutes governing human degradation of the coastal zone. The most serious challenge to the constitutionality of the Act was raised in the case of Zabel and Russell v. Corps of Engineers. The claimants had been denied a permit to fill in 4.4 ha of Boca Ciega Bay, Fla., for the purpose of developing a mobile home community. The case was carried all the way to the Supreme Court, which ruled in 1971 that the Act was perfectly legal, reaffirming similar decisions by lower courts.

The first coastal land-use legislation passed by Congress was the establishment of the National Seashore System, which was begun in 1961:

> ...for the purpose of conserving and preserving for the use of future generations certain relatively unspoiled and undeveloped beaches, dunes, and other natural features... (Fire Island National Seashore Act of 1964)

A different type of federal involvement in coastal areas was taken in 1966 with passage of the National Sea Grant College and Program Act. Under this Program, the state university for each state can qualify for federal funds by providing for approved projects dealing with marine science, engineering, and related disciplines. After three successful years of funding, application can be made to become a Sea Grant College, which assures additional government financing of coastal projects. The Act has provided hundreds of millions of dollars to worthy projects, has resulted in hundreds of publications, and has fostered a new understanding of many facets of the coastal corridor and the important part it plays in resources, economy, and the welfare of citizens.

Two other important pieces of federal legislation were passed in the early 1970s. The Coastal Zone Management Act of 1972 is becoming one of the most significant environmental statutes ever passed by Congress. Its purpose is, "to establish a national policy and develop a national program for the management, beneficial use, protection, and development of the land and water resources of the Nation's coastal zones..." States that fulfill requirements of the Act become entitled to federal funds for both planning and implementation of approved projects. A bonanza of monies has become available for crucial coastal work. The other impact of the law has been to foster comprehensive management strategies by coastal states with accompanying state statutes that provide increased prudent use and conservation of coastal lands and resources. The Flood Disaster Protection Act of 1973 relates to both inland river flooding and to coastal flooding. It requires communities to enforce land-use and building regulations in "erosion-prone areas." In coastal areas, its purpose is to preserve the immediate shorefront, dunes, and upper beach and to protect life and property.

Perhaps the various wetland acts of many states personifies their awareness for the importance of maintaining balance in the land–water ecosystem. Massachusetts, Rhode Island, Connecticut, Maine, and North Carolina all had wetlands acts before 1970. Many of the acts provided for fines if wetlands were altered without first obtaining a permit. In the 1970s, nearly all coastal states had passed wetland legislation, and also, many states additionally passed erosion control acts. For example, Maryland, North Carolina, Delaware, and Florida all have some type of sand-dune protection or beach erosion control statutes. Other important types of state legislation were passed in Delaware with the Coastal Zone Act of 1971, which banned heavy industry from the coast, and the 1972 California coastal law, which controls all development within 1000 ft (300 m) of the shore.

The Coastal Barrier Resources Act became law on October 18, 1983. Barrier islands and bars 1000 km long comprise 188 units that extend along 15 Atlantic and Gulf Coast states. This was the first new land-protection law passed in 14 years. Prior to its enactment, the federal government had spent $750 million during a 3-year period trying to stabilize beaches, building roads, sewers, and other facilities on the island. It is the purpose of the Act "to minimize the loss of human life, wasteful expenditures of federal revenues, and damage to fish, wildlife, and other natural resources associated with coastal barriers along the Atlantic and Gulf coasts by restricting future federal expenditures." It includes such measures as disallowing any new flood insurance after October 1, 1983. It also prohibits construction or purchase of any structure including roads, bridges, sewers, and water-supply lines. Loans or grants of federal funds for development are denied, and limits are placed on stabilization projects. Thus, mankind has finally found an important ally — legal affairs — that can now be used to abate civilization's continuing desecration of this vital resource, the land–water interface that constitutes the coastal landscape.

Finally, it should be mentioned that the so-called Law of the Sea, which has been sponsored by many of the countries in the United Nations, has not received the backing of the United States. This statute does not affect coastal waters, so will be discussed more fully in Chapter 14.

## POSTLUDE

The love of the seashore has proved to be mankind's undoing in this highly vulnerable environment. Throughout history, and especially during the last four decades, many have thought they could win this struggle against what have proved to be unrelenting, unforgiving, and irresistible forces. The arrogance of many coastal engineers in assuming they could conquer nature, has resulted in untold hardships in loss of property and expenditure of huge sums of money in the endeavor. The reason this battle is a lost cause stems from: (1) the awesome power of oceanic forces, and (2) the continuing rise in sea level. Neither of these processes can be mitigated or controlled for long periods. For example, rising sea levels continue to claim more land each year, and there is no end in sight. Thus, those that occupy the coastal zone are hostages to nature, and their gamble is that their short-term enjoyment or amortization schedule will provide sufficient incentive to counteract the hazard and loss they will ultimately face.

The coastal corridor is the proving ground for feedback processes. Changes that occur at one site commonly have repercussions at another site. Groins and jetties are prime examples – sediment that is forced to accumulate on one side creates excessive erosion on the other side; upstream dams that impound sand and sediment reduce river material that nourishes beaches, which subsequently undergo erosion; and removal of coastal area sand for beach nourishment or other use may prove more harmful than beneficial. This certainly proved to be the case at Hallsands, England. Near cities, offshore sand and gravel mining is proving popular because barge loading makes transportation convenient and affordable. Since 1963, New York City has added 165 $km^2$ of land to its five boroughs. The source has been the lower bay of New York Harbor. The resulting holes, however, may lead to problems where organic matter accumulates and reduces the oxygen supply, the tidal range is altered, and coastal erosion is accelerated.

CHAPTER 10

# SOIL DESTRUCTION

"The Farm – best home of the family, main source of national wealth, foundation of civilized society, the natural providence." These words are chiseled into the stone above the entrance to Washington's Union Station. To complete these sentiments, a postscript needs to be added, "To the soil, the backbone of the farm." Just as the farm is indispensable to mankind, so is soil, along with water, the vital ingredient necessary for human existence. Not only does the farmer depend upon the soil, but so do the rancher and the forester, and even the engineer and land manager who must plan structures and all of the elements that comprise the very essence of civilization. Because people live, work, and play on the earth's surface, nearly every activity of the human race causes some change in the soil. These changes invariably cause deterioration of the soil, alter its natural properties, and make it less productive as a resource.

There are numerous ways in which soil is ruined, or lost, by mankind's developmental projects – erosion, siltation, salinization, water logging, deforestation, and urban sprawl. Soil losses throughout the world are staggering, and even in the United States, today's losses are greater than they were during the height of the Dust Bowl years in the 1930s. Improper stewardship of the soil by past civilizations was at times the cause for their downfall. This was true for some of the Mesopotamian cultures, for the Singhalese civilization on Sri Lanka, and probably even played a role in the decline of Rome. Thus, knowledge of the soil and the important role it plays in a nation's health and wealth should be of paramount concern not only to environmentalists, but to those in positions of leadership who make decisions concerning its use and abuse.

## WHAT IS SOIL?

To the engineer, soil is any natural material capable of being bulldozed without blasting. However, it will be more useful in our context to consider soil as those surface materials that have a structure which can support rooted plants. Although there are numerous types of soils, they have in common distinctive horizons that separate dissimilar compositions and structures. The composite of these is called a **soil profile** in which the A-horizon is at the ground surface and contains decayed organic materials. The B-horizon underlies this and possesses leached materials from the top horizon. The lowermost, or C-horizon, is in contact with bedrock and shows partial weathering of the original parent material. Soils are formed by a combination of weathering processes that cause loss of cohesion of the bedrock by physical, chemical, and biological processes. The geologist uses the term **regolith**, which is applied to all other unconsolidated surface materials on top of bedrock, regardless of origin or composition.

Each environmental setting has its own distinctive type of soil. The composition and fabric of a soil depends on such factors as climate, vegetation, rock properties, age, and terrain conditions. For example, in the same region, soils will have entirely different characteristics if one is on a hillslope, and another soil is forming in the valley bottom. In similar manner, tropical soils have vastly different features from arid soils or humid soils. Thickness of soil profiles may range from those that are only several centimeters in thickness to those where the profile and depth of weathering may extend 100 meters or more, as in some places in the Piedmont Physiographic Province.

Soils are so complex that the Soil Conservation Service (U.S. Department of Agriculture) took three decades before they arrived at a classification system that received internal agreement, known as the Seventh Approximation (Olson, 1981). They classify soils into 10 orders, 40 suborders, 120 great groups, 400 subgroups, 1500 families, and 7000 series. Although it is not important for an environmental earth scientist to memorize such a complicated array of types, it is important to know that some soils are more resilient to abuse than others, and some soils are extremely fragile and vulnerable to any manmade changes (Figs. 10.1, 10.2). When forests are cut in the tropics, the soils rapidly change, lose fertility, and can become duricrusted. Soils in permafrost terrane (Chapter 11) quickly undergo degradation when the protective vegetative mat is disturbed. Semiarid soils with high montmorillonite content produce enormous damage to structures and facilities when their expansive properties have not been properly accounted for. The wetland soils in regions undergoing "reclamation" lose their fertility, undergo desiccation and loss of volume, and produce land subsidence (Chapter 4). Obviously, it is therefore important to know which types of soil occur on land parcels before wise decisions can be reached on how to use and manage them.

**FIGURE 10.1** Badland topography near Nahal Besor, in the western Negev, Israel. Photograph (A) shows an overview of the region, and photograph (B) shows the detailed development of piping. This "piped-type" of terrane, according to Aghassy (1973) has had accelerated growth from human impacts in the region.

**FIGURE 10.2** The plastic soil in the Honduras village makes a very poor roadbed for vehicular traffic and is easily degraded. Courtesy of Gerald Olson.

## HISTORICAL BACKGROUND

Throughout history, mankind has engaged in activities that have caused destruction of the soils upon which he is so dependent. As early as 368 B.C., Plato described the denudation that had taken place on Attica, "...Attica... is like the skeleton of a body emaciated by disease, as compared with her original self. All the rich, soft soil has moulted away, leaving a country of skin and bones." Few other writers, however, chronicled contemporary devastation, so our record of past excesses only comes from the deciphering abilities of archeologists and geologists. When the data are combined, it is indeed a very sad record of decline and decay of numerous civilizations. To what degree the human denominator played a major role is uncertain, but in many cases, it undoubtedly was very great.

There are numerous ways that man contributes to the deterioration of the soil and food resource base. These include burning, deforestation, loss of soil fertility, loss of water supply, overgrazing, siltation, salinization, and soil erosion. The ravages of deforestation aided the collapse or migration of such civilizations as the Singhalèse in Sri Lanka and the Inca in Peru. Mesopotamian civilizations from 1700 to 1300 B.C. greatly suffered from loss of soil fertility because of water logging and salinization problems. Silt also clogged many canals that distributed irrigation waters from the Tigris and Euphrates rivers. It has been postulated by Bryson and Baerreis (1967) that the Harappan civilization in the Indus Valley suffered and declined because of a desertification process caused by an increased albedo from farming and deforestation.

The theme of several books emphasizes the magnitude of such losses, such as those by Marsh (1864), *Man and Nature*; Jacks and Whyte (1939), *Vanishing Lands*; Sears (1947), *Deserts on the March*; Osborn (1948), *Our Plundered Planet*; and Eckholm (1976), *Losing Ground.* Some writers such as Simkhovitch (1916) have even attributed the fall of the Roman Empire to a failure to manage carefully the soil resources.

## SOIL DESTRUCTION

In this chapter, we are primarily concerned with the manner of accelerated soil loss due to human activities (Fig. 10.3). Normal or geologic erosion also occurs, and this has been termed "tolerable erosion." For most agricultural soils in the United States, this amounts to about 2 tons per hectare (2 t/ha). It is determined by the rate at which new soil is created by subsurface weathering processes. Modern man amplified normal soil loss with such new ways as industrial fumes and construction activities (Fig. 12.5).

**FIGURE 10.3** Deforestation on these Philippine hillsides has destabilized the soil and led to extensive gullying, landsliding, and aggrading of debris downgradient. Courtesy of John Wolfe.

Even the process of urbanization in the United States contributes to this loss because each year about 1 million ha of cropland are withdrawn from farm use to make way for developmental purposes. I will classify as soil loss, or soil destruction, as occurring on those surface materials that have been eroded from their normal ground position, as well as changes in their chemical composition that inhibit proper plant growth.

The principal method of human-induced accelerated erosion is the reduction in the infiltration capacity of the soil. This increases the erodibility potential because more water is available for surface wash, which results in entrainment and movement of soil particles (Fig. 10.4). Deforestation and devegetation are also important ways that contribute to impoverishment of the soil. On barren ground, raindrop impact and wind deflation of particles cause changes on the land surface. The loss of the vegetative mat also produces a lowering in nutrient levels of the soil as well as destroying the protective root-binding force.

Soil deterioration affects plant productivity when available soil moisture is reduced. This subjects plants to more severe stress, stunting their growth and lowering their nutrient value and palatability. Removal of top soil exposes the less productive and clay-rich horizons to surface conditions. To enrich them and maintain yields, additional fertilizers are needed. These soils are more erodable so that both silt and fertilizers are washed into watercourses, which then become silted and burdened with pollutants. Thus, destruction of soils contains twin problems — erosion ruins the land, and the

**FIGURE 10.4** Accelerated soil erosion in the Andes Mountains near Apartadero, northwestern Venezuela. Horizontal markings on the hillside indicate the steep slopes were formerly intensively farmed. These human activities helped trigger the gullying that is degrading the landscape. Courtesy of Gerald Olson.

concomitant eroded debris in the form of silt ruins rivers and lakes (Fig. 10.5).

## FARMLAND EROSION

Water and wind erosion of farmland, a worldwide problem, has afflicted America's soils since cultivation started in the 1700s. George Washington and Thomas Jefferson both described the problem, and each instituted certain conservational measures in attempts to reduce degradation on their farms. The early tobacco farmers in Virginia and North Carolina constantly moved their farming areas when field after field became infertile after continual use for several years. However, it took the Dust Bowl years of the 1930s for the nation to mount a governmental effort to halt the escalating loss of soils from America's Midwest breadbasket. No other single publication has had such an effect as that written by Bennett and Chapline in 1928 entitled *Soil Erosion – A National Menace.* It catalyzed the government into organizing the Soil Conservation Service with a mission to arrest and reverse

**FIGURE 10.5** Sand deposits in channel of the Canadian River, Okla. Example of the erosion-sedimentation cycle that was accelerated during the Dust Bowl years. Photograph taken by the Soil Conservation Service (USDA) in spring 1938.

the trend in loss of such an important national resource. The soil conservation measures seemed to be largely successful, and the rate of soil erosion was reduced. However, by the 1970s, this was all changed, with increasing tillage for soy beans and expanded farm markets because of foreign grain sales to the Soviet Union, China, and other countries. Also contributing was the demise of federal programs that formerly paid farmers to keep fields idle. Some lands previously considered marginal and fragile were put into production. The former terraces and other structures installed to prevent erosion were impediments to new enormous types of farm machinery, so they were levelled and planted. Thus, once again a giant turn-around occurred, so by 1979, Soil Conservation Service reports showed that soil erosion had become an even more severe problem than it had been during the Dust Bowl years.

In 1975, a report by CAST (Council for Agricultural Science and Technology, a consortium of midwestern universities) showed that more than one-third of the nation's cropland was suffering severe soil losses too great to be sustained and would ultimately be disastrous. The average annual soil loss from cropland has now reached nearly 5 t/ha, three times the level of tolerable erosion. Since it takes hundreds of years for soil regeneration, such loss can be considered the loss of a non-renewable resource. Prior to 1940, more than 80 million ha had been ruined or impoverished by soil erosion, and these areas contained one-third of the nation's rich productive topsoil. Since then, an additional 72 million ha of cropland have suffered from severe water erosion, and 22 million ha are in jeopardy from wind erosion. Only one-third of the farmlands have effective soil conservation measures according to the U.S. Department of Agriculture.

Of course, soil losses are not uniform throughout the country. Iowa has been especially hard hit, where it is estimated that more than $2 billion are necessary to institute proper controls to reduce erosion. Erosion in Ohio has increased 22 percent because of greater farming intensity to produce more crops. To stop wind erosion during the Dust Bowl years, vegetative windbreaks were planted, but many of these have now been dismantled in the rush to increase crop area. In one Oklahoma county, 21 percent of the windbreaks were removed, and according to the Soil Conservation Service, this led to wind damage in the 1970s of 7500 ha, which previously had been protected. The wind storms in 1977 severely eroded 560,000 ha throughout the Great Plains, lands that had previously been fertile.

Each year, erosion in the United States amounts to more than 5 billion tons, of which three-fourths is by water and one-fourth by wind (Fig. 10.6). However, soil erosion is a worldwide problem, and the United States has no

**FIGURE 10.6** Windstorm near Pacific City, Ore. These 80 km/hr winds have blasted the soil and sand and are eroding the soils where protective vegetation had been planted. Photograph taken by the Soil Conservation Service (USDA) on January 29, 1960.

monopoly on losses. Much smaller countries sustain even greater losses per hectare. For example, the annual losses for some countries are: Colombia, 426 million tons; Nepal, 500 million tons; Ethiopia, 1 billion tons. Much of these losses, as indicated later, are by accelerated erosion caused by deforestation. The United Nations has reported that one-fifth of the world's cropland has reached a stage of intolerable soil degradation. Who will feed the world population during the coming decades or centuries?

## DEFORESTATION

Forests of our planet, especially those remaining in tropical and subtropical climates, are fast disappearing. Although in many industrialized nations, accommodation has been reached between timber being marketed and that being planted, in the LDCs, it is an entirely different story. Here, the forests are being rapidly cut down for fuel and to create cropland. Indeed, nearly one-half the world population depends upon wood as their primary fuel source. The **slash-and-burn** type of agriculture is still practiced in many places such as the Philippines and Central and South America (Fig. 10.3). However, the new style of forest destruction is a **cut-and-run** type of logging operation, to market trees as rapidly as possible without further immediate use of the desolated land.

During the 1950–80 period, nearly half of the world's remaining trees had been cut, and yearly losses were about 2 percent of the remainder each year — an area the size of Indiana. Within this short space of three decades, this unusual loss in land resource changed from one of global forest wealth to one of global forest poverty. At these rates, the world's tropical forests will be largely decimated in the next two to three decades. In Thailand, only 37 percent of the forest remains; in Costa Rica, 38 percent; in the Philippines, less than 20 percent; in Java, 12 percent; in Haiti, 9 percent; and in Gambia, 4 percent (down from 56 percent since 1946). Within the next 20 years, total deforestation is predicted for Burundi, the Ivory Coast, Ethiopia, and many others. The problem is nearly equally severe in Kenya, Upper Volta, Panama, Indonesia, India, Burma, and especially the Himalayan foothills. For example, Nepal has 10,000 $km^2$ that are so recently barren that desertification has started.

The Amazon Basin is a special example that points to the severity of the problem, not only on a local level but having worldwide repercussions. This region, with an area larger than the continent of Australia, has one-third of the world's forestland, one-fifth of the world's freshwater, and produces one-half the world's plant-generated oxygen. However, by the 1970s, vast amounts of the forest were being cleared, as in 1975 when 162,000 $km^2$ were denuded. One might think that lands, which can support such lush tropical vegetation, would contain extraordinarily fertile soil. However, just the opposite is true. Tropical soils are extremely fragile, and the nutrients in the system are locked up in the plants with little reserve in the soil. When the lands are cleared, soils are exposed to desiccation by the hot sun, the fine-grained materials shrivel, bake, and are subject to rapid removal with the first rains. Shifting cultivators who resort to slash-and-burn operations formerly allowed such terrains to readjust over spans of a decade or more. Today, with the onslaught of expanding populations, such lands no longer can lie fallow for long periods of time, and become subject to attempts to grow crops within a few years. Soon, their remaining fertility has completely vanished, and they become worthless as support systems for the populace.

The demise of tropical rain forests can have global as well as local and regional effects. The lands lose their fertility for decades. Consequent erosion further devastates the ground and produces excessive siltation in downstream reaches. Perhaps most frightening of all is the possible climatic transformation that might occur. This could be done by changes in the earth's albedo, an increase in atmospheric carbon dioxide, and a loss in water vapor. For example, studies have shown that nearly one-half the rain in the Amazon Basin is generated by evaporation from the forest itself. Thus, deforestation could completely upset the hydrologic balance of the region with possible worldwide implications.

Even in countries that have the ability to harvest timber and replant on a nearly equal basis, the style of forest management and the environmental matters have come into conflict. Rapid expansion in timber use occurred in the two decades from 1951 to 1971 in the United States. Whereas the lumber volume was only 1.5 billion board feet in 1951, it expanded to 8.3 billion board feet in 1961, and to 11.5 billion board feet in 1971. This growth also led to new production methods, namely, clearcutting. The practice gained special momentum starting in the 1960s, when entire tracts of timber were cut at the same time. By 1970, 30 percent of the national forests were being clearcut in the West and 40 percent, in the East. Proponents believe the immediate economic factors outweigh the long-term detrimental environmental impacts. However, loss in soil and nutrients has now been demonstrated at many sites. Runoff is usually increased 20–40 percent in clearcut areas, with accompanying increase in erosion and downslope siltation. In Oregon, clearcut areas produce 100 times more sediment than selective-cut lands. In the Pine Tree Branch Basin in Tennessee, sediment yield was reduced during a 16-year period from 10 t/ha to 0.4 t/ha by replanting the clearcut area. At the Coweeta Experimental Station, N.C., loss in soil nutrients from clearcut areas, such as calcium, potassium, and sodium, ranged from 3 to 20 times higher than in selective cut areas, and nitrate losses were even greater (Figs. 10.7, 10.8). Because forests act as a cushioning force against water erosion and help stabilize the soil and ground surface, special care should be exercised in the manner of timbering and resolving problems that are created by deforestation.

## SALINIZATION

Just as erosion and siltation are two aspects with a single cause, salinization and water logging commonly occur when man decides to irrigate dry lands. This produces destruction of the soil that is equally as severe and maybe even longer-lasting than topsoil removal. Salinization of soils has reached awesome proportions throughout the world so that it now affects one-third of all irrigated lands, and threatens another one-third in the near future. Salinization can also result from deforestation of plains in dryland environments (Figs. 10.9, 10.10).

Nearly all waters used for irrigation purposes contain some dissolved salts. When plants transpire, or soil moisture evaporates, salt is retained on

**FIGURE 10.7** Coweeta Hydrologic Laboratory, N.C. This is a multipurpose site to determine the effects of clear-cutting on timber, wildlife, water yield, and hillslope erosion of soil. Courtesy of James Douglas, U.S. Forest Service.

**FIGURE 10.8** Coweeta Hydrologic Laboratory, N.C. The clear-cut drainage basin in center of the photograph will be compared with environmental factors of the adjacent timbered slopes. Courtesy of James Douglas, U.S. Forest Service.

the land or in the plants. Water not used by plants or evaporated will percolate down to the water table, which will then become saline. If excess water doesn't drain away, the water table will rise. If it reaches the root zone, the crops die because the oxygen supply is choked off. Such water logging is the twin problem of salinization for especially level lands and those near river sources. With repeated wetting and drying of the soil, the salt content creates layers of crust within the soil. Centuries of this cycle left the fields in what is now Iraq as white as snow, and was probably the principal cause for the decline of the Sumerian civilization.

Prolonged irrigation and inadequate drainage systems are principal causes of salinization, but there can also be salt encrustation when too little water is used. Under this condition, the water filtering down is insufficient to flush

**FIGURE 10.9** Ruined soils by salinization process caused by deforestation, Western Australia. Courtesy of Charles Finkl, Jr.

**FIGURE 10.10** This type of salinization of soil is called "salt scalding" and resulted from a rise in the water table due to clearing of vegetation in Western Australia. Courtesy of Karl H. Wyrwoll.

out the inevitable salt deposits. Thus, they too accumulate within the root zone and cause soil destruction and plant loss. Strangely, the paradox of too much and too little water can coexist in nearby areas, such as occurs in Pakistan. By 1960, the twin problems of waterlogging and salinity had reached such proportions that they took a toll of one-fifth the cultivated area of the Indus Plain. In India, 6 million of the total 40 million ha of irrigated lands have now become unusable for crop production.

The United States is not immune from these problems, which now especially afflict the Southwest and California. The rich San Joaquin Valley of California has been a farmer's paradise. The fertile soils, plus the imported water needed because rainfall is only 24 cm per year, have yielded crop values of more than $5 billion each year since 1977. Unfortunately, the very waters that have created this "golden egg" now threaten it with ruination. It is the same old story, salinization and water logging. The perched water table has inexorably risen to depths of less than 1.5 m over more than 160,000 ha. The brackish water is lowering food production, causing some farmers to shift to more salt-tolerant crops, and has forced the installation of extensive subsurface tiling. More than 26,000 km of the tile have been installed in an attempt to drain away the deleterious salts, which must be flushed out of the system with extra heavy water application (see also Chapter 4). The total yearly expenses and losses due to these conditions now approaches $100 million.

Of course, the story does not end here. Just as soil erosion produces downstream siltation, soil salinization projects remove some salts, which then plague downvalley or downstream users. The Wellton–Mohawk Irrigation Project in southwestern Arizona drains the salt into the Colorado River, which has doubled in salinity, and causes grievous losses to farms in Mexico which depend upon the water. To prevent further international illwill, the United States is constructing a desalinization plant near Yuma at a cost of more than $300 million.

## OTHER SOIL LOSSES

The litany of how mankind changes, destroys, removes, and covers soils is seemingly endless. Elsewhere in this book, some of these are further discussed under mining activities, construction of dams and highways, introduction of waste products, and initiation of landslides. For example, what has been called the "quiet theft of soil" occurs from ozone pollution and acid precipitation. These atmospheric changes cause soil and tree losses that are now estimated at $2 billion a year in the United States. The fumes from smelter operations produced drastic soil erosion at such sites as Ducktown, Tenn.; Swansea, Wales; and Sudbury, Ont., Canada (Fig. 12.5). A few of the additional ways that soils are destroyed are discussed in the following sections.

### Overgrazing

Since mankind's domestication of animals, they have often ruined the very lands on which they have grazed. From ancient times to the present, sheep and goats have been important causes for accelerated soil loss on the hillslopes of the Mideast. Sheep and cattle have combined to increase soil erosion in the United States and Australia. Some observers even believe that the enormity of the destruction has been sufficient to induce new cycles of erosion in the American Southwest. And in the Sahel, others feel animals have been important contributors to desertification.

John Wesley Powell, pioneer explorer of the West and geologist who helped lay the framework for the U.S. Geological Survey and U.S. Bureau of Reclamation, wrote the following words more than 100 years ago:

> Though the grasses of the pasturage lands of the West are nutritious, they are not abundant, as in the humid valleys of the East. Yet they have an important value. These grasses are easily destroyed by improvident pasturage, and they are replaced by noxious weeds. To be utilized they must be carefully protected and grazed only in proper seasons and with prescribed limits. They must have protection or be ruined (Crosette, 1970, p. 63).

In terms of total amount of land affected, such improvident pasturage, or overgrazing, has been the most potent force for soil destabilization and desertification in the American West. Its symptoms are declining water tables, increased salinity of soil and water, reduction in surface waters, unusually high erosion rates, and loss of native vegetation.

A U.S. Bureau of Land Management 1975 study found that half the area of 65 million ha it manages had rangelands that were described as being in "only fair condition." On these lands, the valuable forage had been depleted and replaced by less palatable and nutritious plants or by barren ground. Another 28 percent of the range was in poor condition, stripped of much of its topsoil and vegetative cover. Five percent of the total land was in "bad" condition, one where nearly all topsoil was gone. Overgrazing was ruled as the principal cause for the destructive soil changes that now occur on these lands, an area the size of Utah.

In 1978, Congress passed the Public Rangelands Improvement Act and authorized "...an intensive public rangelands maintenance, management, and improvement program involving significant increases in levels of rangeland management, and improvement funding for multiple use values." The mandate was backed with a commitment of $365 million over the next 20 years for a program of intensified rangeland management, but subject to annual appropriations. This was something the Taylor Grazing Act of 1934 was supposed to accomplish but failed. Political pressures from the livestock industry from 1934 to 1976 had effectively blocked the implementation of that Act. Unfortunately, conditions have not improved under the new program, and funding levels have been extremely low.

## Urbanization

By the 1950s, a new phenomenon was being recognized as a process that was quickly changing the landscape, that of a new wave of urbanization. Cities were rapidly expanding by an influx of people from rural areas, and also by the exploding population boom. City governments were not prepared to handle these changes, and in the environs, a new metamorphosis was taking place. Satellite communities and unrestricted housing and other developments occurred, a process described by William Whyte as "urban sprawl." Soils and the environment were usually not considered in such planning, with the result of aesthetic blight, land pollution, and prodigious losses (Fig. 10.11).

Under these conditions, soils were destroyed in two ways. During construction activities in the hurry to complete a project, design engineers rarely undertook methods to safeguard soils or to prohibit the enormous increase in siltation. It was not unusual for housing and road development to increase rates of silt production many thousands of times the normal rate. For example, studies in Maryland showed that at construction sites as much as 55,000 $t/km^2$ of silt were being produced, whereas the pre-construction rate was only a few $t/km^2$. The other type of soil loss occurred when the land was taken out of farm production and covered over by roads, parking lots, housing developments, buildings, utilities, and other types of development in and adjacent to the city as well as the corridors that supply it. Because of the urban sprawl, large amounts of land were being consumed in extravagant fashion. Studies show that the typical urban sprawl that was occurring in the United States up to 1970, used seven times the amount of space of a planned community. The U.S. cities of the 1960s used three times as much land for the same population as those of the 1920s. Is it any wonder that throughout the nation, 1 million ha of croplands per year were being lost to development in the United States by 1974? Since that time, the rate loss of soil has been reduced, however, due to a variety of reasons such as housing costs, gasoline prices, high taxes, construction of more multiple-family dwellings, condominiums, consolidation of industries, etc.

**FIGURE 10.11** Hillside gullying in a new subdivision of Lakewood, Colo. The vegetative ground cover was completely eliminated during construction of these sites. Courtesy of Wallace Hansen.

## Off-Road Vehicles

Another post-World War II phenomenon has added yet another hurdle in the struggle to maintain soil equilibrium throughout the lands — **off-road vehicles** (ORVs). The decades starting with the 1950s have witnessed an explosion in the numbers, types, and mobility of vehicles that travel every conceivable terrain. They range from jeeps and dune buggies, to motorbikes, pickup trucks, and snowmobiles. They are changing the face of the land, on beaches (Chapter 9), in the drylands, in forests, and even on mountainsides. Indeed, few areas seem safe from their ravages.

A 1977 report by the U.S. Bureau of Land Management said that 43.6 million Americans engaged in some type of ORV activity each year, using more than 10 million vehicles in such "recreational pursuits."

The magnitude of environmental damage by ORVs is staggering when realized that they have already ruined about one-half as much land as all U.S. mining activities combined since 1700. A particular problem is that much of the damage is in the dry Southwest and California where new soil generation occurs at a rate of only 1 cm for every 250 to 500 years. Indeed, the soil-forming process is so slow that jeep and tank tracks of General Patton's World War II training manuevers are still visible.

ORVs destroy a wide range of environmental features. The wheels churn topsoil into loose silt that flows downslope with the first rain. With time, the hillsides become pock-marked with gullies and rills that resemble miniature badlands (Figs. 10.12, 10.13). The displaced soils form a new series of sediments at the base of the hills where vegetation is buried and destroyed. On level ground, the ORVs subject soil to degradation by the wind. The traffic pulverizes the topsoil making it more susceptible to entrainment and deflation by the wind. When the topsoil and vegetation are destroyed, the albedo is changed, which further degrades the material by elevating the temperature. In January 1973, dust plumes from heavily used sites in southern California were shown to extend more than 75 km when satellite photographs were examined.

ORVs destroy additional vegetation by undermining the surrounding soil mat and causing the plants to lose strength and collapse by loss of a supporting medium. Compaction from the traffic also injures root systems and prevents germination of new plants. Restoration becomes exceptionally difficult and expensive because of the loss of nutrients, elevated soil temperatures, and slow soil recovery rate. Indeed, many desert plants take decades to grow, such as creosote, which doesn't reach maturity for 80 years.

**FIGURE 10.12** Severe gullying caused by destabilization of the ground by ORVs. This hillside has lost 11 million kg of soil by this manmade type of accelerated erosion. Photograph courtesy of Howard Wilshire, taken in the Red Rock Canyon area, Calif., September 1976.

**FIGURE 10.13** Continuing accelerated erosion on these ORV trails in the Panoche Hills, Calif., after 6 years of closure. Photograph courtesy of Howard Wilshire.

## DESERTIFICATION

Mankind has always lived in drylands and deserts. Indeed, today, more than 630 million still do, 14 percent of the world's population. These fragile environments and their associated nearby lands, constitute a growing problem in today's world. Although the problem is not new (see Bryson and Baerreis, or Sears), a new awakening to the menace of desertification startled the world in the early 1970s when the Sahel area of northern Africa became the site of widespread grazing and other soil-depleting factors, and produced unusual stress on the land and its inhabitants. During the 5-year period from 1968 to 1973, the death rate doubled for the area (which includes the countries of Mauritania, Mali, Upper Volta, Senegal, Niger, Chad, Sudan, and into Ethiopia), the birth rate halved, the average life expectancy dropped by 10 years, soils were devastated, and 100,000 people and 10 million cattle starved and died of thirst.

**Desertification** is the process of making desert-like lands where they should not climatologically exist. As used, the term generally implies that a combination of factors have produced the change, with man being an important component. In the Sahel, a series of good rainfall years had allowed large increases in cattle, which exceeded the carrying capacity of the land. The soil was pulverized by the cattle's hooves and from overgrazing, especially near water holes and wells. Cultivated areas had also increased to meet the rising demand for additional grain, thus exposing more soil for longer periods of time. Deforestation was also occurring at a rapid rate to produce new farmland and fuel. Reduction in shade and loss of root protection in soil triggered increased erosion and albedo changes. Thus, the stage was set for a rapid deterioration of the land with the onslaught of drought conditions.

The southward spread of the Sahara Desert has consumed 650,000 $km^2$ of land, once productive for agriculture and grazing, during the past 50 years. Detailed studies in the Sudan indicate the desert boundary shifted south nearly 100 km in the 1958–75 period.

Thus, desertification ranks with man-induced badlands as the culmination of human despoilation of the environment. Although the extent to which such lands have been desecrated by man is debatable, some observers such as Professor Mohammed Kassas of the University of Cairo, believe that man has contributed to the formation of desert-like conditions of 900 million $km^2$ that once were arable lands. Such warnings, the recent example of the Sahel, and problems in other countries show that mankind should exercise more careful stewardship to such fragile lands. These are not isolated cases. Salinization is likewise producing desert conditions of numerous terrains in Australia, India, Pakistan, and parts of the American Southwest. Other African nations, such as Algeria, Morocco, Libya, and Tunisia are losing 100,000 ha yearly to desert as a result of human activity.

The United Nations has been particularly aware of the desertification problem, which it says bears a close resemblance to guerrilla warfare and states:

> Desertification breaks out, usually at times of drought stress, in areas of naturally vulnerable land subject to pressure of land use. These degraded patches, like a skin disease, link up to carry the process over extended areas. It is generally incorrect to envision the process as an advance of the desert frontier engulfing usable land on its perimeter; the advancing sand dune is in fact a very special and localized case. Desertification, as a patchy destruction that may be far removed from any nebulous front line, is a more subtle and insidious process.

Some of the remedial steps that can be taken to stop drastic losses of soil will be discussed in the next section.

## SOIL CONSERVATION

There are a number of ways the geologist can provide information and design ideas to minimize soil destruction. Cooperation is needed from other disciplines as well, to develop coordinated methods that achieve the best results to save the soil. All techniques for land management of the soil resource have the following general objectives: (1) retain soil moisture on site, (2) reduce movement of water over the land, (3) minimize length of time barren soil is exposed, and (4) assure that carrying capacity of soil is not exceeded. Toward these goals the procedures of soil conservation can be classed into soil-bearing capacity controls, biological controls, and structural-engineering controls.

### Soil-Bearing Controls

This group of techniques depends on some type of land management whereby the numbers of animals, people, and crop density are controlled. All soils have a capacity that can be in equilibrium with certain environmental factors, but when exceeded, a threshold occurs that starts the deterioration process in motion. When the rate of tolerable erosion is exceeded, those activities that have produced the excess are then prohibited. Thus, only a prescribed number of cattle, for example, would be allowable for a certain size of grazing area. To achieve further erosion control, damaged land can be retired from production.

### Biological Controls

These methods depend upon the type and manner in which crops are planted. With sloping terrain, crops should be planted along the contours of the land to minimize the flowlines and velocity of slope wash. Strip cropping can also be advantageous with planting of different crops that mature at different times, so some part of the slope always has a vegetative cover (Fig. 10.14). Rotation of crops is another practice that over the long run can produce greater yield and prevent continuing loss of soil nutrient. The same crop, such as corn, should not be planted in the same field year after year, but should be interspersed with those crops that can restore important elements to the soil. For example, a legume such as winter vetch can add as much as 25 kg of nitrogen per hectare during a single season. A study of soil losses on 16 percent slopes at La Crosse, Wis., showed that after six years of

**FIGURE 10.14** This is a combination of strip-farming and contour-farming methods in Illinois. Courtesy of Soil Conservation Service (USDA).

growing corn 36 t/ha were lost, but when one year of crop rotation with hay was introduced, the losses were reduced more than 100 percent.

No-till agriculture became feasible with the development of herbicides, which could then be used to replace plowing as a method to control unwanted plants. The only manipulation required is to open a slit or trench just wide enough to receive the seed, and then cover it with soil. No tillage leaves nearly all the previous crop residue on the surface thereby reducing wind and water erosion. Efforts to reduce tillage was almost nil until the publication of *Plowman's Folly* in 1943 by Faulkner. He regarded the plow as an impediment and an artificial method for plant growth. With the introduction of the herbicide 2,4-dichlorophenoxyacetic acid (2,4-D), a revolution occurred with weed control much the same way that DDT had completely altered techniques for insect control. The further invention of atrazine in 1957 made no-till corn possible and practical because it kills most weeds without harm to the corn. The introduction of paraquat in 1966 was another significant advance because it kills existing vegetation and then becomes inactive toward subsequent crops which are not harmed. Even with these advances, no-till farming spread very slowly, so by 1977, only 2 percent of the total cropped lands in the United States were farmed by this method. However, minimum till methods had become more widespread as the next best alternative. By 1981, some form of no-till or minimum tillage was used on 28 million ha compared with only 1.6 million ha in 1964. Furthermore, between 1980 and 1982, no-till farms doubled.

No-till methods work best in well-drained soils, on hilly terrain, and for row crops. Its advantages include:

1. Increased production;
2. Reduced capital investment and maintenance expenditures;
3. Reduced pre-harvest labor by 70 percent;
4. Reduced growing season labor by 300 percent;
5. Reduced fuel costs as much as eight times;
6. Reduced soil erosion. The crop residue inhibits raindrop impact, wind deflation, evaporation, and water erosion. Studies in Ohio showed water erosion was reduced 10 times and wind erosion 65 times.

However, no-till agriculture is not a panacea everywhere. Herbicides are an environmental pollutant and can affect the land and the waters draining from them. The method is not practical on soils or land with poor drainage. Insects may also thrive on the mulch, and disease-forming organisms may prosper. This requires the application of pesticides, which again have harmful environmental effects. Paraquat is extremely toxic and can be a health hazard to those who handle it and in soil waters that move from the site. Such methods may not be possible or practical in many Third World nations where funds are unavailable to purchase the proper pesticides required for pest control.

Revegetation and reforestation can be used as measures to further reduce the ravages of water and wind erosion.

## Structural-Engineering Controls

These methods require rearrangement of the topography and introduction of non-crop materials for the purpose of stabilizing soil and reducing siltation. When slopes are too steep for conventional farming, terraces can be constructed. These have the effect of reducing the slope and allowing for greater water retention, or its channelization to sites where it will not be destructive. Terraces need to be carefully designed to reduce maintenance and to lessen farming difficulties during plowing, planting, and harvesting operations. They have been used extensively in numerous countries, such as Nepal, the Philippines, China, Malaysia, and others, where there were no alternative ways the hillside soils could be farmed (Fig. 10.15).

To prevent expansion of drainageways by gullying or other channelized flow, a variety of structures, check dams, and impediments for incision can be emplaced (Fig. 10.16). These serve to control flow into designed positions

**FIGURE 10.15** Rice production on terrace farming in the Philippines. Courtesy of John Wolfe.

FIGURE 10.16 This check dam and overflow structure was constructed in the summer of 1969 near Hanover, Ind. This structure helps control runoff of 54 ha. Courtesy of Soil Conservation Service (USDA).

where soil damage cannot occur. A further aid to restricting water erosion is the emplacement of dikes to divert water flow from the fields, and the vegetation of waterways to cushion and impede the development of high-flow discharge.

Trees can be used as a designed engineering plan to control acclerated soil deflation by the wind. Such a method was first used in the United States more than 100 years ago. The Timber Culture Act of 1878 offered to home-steaders 160 acres with the sole provision that trees be planted on 40 acres. Unfortunately, the law was repealed in 1891, and new governmental support for windbreaks did not occur until the 1930s. The Dust Bowl storms showed only too well the devastation that wind erosion could cause. During the 1930s, the planting of 218 million trees on 30,000 farms on the Great Plains totalling 32,000 km in length was a powerful deterrent to wind erosion.

Vegetation shelter belts are not only effective in moderately drylands such as Nebraska, but have even been used in such arid regions as Algeria, the Gobi Desert, and Saudi Arabia. Additional benefits, besides diminishing wind erosion, occur, such as increasing nearby humidity and lowering the temperature of nearby soils.

## THE WORLD SITUATION

Soil is not a replaceable resource, and its loss has reached worldwide proportions. Although such losses in the United States are severe, they are infinitely more tragic in most other countries. Threats common to such losses include: (1) conversion to other uses, (2) soil erosion and loss of fertility, (3) undesirable patterns of farm ownership and management, (4) drought and climatic change, (5) water scarcity, (6) energy shortages and high costs, (7) genetic overspecialization, and (8) declining yield with soil impoverishment. Although erosion in Third World nations is poorly documented, available studies indicate it is of mammoth proportions. In the tropics, violent rains rip away vulnerable topsoil left unprotected. Awesome manmade gullies fill the hills of much of Asia, Africa, and Latin America (Fig. 10.4). In drylands, desperate farmers plant areas that can only sustain pasturage, and forests and woodlots are decimated to obtain fuelwood in other areas. The United Nations has estimated it would take $48 billion in developing countries over a 20-year period to rehabilitate their damaged irrigation lands, half their rangelands, and 70 percent of their impaired rainfed croplands, and to stabilize sand dunes on 2 million ha (Eckholm, 1982).

Lack of understanding such worldwide problems, and lack of action to substantially address them, may lead to unrest of such proportions that renewed conflicts will result. By 1990, the developed nations will account for only 24 percent of the world population, but contain 85 percent of the world economic activity and 50 percent of the world grain consumption. Low income countries will need massive infusion of capital investment, research support, and education to help them build an infrastructure with the capacity to produce, distribute, and market food supplies. Continued or expanded disparities will only lead to the spector of worldwide conflict (Woodwell et al., 1983).

## POSTLUDE

At the present time, Americans are in an exceptionally complacent mood regarding farms and the institution of any programs to substantially address soil degradation problems. Instead, they cite the extraordinary efficiency of the American farm unit to produce such extravagently bountiful crops that they are a glut in the marketplace. The huge surpluses cannot be sold, so farmers are encouraged to plow under their crops and to eliminate many lands from growing any crops. Furthermore, figures are cited showing that only 4 percent of the American population grow these astounding amounts whereas in Russia even 30 percent of the population cannot grow enough grain to feed the country. However, this is like a time bomb and has currently provided a false sense of security. It is also a paradox because American farms are losing more land and soil than during the heart of the Dust Bowl years in the 1930s. What would happen if severe drought conditions become superimposed on other conditions that now threaten many farms? These include:

- Large farm equipment makes working hillsides difficult, so terraces and other conservation structures have been removed. Replacement of structures has also become too costly to implement construction.
- The increased costs of fuel, fertilizers, pesticides, and herbicides have caused some farmers to use marginal land that should be set aside for conservation.
- Absentee landlords rent land to leasees and are unwilling to make appropriate long-term investments in conservation methods. One-third of farms are not owned by those working the land.
- There has been a lag in research at universities and agricultural schools, and inadequate communication with farmers. For example, the General Accounting Office in a report documented that major attention was given to farms to increase production instead of those farms that were having significant erosion problems.

Thus, we are living on borrowed time. Soil, along with water, constitutes our most precious resource. However, soil, unlike water, is not renewable. It takes decades to hundreds of years to develop a fertile soil, and with the world's expanding population, there is not sufficient time to produce a second soil crop. In the United States, there is a growing concern that new conservation strategies must be imposed on farmers. Some specialists advocate that farmers who do not install appropriate conservation measures should be denied price supports and other forms of farm aid. Such an approach is called "cross compliance," but is viewed as too drastic by others. They urge, instead, that a greater education effort will produce a sufficient volunteer solution to erosion problems. In the Uttar Prades State of India, an unusual approach was taken to stop erosion and prevent deforestation. Every member of the Chipko Andolan ("to hug the trees") movement was assigned a particular tree to guard. If a poacher or timber cutter appeared, the member literally entwined his body around the trunk they had agreed to protect. Both governmental wisdom and personal ardor are required if soils of the world are to be retained as a heritage for future generations.

CHAPTER 11

# GEOENGINEERING

By now, it should be obvious to the reader that geology is a science that can become thoroughly immersed in the affairs of man. When these involvements are directly related to construction activities, such application of his knowledge falls within the realm of geoengineering. Practioners of this science are known as engineering geologists or geoengineers. It is their job to bring geoscience knowledge into the service of man, whereby dams are built, highways constructed, developments sited, and the entire range of operations that occur when mankind alters the land–water ecosystem by his activities.

Geoengineers should be experts that understand how the materials of the earth, the processes that operate, and the terrain interact when developments are designed and emplaced. It is especially necessary to determine the character of feedback systems that may result because the earth is an open system, and changes made at one site will influence contiguous areas. Thus, developments cannot be installed in isolation from considerations of impacts to neighboring environments. This is especially the case for such structures as dams and highways.

Engineering works are requisites of civilization by which materials are mined, processed, transported, manufactured, used, wasted, and disposed. All of these steps involve geoengineers at some point. This is demonstrated by many of the other chapters:

- The mining of minerals and fossil fuels (Chapters 2 and 3);
- The development of water retention and supply systems (Chapter 4);
- The remedial steps to control landslides (Chapter 7);
- The structural works to control floods (Chapter 8);
- The structural installations to preserve beaches (Chapter 9);
- The measures emplaced to minimize soil erosion and siltation (Chapter 10);
- The proper siting of waste disposal systems (Chapter 12).

The geoscience disciplines that are especially useful in such planning are those in soil and rock mechanics — the mineralogists and petrologists, structural geologists, and geomorphologists. These are the people who are versed in knowledge of materials and processes of the earth — the environment in which the geoengineer works.

The need and market for geoengineers is continuing to expand rapidly. Their technical skills are required in such a variety of instances that only some of the broader categories will be mentioned in this chapter. During the last decade, there was an especially large rise in the number of consulting firms and scientists employed by them. One reason for this growth has been environmentalism, as mandated by the National Environmental Policy Act of 1970 (NEPA). Now more than ever, geoscientists are required to look at all aspects of their work and to predict repercussions that might occur throughout other parts of the environmental system. We should make one point clear: the environmental imprint created by some engineering works is worse for some developments than others. Furthermore, not all engineering projects have the blessing of geoengineers. Society may demand a particular construction program, and work progresses and is completed. The role of the geoscientist, who becomes involved in such matters, is to make people aware of what the impacts will be and to try and minimize them whenever possible.

## CONVEYANCE SYSTEMS

One of the biggest jobs in engineering is to develop the transportation media whereby goods and services are brought to the consumer. Thus, highways and railroads need to be properly sited for the transport of materials by vehicular traffic. When the resources are fluid, canals, pipelines, and aqueducts must be employed to allow their movement from site to site. Conveyance systems have become increasingly important as either the source area or the consumer areas become farther and farther apart.

### Highways

Although roads have always been important arteries linking trade and commerce from place to place, the size and number of highways greatly expanded throughout the globe during the post-World War II period. In the United States, the 64,000 km of the Interstate Highway System added an entirely new dimension to travel in the country. Similarly, such massive construction changed the environment of numerous localities throughout the nation (Fig. 11.1). Proper highway design should consider possible changes to the hydrology and slope stability of the sites. Three groups of damaging impacts can result from improper location and construction.

**FIGURE 11.1** Impact of highway construction in the ridges of the Folded Appalachian Physiographic Province. Courtesy of John Markham.

1. Hydrologic systems. A highway cut may intersect the water table and thereby behead water as a supply to wells and users downslope. Drainage from the highway may be diverted to areas that cannot properly handle such water increases. If the roadcut is deep, the excavation may act as a groundwater sink and lower the water table, which can, again, affect nearby homesites.

2. Slope stability. An artificial cut into a hillside always increases the steepness of the terrain at that spot. If earth materials are weak, such an incision may exceed the stability index of the materials and trigger slope movements from creep to landsliding (see Chapter 7). Obviously, accurate studies in the field of soil and rock mechanics are required to determine the safety factors when highways are built (Fig. 11.2).

3. Pollution. Highway construction is notorious for accelerating erosion and siltation. Studies show that such processes are increased several thousand fold from those of undisturbed landscapes. For example, four-lane highways require the exposure of 3–10 ha of land per kilometer during construction. It is not unusual for 2000 tons of sediment per kilometer to be washed away during such exposure. In addition, the cut may contain rocks of a type that are quickly weathered and produce acid water when they drain. Salting of roads, exhaust fumes from vehicles, and spraying of pesticides and herbicides along the right-of-way all provide additional contaminants to the environment. Many studies have shown, as in New Hampshire and Massachusetts, the potential danger to plants and people from ingesting the saline waters that accompany road sites in those states.

**FIGURE 11.2** Roadcuts for the Southern Tier Expressway, Route 17, Catskill Mountains, N.Y. The Delaware River valley was largely followed by this new construction. Benches have been cut in the sandstone bedrock to help stabilize this road cut, which was the deepest road cut east of the Rocky Mountains when constructed in 1965.

## Aqueducts, Canals, and Pipelines

This family of features provides flow media for vital resources such as water, oil, natural gas, and other potential fluidized materials such as coal slurry. Canals by definition are constructed on the land, whereas aqueducts and pipelines are conduits that may be either underground or aboveground. Again, their emplacement requires geoengineering consideration to determine the most feasible route, and to bypass terrain that would prove damaging to the installation.

Canals are used for a variety of purposes including water supply, wetland drainage, navigation, and flood control. Their development dates back to antiquity. Some of the earliest and best known are those that brought water from the Tigris and Euphrates rivers to the early civilizations of the Fertile Crescent. Their construction dates back more than 4000 years. Unfortunately, their design made back-breaking work for the populace because of their propensity to silt. Indeed, some historians report that one-half their time was spent by those early agrarians in cleaning the silt from the canals or repairing breaks caused by erosion. The science of hydraulics and geomorphology are vital in canal construction because when gradients are too steep erosion results, and if too gentle, sedimentation occurs.

Of course, the Panama Canal and the Suez Canal are premier examples in the world. Digging the Panama Canal was particularly gruesome because of disease and inability to control earth movements and landsliding. The solution to that problem of accelerated erosion and sedimentation was strictly a better understanding of the geoengineering of the materials and knowledge of how to deal with them. The U.S. Bureau of Reclamation has dug many mammoth canals in the arid West, transporting water from dams they also constructed. (Figs. 11.3, 11.4). Some extend hundreds of kilometers with huge channels. For example, the San Luis Canal in California is 11 m deep and 78 m wide at the top and is composed of a concrete liner. A rather mind-boggling enterprise is currently under construction in Africa. The Jonglei Canal through the Sudan will be 350 km long, twice the length of the Suez Canal. It will transport 20 million $m^3$ of water a day, or about one-

FIGURE 11.3 U.S. Bureau of Reclamation canal in California, constructed in a meandering pattern that is most consistent with the amount of water and topographic slope so as to prevent bank erosion and sedimentation.

**FIGURE 11.4** Sun River Irrigation Project, Mont. The Greenfields Main Canal delivers water to the nearby grain and alfalfa fields. U.S. Bureau of Reclamation photograph taken July 20, 1970.

fourth the flow of the White Nile. Its purpose is to save water, which otherwise becomes lost in The Sudd, a giant wetlands.

Aqueducts and pipelines are also of ancient vintage. Of course, the Romans were known for their extensive aqueduct system, some of which is still in use today. The earliest large aqueduct in the United States brought water from the Catskill Mountain region to the New York metropolitan area. It was designed by the reknown engineering geologist, Charles P. Berkey in the early 1900s and became a classic of inventiveness and careful design.

Extraordinary mapping of the subsurface rock types and structures was required to make construction safe and maintain the integrity of the system.

Another marvel of recent times has been the completion of the Trans-Alaska Pipeline Systems (TAPS). Its purpose is to move oil from the Prudhoe Bay Field on the north coast of Alaska to southern Alaska where it can be transported to the continental United States by tanker. The pipeline extends 1270 km and traverses three mountain ranges, more than 800 streams, and is in permafrost soil for 950 km (Fig. 11.5). Numerous environmental safe-

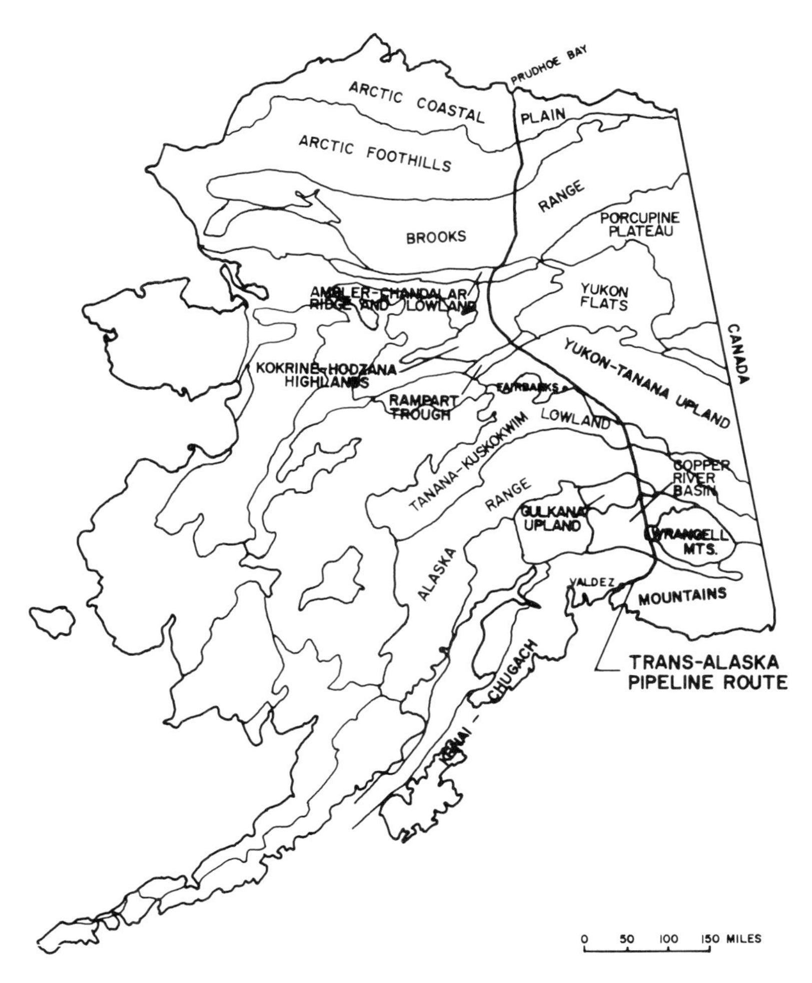


**FIGURE 11.5** Trans-Alaska Pipeline route from Prudhoe Bay to Valdez, a distance of 1270 km. The pipeline was completed in June 1977, crossed more than 800 streams and is in permafrost terrain for 950 km (Kreig and Reger, 1976).

guards were taken to assure the minimum of ecological damage and still maintain the structural solidarity of the pipeline. For example, there are more valves per kilometer than any other long-distance pipeline. The purpose is to isolate any leakage that might occur. For 680 km, the pipeline is above ground, resting on 78,000 vertical supports – the longest bridge in the world. At hazardous spots, flexible pipeline was installed, capable of withstanding damage from major earthquakes. Pump stations were located at sites that would not interfere with wildlife, and extensive plantings were made to provide habitat for big game grazing. As this goes to press, another major pipeline is being planned, to transport natural gas from the Arctic through Canada and into the United States. It will be interesting to see the final figures, data, and impacts for this huge venture.

## DAMS

Dams seem to have so many benefits that construction of them was not seriously debated until the last 25 years. Nearly everyone seemed to favor them – except, of course, those people whose homes and livelihood would be lost because of the inundation. Originally, dams were built for a single purpose, but to justify their construction today, dams have almost universally become multipurpose in design and justification. Development of a new water supply is the foremost reason for many dams. Water that might otherwise rush unused to the ocean is temporarily detained and then diverted to surrounding regions for irrigation, industrial, and municipal uses. When the imponded water flows from the reservoir top to the dam base, the fall in the water, when diverted through turbines, produces hydroelectric power that is pollution-free. The new lacustrine environment can become a haven for wildlife and fish. The site can also be developed for recreational purposes – fishing, boating, swimming, camping, etc. New industry and communities may also develop nearby. The increased food supply grown by irrigated farming may be a necessity to sustain life. Indeed, why should anyone question the value of dams? But read on.

To justify dam construction, all the aforementioned values are used to show that the benefit/cost ratio is highly favorable for building the dam. Such rules must be applied for all federally financed dams, which are now built by such agencies as the Corps of Engineers, Bureau of Reclamation, and TVA (Tennessee Valley Authority). Even the SCS (Soil Conservation Service) can build some dams up to a certain size not to exceed the impondment of 4000 acre-feet of water.

## Case Histories

The Tennessee Valley Authority reclamation and dam project is the most spectacular and comprehensive environmental program in the world (Figs. 11.6, 11.7). In the Tennessee Valley region, 106,000 km$^2$ are managed with a massive environmental approach to design, construction, and resource use. Here, the TVA has constructed 20 dams that make the watershed one of the most highly regulated systems in the world. At the time of construction, however, other energy resources were not sufficiently available to make

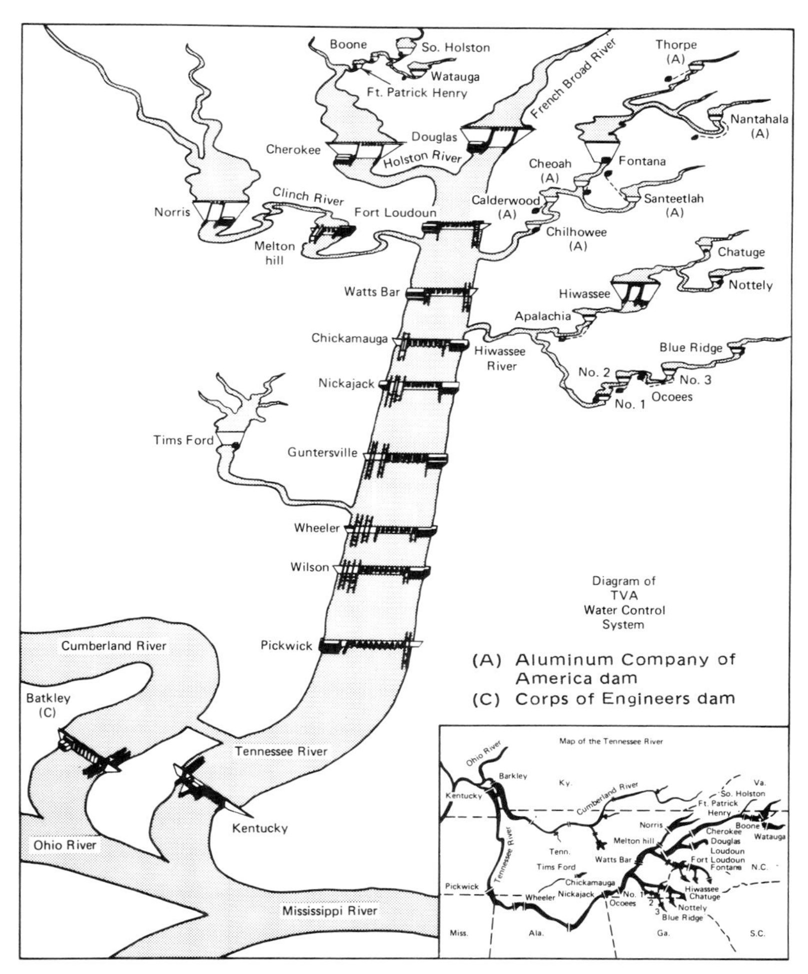


**FIGURE 11.6** Maps showing some of the essential features of the TVA Project. Courtesy of TVA.

**FIGURE 11.7** Fort Loudon Dam (built 1940–43) is the farthest upstream main river TVA dam. It brings the navigation channel to Knoxville, 129 km distant. This dam is 37 m high and 1277 m long, and its reservoir covers 5900 ha. Courtesy of TVA.

such an immediate impact on the socioeconomic welfare throughout the region. The total hydroelectric network affects even a larger area, 209,000 km$^2$. More than 150 municipal and cooperative electric systems are served that have nearly 2 million consumers. The programs have also been responsible for extensive reforestation of the region, influx of tourism, and development of mineral resources.

The Snowy Mountain Hydro-electric Authority has constructed an immense project in the mountains of eastern Australia. In an area of 5200 km$^2$, there are 80 km of aqueducts, 144 km of tunnels, 16 large dams, pumping stations, and 7 power stations (Fig. 11.8). This grandiose scheme has diverted waters of east-flowing rivers westward under the drainage divide to the parched lands of the southeast part of the continent. Although con-

FIGURE 11.8 Murray 1 Power Station and Pipeline, part of the Snowy Mountain Hydro-electric Authority Scheme, Australia. Courtesy of the Snowy Mountain Hydro-electric Authority.

structed in a remote area, many conservation measures were undertaken, for example, planting of hundreds of thousands of trees and shrubs. These along with prohibition of domestic animals have helped stabilize hillslopes.

### Adverse Impacts

The loss of property and life rank high on the deleterious impact side of building dams. Property loss occurs when people are displaced because their lands are within the construction and reservoir area, or when dams fail and destroy downstream areas. During the last 25 years, more than 10,000 people have been killed due to dam failure. The only good thing that emerged when the St. Francis Dam, Calif., dam burst on March 13, 1928, was that it prompted State legislation whereby geological investigations would be required for all future dam construction. Unfortunately, a geologic study had not been conducted of the St. Francis site because it could have revealed the presence of dissolvable minerals, which produced the crack in the dam abutment.

Prior to 1976, the major federal agencies charged with construction of large dams had a spotless safety record. That was all ruined, however, on June 5 of that year when the Teton Dam in Idaho disintegrated, releasing 302.8 million $m^3$ of water that devastated downstream areas. Here, the geoengineers had not sufficiently predicted the character of the abutment bedrock and its reaction to grouting as being likely to cause failure. The disaster did, however, prompt strong political action. A National Dam Inspection Act had been passed by Congress in 1972, which authorized the Corps of Engineers to inventory and inspect U.S. dams, but no funds were allocated. Therefore, a Presidential Directive of December 2, 1977, had the force of providing funds for a 4-year investigation to examine the 9000 dams that were listed as "high hazard."

The reservoirs behind dams become siltation basins that trap upstream erosion products. The buildup of these sediments shortens the effective life of the dam. This was noted as early as 1941 in a U.S. Geological Survey study that reported 39 percent of the nation's reservoirs would be largely filled with sediment before the end of the century. The problem is worldwide. The Tarbela Dam in Pakistan was completed in 1975 as the world's largest earth and rockfill dam. Unfortunately, it will be made nearly useless by sediment filling in less than 50 years. Similarly, Anchicaya Dam in Colombia was completed in 1955, but it took only two years to be 25 percent sediment-filled. In another 7 years, it had lost capacity to store water or generate electricity.

Siltation in reservoirs produces an entire chain reaction of environmental damages. The sediment plume that discharges into the standing water eventually forms a delta that grows headward and can engulf upstream properties adjacent to the reservoir. Below the dam, the released waters are clear (having been deprived of their sediment load) and thus, have increased ability to erode. Downstream erosion extending from the dam is a common occurrence. Hoover Dam on the Colorado River has caused more than 3 m of erosion for a distance of several kilometers, and Glen Canyon Dam has caused 6 m. Other dams on the River have had similar histories. However, the story does not stop there. Tributaries that formerly entered the river at grade also erode their channels to keep pace with the master river. The erosion products become deposited farther downstream, such as at Needles, Calif., which has flooding problems due to sediment buildup in the channel from the new erosion cycle. And finally, even coastal areas become impacted. Nearby beaches, which formerly received sediment from now-dammed rivers become sediment-starved and undergo accelerated erosion. For example, a series of impondments caused a 71 percent decrease in the sediment load of the Brazos River, Tex. By 1958, coastal erosion where the river empties into the Gulf of Mexico had reached a rate of 4 m per year.

There are still other impacts created by dam construction. The Aswan Dam in Egypt has deprived downstream farmers of the usual nutrients on their land, which formerly were supplied by river flooding. Consequently, much money must now be spent to artificially add nutrients to their fragile soil. Because the reservoir area is now a lake instead of a free-flowing river, new aquatic and wetland organisms have intruded the region carrying diseases that have affected much of the local populace. Even the quality of water changes, such as in the delta region of the Nile. Here, the Mediterranean Sea has become more saline because of diminution of the fresh river water and caused a reduction in fishing and other aquatic life that had depended upon the original lower salt content.

One of the most frightening aspects of large dams and reservoirs is that the additional load superimposed upon rocks below is capable of producing earthquakes. This discovery was first made in 1931 studies of the Marathon Dam, Greece. To date, the most severe disaster from this cause was from the Koyna Dam, India, in 1967. On December 10, a 6.3 M earthquake killed 200 people, injured 1500, and left thousands homeless.

## CHANNELIZATION

Channelization is the engineering practice of deliberately rerouting streamflow to another site. There can be many reasons for this, including drainage, flood control, navigation, erosion control, production of more usable land, and protection for other structures. The process of channelization consists of such modifications as straightening, deepening, widening, clearing, or lining the natural channel (Figs. 11.9, 11.10). Thus, the normal hydraulic properties of the stream are changed because it is shortened, the gradient steepened, the bed changed to smoother materials, and temperature

**FIGURE 11.9** Channelization along the newly constructed highway corridor of Interstate 88, Broome County, N.Y. Note bank destabilization and newly increased sediment load in the stream.

**FIGURE 11.10** Channelization, showing displacement of stream and straightening of the channel. Courtesy of John Conners.

increased with diminution of vegetation. All these factors result in higher velocities, thereby increasing the discharge and erosional ability of the water (Fig. 11.11).

The Mississippi River has undergone channelization by the Corps of Engineers at various sties since the 1870s. A new wave of U.S. channelization was ushered in by the Watershed Protection and Flood Prevention Act of 1954. Under this statute, the Soil Conservation Service has participated in

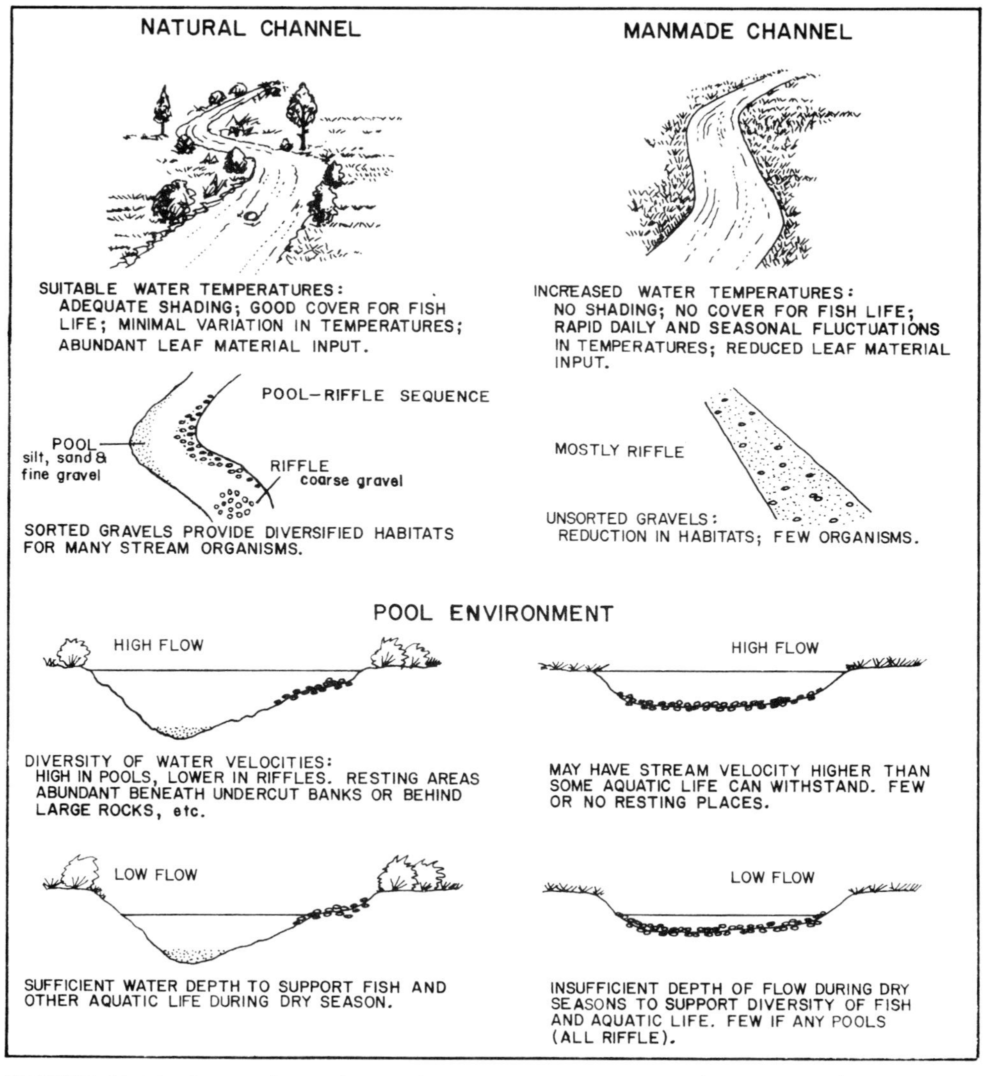


**FIGURE 11.11** Comparison of natural and channelized streams (Keller, 1976).

channelizing more than 13,000 km of streams throughout the nation. The Corps of Engineers channelled an additional 2400 km of larger streams during this same period. Although congressional approval has been received to channelize an additional 20,000 km of waterways, the pace has greatly slowed in the past five years because of the mounting torrent of environmental objections.

The Arthur D. Little Company made a survey of 42 channelization projects in 1973 wherein 3700 km of streamways had been altered. They concluded that 11 had produced significant wetland modification; 5 created dramatic loss of hardwoods; 6 adversely affected water tables and recharge capabilities; and 19 produced unsightly landscapes. Plant and animal aquatic life was invariably destroyed or highly stressed.

### Case Histories

Channelization effects of the Willow River in Iowa have been documented by several geomorphologists (Daniels, 1960; Ruhe, 1971). This Missouri River tributary began to be straightened and modified in 1906, and the project was completed by 1920. Since that time, there has been deepening of the main new channel up to 12 m, and a new cycle of erosion and gully development has occurred in the adjacent watershed as tributaries become entrenched in order to join the master stream.

The Blackwater River, Mo., was channelized in 1910. The 53.6 km length of the original channel was shortened to a 29 km length, thus increasing the gradient from 1.67 m/km to 3.1 m/km. This resulted in serious erosion along the banks, caused new gullies and headward erosion of tributaries, and required replacement of bridges.

Of course, major rivers have also suffered through channelization, and again, there are environmental repercussions. For example, many argue, such as Charles Belt (1975), that channelization projects on the Mississippi River system have actually increased flood stage. During the disastrous 1973 Mississippi River flood, the highest previous flood at Cape Girardeau, Mo., had occurred in 1884. Although the 1973 flood was only 0.6 m higher, it accomplished this with discharge that was 35 percent less! The Sacramento River, Calif., has undergone several channelization projects, some which date back to the late 1800s. Throughout these times, the river continued to flood, and one particularly disastrous flood claimed 40 lives in 1955 and destroyed $48 million in property.

### Benefits and Design

In some instances, channelization is a must. In urban settings and areas of large development, there may be no alternative. When such geoengineering is required, it is important to design with nature as much as possible (Fig. 11.11). The hydraulics of the stream must be understood, and also the flows it must carry under any new circumstances brought about by development. For example, natural streams have predictable pool and riffle sequences, usually they alternate with length, six times the channel width. Thus, accommodation should be made to allow the stream to obey such a geometry. If not built into the design, the stream will continue to attempt to erode and deposit sediment in accordance with such hydraulic laws.

## ENVIRONMENTAL ENGINEERING

Certain environments and geological settings require special care in the design of structures, and it is necessary that geoengineers provide the necessary data base and expertise upon which the plannning and building can be consummated. For example, ground conditions are especially fragile in those lands with climatic extremes – the drylands, the arctic lands, the tropical lands. Such areas have more than the usual type of engineering problems, and the following sections provide illustrations of this.

### Expansive Soils

Expansive soils are earth materials that greatly expand when they become wetted. Also, if they contain original moisture, when dried, they can also lose much volume. Thus, they have both shrink and swell properties. Certain clays, such as montmorillonite, are especially prone to these changes and are 10 times more reactive than other clays. Although the majority of damage from this phenomenon occurs in dry climates, soil changes are not necessarily restricted to such regions. About $3 billion in property and development damage occurs each year in the United States, and many other countries also suffer grievous losses such as Argentina, Australia, India, Mexico, South Africa, and Spain. The problem was not even identified by the federal government until 1938 when damage to some installations of the U.S. Bureau of Reclamation was recognized as being caused by expansive soils (Fig. 11.12).

There is nearly an endless list of types of structures damaged by this shrink-swell process of expandable soils. These include highways, building

**FIGURE 11.12** These 10 cm thick concrete linings have suffered severe damage from hydrocompaction of the underlying soil. This location is at the "Mendota Test Site," a facility of the U.S. Bureau of Reclamation for calculation of hydrocompaction effects prior to construction of canals in the San Joaquin Valley, Calif. Courtesy of Nikola P. Prokopovich.

foundations, underground pipelines, pilings, retaining walls, hillslopes, swimming pools, sidewalks, etc. The amount and type of damage is a function of such factors as water-table depth, antecedent soil conditions, magnitude of new moisture, vegetation, topography, drainage conditions, and character of the construction (Fig. 11.13). Mathewson and Clary (1977) describe landsliding in Texas in new highway roadcuts caused by the expanding clays. Jones and Holtz (1973) and Font (1977) provide other case histories of damages by this process.

It is important for homeowners to be aware of such potential problems. When building a new house, preventative measures should be employed, and for older structures, remedial measures may be necessary to alleviate and diminish the problem. Steps that can be taken include:

- Location of vegetation away from the house;

FIGURE 11.13 Deterioration of a road in Waco, Texas, from expansive sediment. The swell-shrink properties of these montmorillonitic soils produce extensive structural damage in the region. Courtesy of Robert Font.

- Provide sufficient weight to the foundation to withstand expansive force;
- Elevate structure so that moisture cannot collect near or under it;
- Replace potential swelling soil with more stable earth material;
- Provide physical barriers in the ground to prevent moisture from collecting underground;
- Pre-wet the terrain prior to construction to assure all stresses are relieved (Fig. 11.14).

By taking such measures, the possibility for creating structural damage by the end lift and center lift stresses that commonly develop in such sensitive sites can be largely eliminated.

**FIGURE 11.14** Fissures and accelerated erosion produced in dryland soils of alluvial fans in California. Artificial wetting of these materials has resulted in such hydrocompaction features as this throughout large areas of southern California. These might be prevented by prewetting. Courtesy of Nikola P. Prokopovich.

## Permafrost Terrane

**Permafrost** is perennially frozen ground. The term is used to describe thermal conditions of earth material whose temperature remains below freezing for at least two years. Such conditions are common in arctic areas where 20 percent of the earth's land surface is covered with such material. Depths of the frozen ground vary, but in Alaska and Canada it can reach depths of 600 m and be twice that amount in Siberia. As the world's population continues to expand, more and more people will need to settle in such regions.

Although permafrost provides excellent bearing strength for structures, its strength becomes greatly reduced with a temperature increase, and when thawed, may not have the ability to support even light loads. Serious engineering difficulties arise with fine-grained soils, those with high moisture content. When thawed, these materials turn into a slurry with little strength and cause failure of the structure emplaced on it. During World War II, extensive military construction was undertaken in Alaska to safeguard the area from possible invasion by Japan. The engineers had little experience in dealing with permafrost and its accompanying problems, so during the early construction work, numerous failures occurred to installations. The protective and insulating mat of vegetation was removed, and structures were placed on top of the barren ground. The airplane landing fields became quagmires with enormous ruts and washboard-type surfaces. Quonset huts, which housed troops and workers, sank deep into the ground when the heat from the buildings melted the ice between the underlying soil grains. Such construction failures finally prompted the U.S. Army to organize a special group of scientists, engineers, and technicians to solve these problems. This group is now known as CRREL (Cold Regions Research Engineering Laboratory) and is headquartered at Hanover, N.H. Their work has provided the necessary expertise to solve many difficult permafrost problems.

Unless proper precautions are taken, it is not unusual for roads cut into permafrost to completely disappear within the ground and sink to depths of 1-2 m within a single season (Figs. 11.15, 11.16). Therefore, in all engineering design, it is vital to maintain the integrity of the natural conditions as much as possible.

It is almost platitudinous to state that geoengineering investigations are necessary before proceeding with construction in permafrost terrane. It is fundamental to know that such materials are sensitive to thermal changes, are nearly impenetrable to moisture, and that fine-grained sediment and

**FIGURE 11.15** Severe thermal and water erosion of a winter trail after one thaw season in Alaska. Courtesy of Richard Haugen.

**FIGURE 11.16** Massive ice lens in permafrost terrane is exposed in this fresh road cut near Livengood Alaska. Courtesy of Richard Haugen.

organic matter are especially fragile. Site investigation is an absolute necessity, and for larger projects, geological mapping must be conducted. The thickness and distribution of permafrost must be delineated, and borings, test pits, and other exposures of the material must be scrutinized. The depth of summer thaw and drainage problems must be carefully assessed. The total environmental setting must be analyzed — the slope, kind of materials, thermal regime — and then linked to the type of construction to be undertaken. For example, Kreig and Reger (1976) reported on the character of the pre-construction terrain evaluation in preparation for building the Trans-Alaska Pipeline (TAPS) (Fig. 11.5). The route was selected after careful air-photo analysis, landform classification, field surveys, and examination of data obtained from more than 3500 boreholes.

When dealing with permafrost terrane, and indeed many others, there are four options that may be followed in construction: the conditions can be neglected, the frozen conditions can be preserved, the materials can be eliminated prior to building, or the thawing effects can be accounted for by special design of the structure. For example, fine-grained materials can be replaced with coarse-grained materials, which do not have such a potential for volume and water changes. Proper ventilation and insulation of buildings can prevent heat leakage that otherwise would alter the frozen ground regime. Pilings can be installed that are designed to "breathe" or accommodate stresses. Thus, geoengineers can provide the data base to aid planners in building structures that can withstand destruction from these capricious materials.

### Wetlands

Few other terrains have been so devastated by engineers, generally acting for developers, as wetlands. These environments, where the topmost soil is saturated, have only belatedly received the attention they deserve during the last 10–15 years. Such sites were formerly tagged as swamps, bogs, or marshes, and were looked upon as entirely undesirable and worthless. They were considered only good areas after they had been dredged, filled, or drained and put into "productive use by man." Too late for many regions, and almost too late for others, was it discovered that wetlands serve many vital functions and are necessary for environmental and ecological equilibrium. For example, the name "Dismal Swamp" covering a 2000 $km^2$ area in Virginia and North Carolina typifies the low esteem in which such environments are held.

The scope of wetland desecration and destruction is appalling. In the coterminous United States, only 28 million of the more than 50 million hectares of original wetlands remain, and these are being lost at a rate that exceeds 1 percent per year. In the northeast, nearly 300,000 ha were destroyed in the 1950–69 period. Fortunately, the states now have wetland legislation, which largely preserves the remaining areas. Wetlands have also become an "endangered species" throughout the world. Historically, England and the Low Countries such as the Netherlands, were the first to reclaim coastal wetlands. In England, coastal wetlands started to be changed during Caesar's rule, and by 1885, the Romney Marsh area had 20,000 ha of reclaimed land. As early as 4000 B.C., coastal areas were being placed into some type of production in what is now the Netherlands. Since that time, the country has continually added new land at the expense of the sea, the deltas, and the wetland terrain. Nearly one-third of the country was formerly in a non-productive state so far as the inhabitants were concerned.

Wetlands can be divided into whether they are freshwater or saltwater, but they both perform the same range of environmental–ecological benefits.

- They act as a sponge reducing the severity of floods and release water slowly to retain the hydrologic balance of the area.
- They trap sediment and pollutants, and act as a living filter to absorb detritus and to release oxygen.

- They are breeding grounds for animals, such as waterfowl, and most important of all, initiate and sustain the food chain of living organisms.

Many of these valuable assets were lost during the installation of canals in 1906 that began to drain the Everglades wetlands in Florida. Agricultural interests, and later the burgeoning metropolitan areas, coveted the 10,000 km$^2$ region. The diversion of water from the region caused a large range of undesirable impacts. The native vegetation died when deprived of natural conditions. The lowered water table permitted more fires. The drying out of the organic-rich soil caused its desiccation. This diminution of volume produced subsidence, and the land lowering has amounted to more than 1 m in many places. And finally, wildlife and fisheries have been depleted when deprived of their normal habitat.

On a cheery note, the example of the Thomas Pell Wildlife Sanctuary can be cited. It occurs on the outskirts of the Bronx, N.Y., behind one of the largest housing complexes in the country and across from a recently developed New York City landfill. It encompasses a mere 20 ha. Thus, it is too small for anyone to argue that its disappearance would jeopardize the fishing or shellfish industry. Neither can it significantly affect pollution diminution. Yet this seemingly worthless piece of land was sufficiently valuable to the people of the Bronx, that when New York City tried to turn it into a landfill, the opposition was so strong that the plan had to be abandoned. Another happy wetland preservation is that of the Bashakill wetlands adjacent to the southern Catskill Mountains. Here, the 1972 New Yrok Environmental Bond Act allocated $2.4 million to purchase the 900 ha area. Its nearness to a major metropolitan area will permit its use as an important recreation resource. Not only is the Bashakill a spectacular wildlife site, but it contains a range of topography that can be used for hiking, bicycling, fishing, and other outdoor sports.

## OTHER CASE HISTORIES

Unfortunately, many engineering operations do not employ geoengineers. Legget (1974) cites the case of the excavation for the St. Lawrence Seaway. Here, a series of lawsuits were filed by contractors who claimed more than $27 million because the materials excavated were more difficult to remove than predicted. Their workers had misinterpreted the drilling records and failed to realize the material was a dense glacial till. It was apparent that they had not read the geological literature of the area, which contained sufficient information and described the materials accurately.

Kaye (1976) described another engineering lapse in Boston, Mass. Although boring records were available to determine the character of soil that underlay building sites, such information was not used in the planning of the excavations for buildings on Beacon Hill. Here, at least three major projects encountered construction problems – the Boston Common Garage, the Leverett Saltonstall State Office Building, and an apartment building – because of misinterpretation of information. Workers anticipated materials would be glacial till because they thought the landform was a drumlin. Instead, the sediments were a complex of materials with prominent sand and gravel deposits that formed a glacial end moraine (Fig. 11.17). This mistake was very costly and delayed the projects many months because the materials had to be excavated in a different manner than planned. Groundwater flow produced additional problems.

A study of well yields on the Papago Indian Reservation in central Arizona revealed several important wells had lost much of their production capacity. To increase the yield (Coates, 1981), a series of dry-icing injections was used. This technique depends on insertion of dry ice into the well bore and closing the casing to build gas pressure from the decay of the solid carbon dioxide. When the plug is released, tremendous gas pressure surges through the system, creates a flushing action – along with a manmade

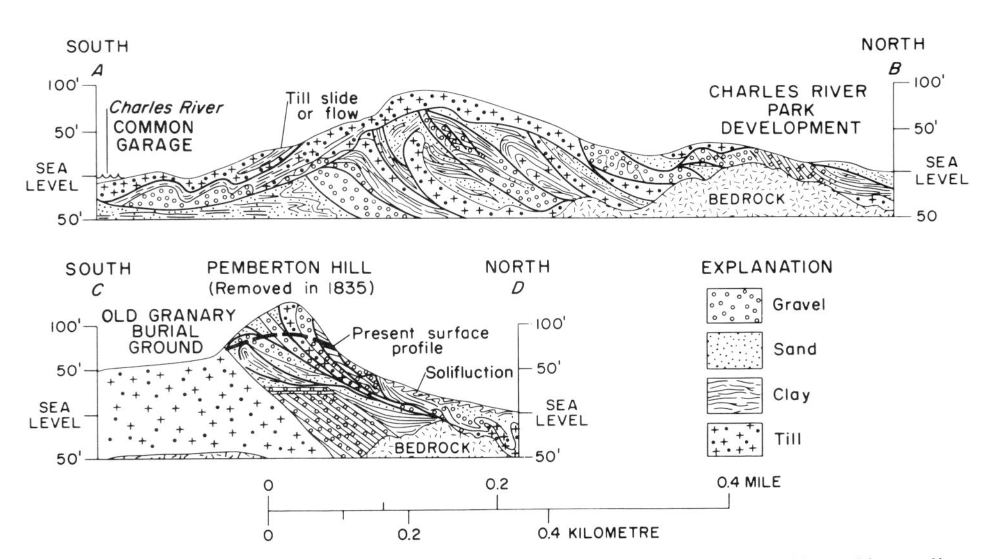


**FIGURE 11.17** Geologic cross sections across Beacon Hill, Boston, Mass. Heavy lines indicate faults. The drill cores had been misinterpreted as glacial till in a drumlin. Instead, the underlying structures were part of a series of imbricate thrust sheets in a glacial moraine. This was a costly mistake in several construction projects of the area (Kaye, 1976).

geyser — and the debris that had clogged the well screen is loosened, thereby permitting greater groundwater flow into the bore of the well. Applications of this method were able to increase yield in some wells as much as 100 percent.

The siting of nuclear power plants is probably the most crucial engineering decision in today's world. At Nine Mile Point, N.Y., on Lake Ontario, several faults were discovered near the construction site of a new reactor. Here, it was important that we assure the plant's integrity by making exact determinations on the characteristics of the faults as well as the stress field within the rocks. This was necessary to plan the structural security of the plant's foundation and services. It was also necessary to install a highly sophisticated battery of monitoring equipment to detect the slightest movement or change in the rocks.

And finally, one part of geoengineering falls within the realm of geotechnology, that subdivision whereby geologists help design equipment and apparatus that is needed for their work. For example, they can assist in determining which materials are best suited for a specific project. Gabion (wire mesh crates filled with rock) may be the most appropriate materials for some jobs, whereas fabriforms (porous nylon netting secured by injection of high-strength mortar) may be preferable for other types of bank stabilization (Figs. 11.18, 11.19). The design of monitoring and measuring devices may also benefit from specific input by geotechnologists.

**FIGURE 11.18** Construction of the Humacao River Channelization Project, Puerto Rico. Here, gabions are being installed to stabilize the river bank. Courtesy of Bekaert Steel Wire Corp.

**FIGURE 11.19** Construction of a fabriform plastic revetment at Kinzua Dam, N.Y.-Pa. Courtesy of Prepakt Concrete Co.

## POSTLUDE

In the next decades, geoengineering will become even more crucial than it is at the present time. Mankind is rapidly running out of safe, convenient, suitable, and choice building sites. An expanding population also places more constraints for location of structures necessary for all of societal needs and activities. Thus, siting problems become more difficult, and the assurance that structures will not fail or produce their own brand of degradation on the environment becomes more problematical.

Geoengineering requires a large blend of other geological subdisciplines. Furthermore, it needs to be both inter- and intra-disciplinary because other aspects of a planning matrix are required. For example, cost, time, political boundaries, and public preferences are all part of the process whereby final decisions are reached. Thus, the geoengineer needs to be tuned to a variety of factors that will all affect the degree of freedom with which he can ply his trade. Furthermore, it is necessary that environmental managers realize that structures cannot be placed anywhere they may choose. If there is not a proper line of communication and understanding among the scientist, engineer, and planner, we can expect dislocated buildings, and more, to spring up all over the planet (Fig. 11.20).

**FIGURE 11.20** Leaning Tower of Pisa, Italy. This structure was placed on materials with low soil strength and began tilting prior to completion of the building. This engineering mistake proved to be a blessing in disguise because of the large amount of revenue brought to the area by sightseers. Courtesy of David Alexander.

CHAPTER 12

# ENVIRONMENTAL CONTAMINATION

The despoilation of his own habitat by man is nothing new – only the magnitude and variety of degradation is different. Early man was highly mobile and few in numbers, and his refuse was scattered along the terrain in his wanderings. Once he settled down with first the Agricultural Revolution, then the Industrial and Urban revolutions, however, the waste products became concentrated in smaller areas. Some of the early civilizations simply rebuilt their cities on the rubbish of earlier regimes. In today's world, this is no longer a possibility. The disposal of waste material has become a monumental problem, but is the price we pay for our life style.

Although natural processes produce their brand of deleterious materials, such as noxious fumes in volcanic areas, we will emphasize in this chapter the manner in which mankind is contaminating the land, water, and air of our planet (Fig. 12.1). Indeed, nearly every action by humans alters the land–water ecosystem in some way. The awareness of contamination, which includes all facets of pollution and waste materials, was very slow in coming. The conservation–environmental movements in the United States that first occurred in the early 1900s, and again in the 1930s were strangely silent about contamination. Instead, they were largely resource-oriented. The beginning of the new wave of environmentalism dates from the 1962 publication of Rachel Carson's dramatic book *Silent Spring.* For the first time, an international authority showed our nation, and the world, the perils of unrestricted and saturated use of manufactured chemicals on the quality of life. Thus, this new and third wave of environmentalism was founded on an original core of attempting to contain the rampant desecration of the earth by human-induced contamination.

Pollutants and waste products affect the land, water, and air and become incarcerated in these media. The geologist becomes involved because these domains influence geologic processes and materials. Furthermore, his input is vital in providing a data base upon which wise decisions and environmental management can be reached. Each segment of society produces its own type of contaminants – the mines, the farms, the cities, and the industries. The emerging subdiscipline of geomedicine is starting to deal with the character of earth materials and how they influence human health.

**FIGURE 12.1** Waste disposal in the open canals of Venice, Italy.

Many factors have contributed to the escalating mountains of trash, refuse, and fumes, which come in all sizes, shapes, and forms — the huge number of Earth inhabitants, the inventiveness of man in creating thousands of new products and chemicals, the inability of people to employ proper safeguards, the deliberate act of using the least expensive production methods regardless of environmental impact, the psychology of a throw-away society, the packaging industry, planned obsolescence, and many more (Fig. 1.8). How to deal with these manifold contamination problems should cause sleepless nights for environmental managers, give headaches to budget directors, and frighten citizens who must try to live in an ever-deteriorating homeland.

## ROLE OF GEOLOGY

Geologists play a crucial role in those aspects of environmental contamination where the land-water ecosystem is involved. Their knowledge of earth materials and structures, and the characteristics of surface and ground waters is essential for environmental managers. The location of proper disposal sites is entirely the province of the geologist because this is his domain of expertise. Determining the type of incarceration of waste material requires the skills of geoengineers, and the character of materials to be used and handled needs the research and data compiled by those in geotechnology. Thus, problems of environmental contamination cut across many disciplines in society, but the earth sciences are vital for the successful location and operation of disposal systems that have high integrity and provide safety and well being for society.

My own work in this field provides illustration of still other ways that a knowledge of geoengineering can aid planning and designing of structures. When it was discovered that Deltown Chemurgic Corporation's Delhi, N.Y., creamery was polluting the West Branch Delaware River, they were required by the Delaware River Basin Commission to cease such contamination. An engineering firm was hired that installed a settling basin for wastes (Fig. 12.2). These were then pumped to a spray irrigation plot where alfalfa was attempted to be farmed. Such technique had proved successful in studies at

**FIGURE 12.2** Deltown Chemrugic Corporation dairy plan, Delhi, N.Y. A holding pond is in the foreground for temporary storage of effluent pollutants. Note the white-appearing scum on far side of the lagoon. These materials were formerly sprayed on soil in an attempt to develop a spray-irrigation type of alfalfa crop. It failed, and the company was required to install a sewage plant.

Pennsylvania State University. Unfortunately, the river contamination continued, and I was consulted as to why the system failed. The answer was simple: the excessive moisture on the land built up a "groundwater high." This allowed for increased flow velocity of the water through the coarse valley-fill sands and gravels without sufficient time for cleansing. These "natural" conditions in the field were at odds with engineering handbooks that indicated such materials should act as an effective pollutant filter.

## AIR POLLUTION

Historically, air pollution takes precedence as the first realm where man became aware of contamination. As early as 100 B.C., Horace wrote very unfavorably of the "smoke-blackened temples of Rome." By 1306, London's air had become so fouled by smoke that a royal proclamation was issued that curtailed the use of coal in the city. Violation of the proclamation was punishable by death. The public Nuisance Law of 1536 in England was primarily prompted by air pollution, and it became the cornerstone of English Common Law in the field of environmental nuisance (see also Chapter 14). The initial legislation in the United States dealing with air pollution was the Air Quality acts of 1963, 1965, and 1967. These acts empowered the government to monitor and set air-quality standards. By 1968, such states as Maryland and New Jersey, as well as New York City, and several northern Virginia counties had set maximum sulfur levels that would be permitted in the burning of fuels (Fig. 12.3). The 1970 Clean Air Act had as a major goal a healthful air over the entire country by 1977. However, it was continually beset with administrative problems and industrial inertia and conflict. Even today, it faces softening of the guidelines it originally established.

The most common pollutants are particulate matter, sulfur dioxide, carbon dioxide, carbon monoxide, oxides of nitrogen, and various hydrocarbons. Such contaminants are produced in many ways. About 60 percent comes from motor vehicles, 17 percent from industry, 14 percent from electric generating plants, and 9 percent from space heating and incineration. Additional burdens, thrust into the atmosphere, come from such activities as pesticide use, construction, agriculture, mining, and others.

### Human Health

Modern air pollution is more insidious than earlier air contaminants because often some materials have been removed at the source, and the

**FIGURE 12.3** New York City pollution before the Clean Air Act of 1972.

remnant matter entrained into the atmosphere is not visible, thus lulling the populace into false security. Although the bulk of contaminants are oxides of sulfur, carbon, and nitrogen, numerous exotics, especially harmful, are released by some operations. These include the bewildering array of such elements as arsenic, copper, beryllium, fluorine, lead, mercury, and even aluminum, chromium, iron, and vanadium.

In December 1930, the smoke stacks of blast furnaces, steel mills, and power plants were belching forth as usual in the Neuse River valley of Belgium, but the thick upper air did not allow for dissipation of the fumes. As a result, thousands of people were sickened during a three-day period and more than 60 died. A similar fate befell Donora, Pa., during the period of October 27-31 in 1948. Smoke from heavy industry and the temperature inversion of the air produced the same effects on people – irritation of the respiratory tract, eyes, nose, and throat. A total of 5910 inhabitants became very ill, and 17 died (usually those with a pre-existing disease of the cardio-respiratory system). However, the greatest tragedy of all occurred December 5-9, 1952, in London. A continuous heavy blanket of air (really smog, the mixture of fog and smoke) retained all the harmful contaminants within the city, resulting in many thousands of illnesses and the deaths of 4000 people.

As early as 1970, Lave and Seskin made a detailed study of the relationship of air pollution to human health in the United States. In urban and industrialized areas, they showed a remarkable correlation with many diseases, including certain types of cancer, bronchitis, and cardiovascular mortality. In terms of human health costs, the study indicated savings of more than $2 billion would occur if air pollution could be reduced 50 percent (Fig. 12.4). A 1977 study by the National Academy of Science concluded that 21,000 people east of the Mississippi River die each year as a result of power-plant emissions. Another 4000 deaths nationwide, and 4 million sick days are caused annually from auto exhaust. Clearly, humans are highly vulnerable to contaminated air.

## Acid Precipitation

Oxides of sulfur and nitrogen are created by combustion of fossil fuels. When released into the atmosphere, they react with oxygen and water to

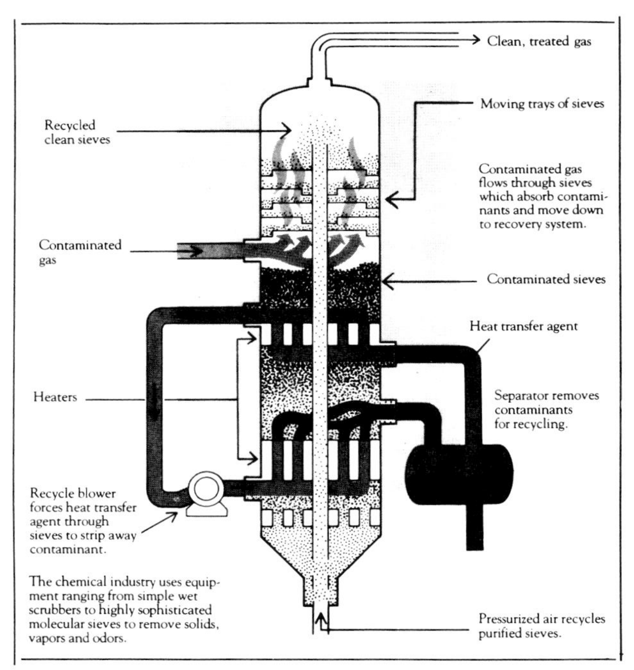


**FIGURE 12.4** Schematic drawing of the molecular sieve process to reduce air pollution. A variety of techniques are now in use in attempts to largely eliminate the source of "acid precipitation." Scrubbers can trap as much as 1300 tons of particulate sulfur each year. Electrostatic precipitators hold fly ash emissions down to nearly 100 percent. Courtesy of Chemical Manufacturers Association.

form sulfuric acid and nitric acid – the principal ingredients of "acid rain." Immense environmental damage may ensue when such materials fall from the sky and interact with the soils, wildlife, and waters of the land.

The cause and effect relationship of acid rain and the disappearance of fish in Scandinavia was first suggested by a Norwegian fisheries inspector in 1959. The decline of salmon catches became unprecedented in the late 1920s, but it took later documentation to show their low tolerance for acidic waters. Also in the late 1950s, episodes of highly acid precipitation were noted in Belgium, the Netherlands, and Luxembourg. By the late 1960s, additional regions in Germany, northern France, and eastern British Isles were receiving abnormally heavy acid rain. This phenomenon was first recognized in North American at Hubbard Brook Experiment Station, N.H., which showed a rainfall pH of 4.1 during the 1963–69 period. (On a scale of 1 to 14, neutral chemicals are 7, and lower numbers become increasingly acidic.)

Since the 1960s, there has been a rapid increase in American acid rain production. In 1940, about 28 million tons of sulfur and nitrogen oxides were released into the air, but by 1976, the amount has doubled to 57 million tons. The principal culprits are smelters and coal-fire electric plants. Although there are numerous local problems, the one of greatest magnitude emanates from the industrial complex of the Ohio Valley and contiguous area. With the advent of higher smokestacks, first recommended so that the local areas would not be contaminated, the entrainment of pollutants soared into the higher winds, thus spreading the plumes of acid rain to ever greater distances. For example, it has been reported that the 400 m high smokestack at Sudbury, Ont., is responsible for 1 percent of all sulfur dioxide emissions throughout the world!

The environmental effects of acid rain are in part determined by the chemistry of the water, soils, and rocks upon which they fall. The reason the effects are so devastating in Scandinavia, the Adirondacks, and parts of eastern Canada is that the rocks and soils are crystalline types, already high in acidic minerals. Lacking a carbonate content, such as limestone, which can act as a buffering agent against acidity, such terranes are unusually sensitive. The tolerance or threshold level for many fish and plants is soon reached, thereby causing drastic changes in the ecology. In 1975, a Cornell University study of 217 Adirondack lakes at elevations above 2000 ft (600 m) showed that 51 percent of the lakes had a pH of 5 or less, and 90 percent had become devoid of fish. Recent studies have shown 256 other lakes are in danger of losing their sportfish. Similar studies from southwest Ontario, Canada, eastward to Nova Scotia, also showed decimation of fish populations. This is becoming a point of great international aggravation to the Canadians because the cause of the drastic increase in acid can be traced to the United States.

There are many other environmental disruptions from long-travelled acid rain in addition to fish kill. Terrestrial ecosystems are adversely affected, and the food chain is disrupted. Alteration in the soil chemistry upsets the normal cycling of metals in soils and accelerates the leaching of minerals, ultimately influencing forest productivity. Most bacteria cannot tolerate a pH of less than 5, so instead, their place is taken by fungi. Because bacteria are the major decomposers of dead organic material, decomposition slows along with the nutrient cycle. A type of self-accelerating oligotrophication begins. In this state, mosses and aquatic plants, which are more resistant to acidity than aquatic animals that feed on them, also increase. Decay continues to slow, and the filling-in stage of the lake becomes faster. It ultimately dies. Groundwaters also become polluted by these processes. A recent report by the New England Interstate Water Pollution Control Commission estimated yearly damages due to acid ran in New York and New England at $250 to $500 million for both aquatic and terrestrial systems. If secondary human costs are considered, the figure would inflate to $2.5 billion a year! The principal losses include recreational fishing and related tourism; drinking water supply treatment; lumber, paper and related industries; crops and vegetation; and manmade structures such as buildings and monuments.

When long-range air pollutants are added to short-range exhaust impacts, the amount of damage to soils and property becomes staggering. Smelters at Ducktown, Tenn.; Smelterville, Idaho; Sudbury, Ont.; and Swansea, Wales, have destroyed soils and vegetation for several kilometers around the plants (Fig. 12.5). For example, the damage was so severe at Ducktown that an entire new erosion cycle commenced because there was no plant life to protect the barren soil. The combination of air contaminants has defaced and ruined buildings throughout the world. The priceless Acropolis and Parthenon in Athens have been greatly damaged by auto fumes, as has the Taj Mahal in India. The sculptured treasures on view in Venice have had to be taken indoors, and substituted by replicas, because of industrial air pollution from the adjoining mainland. In the United States, the Statue of Liberty and the hierographics of Cleopatra's Needle, a giant obelisk in Central Park, New York City, have suffered deterioration. Cleopatra's Needle had resided unchanged in Egypt for 3500 years, but within 50 years after having been installed in the City, the ancient writings had been ruined by air contamination. When all such losses are added for the eastern part of the country, the New England Commission estimates an annual loss of $13 billion in degradation of buildings and materials — mortar, stone, concrete, textiles, paper, paint, and leather.

**FIGURE 12.5** Accelerated manmade erosion. Fumes from the smelter at Ducktown, Tenn., killed all nearby vegetation and helped usher in a new cycle of erosion (Glenn, 1911).

## Weather and Climate Effects

Man's influence upon local weather conditions, and possible global impacts, is still subject to much debate. In urban areas, there is now good evidence that the combination of buildings and pavement, which change the radiation heat balance, and the air contaminants do influence local weather. This "heat island" effect causes many cities to be up to 10°C warmer at night, and several degrees warmer at other times than adjoining rural areas. Rainfall is also increased. For example, when the urban parts of the following cities are compared with the surrounding rural countrysides, Chicago receives 17 percent more rainfall; St. Louis, 15 percent more; Detroit, 25 percent more; and Cleveland, 27 percent more.

Worldwide effects of man's intervention and the entrainment of soil particles into the atmosphere (see Chapter 10) may induce desertification by the change of albedo on croplands, rangelands, and forests. Another pressing problem is whether long-range climate influences are occurring due to the "greenhouse effect" (the accelerated introduction of mostly carbon dioxide into the atmosphere, which allows for incoming solar radiation, but the retention of heat energy within the planetary system, thereby elevating the temperature). Although carbon dioxide constitutes only 0.03 percent of the gases in the atmosphere, it contributes to an increase of 10°C for near-surface temperatures. Atmospheric carbon dioxide was about 285 ppm (parts per million) prior to the Industrial Revolution, but has increased to about 335 ppm today and grows 3–4 percent each year. If present trends continue, by the year 2000, such additions could conceivably cause at least a 1°C annual change in temperature, which would have significant global repercussions – effecting a rise in sea level and shifting the boundaries of arable lands.

## SOLID-WASTE DISPOSAL

Mankind is almost being drowned in the debris from his own activities. All human endeavors produce solid waste – households, mines, businesses, and governments. In 1930, city inhabitants discarded an average of 2 pounds (0.9 kg) of trash (refuse, garbage, and other debris) per day. By 1970, the figure had grown to 5 pounds (2.2 kg), whereas today it is nearly 10 pounds (4.5 kg)! Such astronomical amounts are placing a huge burden on collection and disposal of these wasted materials. The United States generates more than 4 billion tons of solid waste each year. About 90 percent comes from agricultural and mining wastes, and about 3 percent from industrial processes (some of which are recycled). However, the greatest problem seems to be what to do with the nearly 300 million tons of garbage and trash thrown away from homes, schools, office buildings, stores, hospitals, and municipalities. This might be called the "smothering of society," because a recent study showed that half of the cities will run out of their disposal sites within a 5-year period.

The earliest pollution law in the United States was passed in 1899, known as the Refuse Act (also Rivers and Harbors Act). It forbade dumping of refuse into navigable rivers without prior permission from the U.S. Corps of Engineers. However, it was almost totally uninforced until the 1960s at which time the term "refuse" was interpreted to include pollutants and also cover acts of dredging. By 1965, the federal government had become further alarmed at the mounting waste problem, and Congress passed the Solid Waste Disposal Act of 1965. Although this was the first national recognition of the problem, the act was incomplete. Its purpose was to provide assistance to local governments to plan disposal programs, but it failed to mandate any regulatory authority and overlooked the problems of collection and transportation of waste. However, the Resource Recovery Act of 1970 amended the original 1965 act, and shifted federal involvement from disposal to recycling, resource recovery, and conversion of waste to energy. It also recognized the importance of proper storage of hazardous waste, which then became the focus of the Resource Conservation and Recovery Act of 1976. Meanwhile, many states were passing stricter and stricter legislation about solid waste. One example, is the Solid Waste Management Act of 1969 in Pennsylvania. This law requires the involvement of geologists on the siting of landfills, which must receive geological approval, study, and reports before they can be so designated by the state.

The twin problems of disposal cost and which disposal method to use plague all elements of society. It has been estimated that more than $15 billion is spent annually to dispose of all solid waste, and collection alone of waste in cities is nearly $5 billion per year. The following sections briefly describe some of the methods that are used for solid-waste disposal.

### Open Dumps

Throughout history, this has been the most common disposal method, and unfortunately, still is today. In the United States, 95 percent of total wastes use this "method." The best rational for such a practice is economics. Open dumps require little preparation, maintenance, or supervision. They are the cheapest procedure, but their environmental effect is a disaster. Surface water and groundwater contamination become rampant. Disease can easily result because the breeding of pathogenic organisms is uninhibited. They are an aesthetic blight to the landscape, in sight and smell.

### Marine Dumps

Many coastal cities employ this disposal method. New York City is a prime example where offshore garbage dumping has been a principal waste elimination procedure used for more than 50 years. The dictum of "out of sight, out of mind" and the cheapness along with space limitation have caused continuation of this practice. However, garbage is sometimes washed ashore, and commercial fishing and shellfish harvesting suffer financial loss from such operations.

### Incineration

Burning is also a method to eliminate bulk in many waste products. Although normal burning can reduce the volume of wastes, the resulting fumes can cause severe air pollution. A recent 2-year study by the Danish Environmental Protection Agency showed the hazardous dangers that result from refuse burning by increasing cadmium (harmful to human lungs, bone tissue, and kidneys) in the soil.

The installation of high-temperature incineration is a great improvement on previous burning techniques. In such plants, the weight and volume of what needs to be finally buried is greatly reduced. For example, only 3 percent of the original volume remains. Potential air pollution can be greatly subdued with the addition of electrostatic precipitators. The pyrolysis method is a special technique whereby garbage is baked in a low oxygen atmosphere. In the process, methane gas is produced, which can then be either recycled as the fuel that burns more garbage, or diverted off by pipelines for use as a resource.

### Sanitary Landfill

When land is abundant, and the environment setting appropriate, waste disposal in a "sanitary" landfill can be an effective disposal method. This technique has become especially popular in many communities during the past 20 years. To assure integrity, and thereby minimize leachate from getting into water systems, landfills should be located at sites where geologists have given the okay. The site should have a predictable water table, be free from flooding, have suitable cover material, and be sufficiently remote from important downgradient water sources.

The name sanitary is applied to indicate the attempt to contain all waste and possible pollutants on site. This is done by sealing the landfill off from surrounding processes that could affect it adversely, and containing the undesirable leachate from penetrating into water systems, either surface water or groundwater. It is important that the base of the trenches of excavations that receive the waste are impervious so that drainage through the waste will not reach the water table or nearby streams (Fig. 12.6). Natural sealants, such as clay can be used, or the site can be lined with plastic sheets that are non-degradable. During the day, waste is placed into the landfill and compacted as much as possible with bulldozers or other heavy equipment. At the end of each day's operation, the waste is covered by a layer of impermeable dirt to a 15 cm thickness. This prevents trash from blowing, infestation by animals, and reduces infiltration from rainfall.

There are three main methods for development of a sanitary landfill, each depending upon the local conditions and terrain. The trench method is used on relatively level ground, and when one trench is filled, another is excavated at its side (Fig. 12.7). The area method can also be used on level

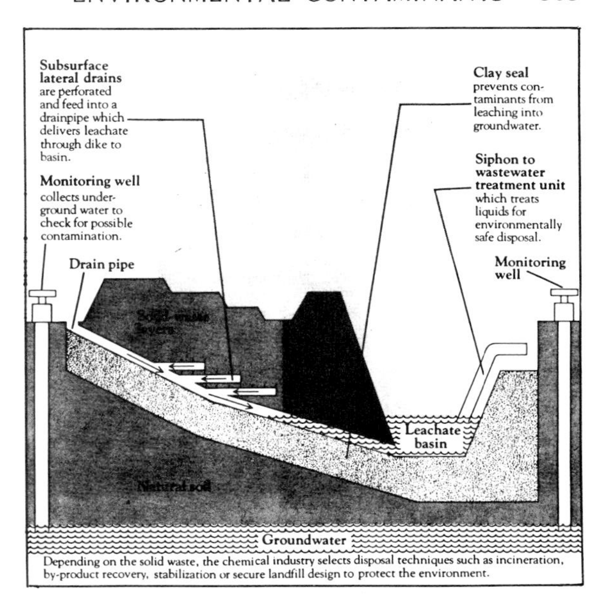


**FIGURE 12.6** Design for the type of engineering that can help produce secure sanitary landfills. Courtesy of Chemical Manufacturers Association.

ground, but prior excavation is not required. Instead, the waste is spread row by row from the boundary of the landfill, and the surface is gradually built up (Fig. 12.8). The ramp method is a hybrid technique that combines parts of both the trench and area methods. It is especially suited for sloping terrain. Here, refuse is spread over the hillslope, compacted, and then covered with earth from a cut at the base of the ramp.

Sanitary landfills have several advantages: the financial investment is low, the operation can begin within a short time, all wastes can be incarcerated, and when filled, can still be used for other purposes. For example, completed landfills have been used as recreation sites for ball fields, ski slopes, raceways, green space, agricultural uses and even some light building projects.

## HAZARDOUS WASTE

As if there were not enough problems to beset environmentalists, a new villain seemed to suddenly appear on the scene in the late 1970s – hazardous waste. Although the dangers of radioactivity from nuclear processes had been known for decades, surprisingly, other components of man's tampering with chemicals had largely been ignored. In a variety of studies, the Environmental Protection Agency concluded in 1983 that the United States produces 20 billion kg of hazardous chemical waste each year, that there

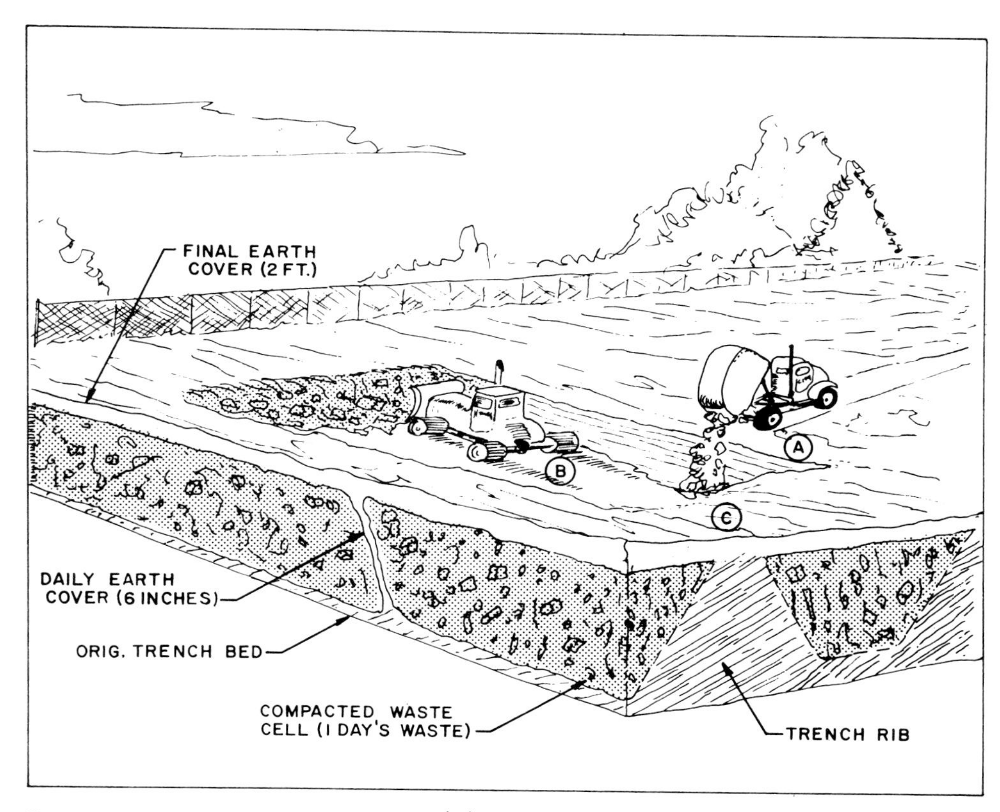


**FIGURE 12.7** Trench-type landfill. (A) Collection trucks discharge refuse into excavated trench. (B) Shredder-compactor spreads and compacts material. (C) A final 6 in (15 cm) cover is applied between cells, and a 2 ft cover is placed on top before seeding (Foose and Hess, 1976).

were at least 15,000 disposal sites that presented severe health hazards, and that toxic waste is the most important environmental problem.

Under the federal Resource Conservation and Recovery Act of 1976 (RCRA), hazardous waste is defined as "a solid waste, or combination of solid wastes, which because of its quantity, concentration, or physical, chemical characteristics may cause an increase in serious mortality or an increase in serious illness or pose a substantial present or potential hazard to human health or the environment when improperly treated, stored, transported or disposed of, or otherwise managed." The U.S. Environmental Protection Agency (EPA) has identified four principal characteristics of what they classify as hazardous waste: ignitability, corrosivity, reactivity, and toxicity. Also, in 1976, Congress passed the Toxic Substances Control Act, which authorized EPA to require testing of certain products by manufac-

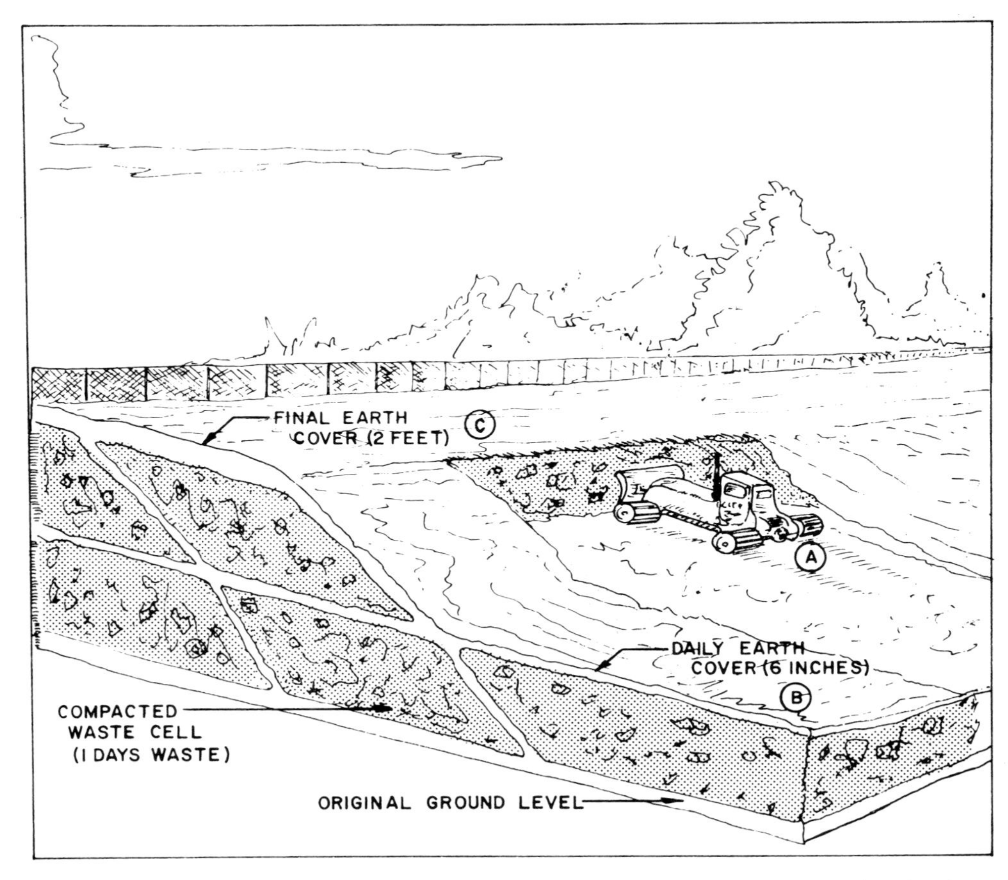


**FIGURE 12.8** Area-type landfill. Here, the waste cells are placed on top of each other, sufficiently separated by cover material. (Foose and Hess, 1976).

turers, and to provide guidelines and monitoring surveillance systems for specific materials. The federal Comprehensive Environmental Response, Compensation and Liability Act of 1980 is the newest legislation that deals with hazardous waste problems. The Act is commonly known as the "Superfund" because it authorized expenditure of $1.6 billion over a 5-year period to clean up the worst hazardous waste dumpsites in the country.

"The presence of toxic chemicals in our environment is one of the grimmest discoveries of the industrial era." So spoke President Carter as he attempted to marshal the involvement of the country on this critical problem. A recent congressional report further noted, "The hazardous waste problem cannot be overstated." The first estimate of the quantitative magnitude of the problem was made in the 1979 Hart Report of the EPA. It was estimated that between 32,000 and 50,000 sites in the United States contained hazardous wastes, and that 1000 to 2000 pose a significant risk to human health and the environment. To alleviate the dangers from such sites would cost an average of $25.9 million per site. The Love Canal problem at Niagara Falls, N.Y., prompted a 1980 survey in the State that showed hazardous wastes occur in 852 disposal areas. Of these, 157, according to N.Y. State Department of Environmental Conservation, pose "substantial environmental and public health concerns."

The villains in the hazardous waste scenario are the vastly expanding number of synthetic chemicals on the market, trace elements, and inadequate handling and disposal methods throughout their use. Chemists have a register of 4.3 million distinct chemical compounds, and currently 70,000 are in production in the United States. Although many are harmless, the Occupational Safety and Health Administration (OSHA) has identified 25,000 compounds that are toxic to human health, and more than 1900 that are carcinogenic. These occur in such materials as drugs, food additives and preservatives, ores, pesticides, herbicides, fungicides, dyes, detergents and soaps, plastics, animal and plant extracts, and waste by-products. There are 115,000 industries involved with manufacture, distribution, and disposal of chemicals. But the really sobering statistic is that the National Institutes of Health estimates 90 percent of human cancer is environmentally related.

Not only the prodigious amount of chemicals, but their widespread use and the large number of dangerous ones, have caused severe problems all along the manufacturing and utilization line. U.S. industry generates about 60 million tons (wet base) of proved hazardous waste each year. The rogue's gallery of toxic materials is enormous, and includes the following: (1) chlorinated hydrocarbons, (2) polychlorinated biphenyls (PCBs), (3) metals such as antimony, arsenic, cadmium, lead, mercury, selenium, and thallium, (4) halogenated aliphatics, (5) phenols, and (6) other groups of aromatics, ethers, polyvinyl chlorides, dioxins, etc. The manner of entrainment of hazardous wastes can come from either a point source (a restricted site such as a plant, a lagoon, or a landfill), or a non-point source (a large area where wastes are disseminated, as in pesticide application of entire fields or orchards).

## Dioxins

There are 75 related chemical compounds that are dioxins. They come in a variety of forms, but the most toxic is 2,3,7,8-tetrachlorodibenzo-p-dioxin (or simply 2,3,7,8-T, or TCDD). Dioxin is colorless, odorless, and tasteless. It develops as an unintentional by-product in the manufacture of herbicides (including the defoliant Agent Orange), detergents, pesticides, disinfectants, and wood preservatives. By itself, it has no known use or value. However, it is so toxic and lethal that 5-thousandths of a gram can kill a hamster, and 1-millionth of a gram can kill a guinea pig. In humans, according to the EPA, 1 ppb is hazardous to human health. For example, forest workers in Sweden exposed to herbicides with dioxin developed soft tissue cancers six times the normal rate. In Seveso, Italy, a malfunction in a plant manufacturing chemicals that contained dioxin as a by-product killed thousands of domestic and wild animals. Residents throughout the region had to be evacuated. In Times Beach, Mo., three dirt-floor horse rings were sprayed with waste oil (unknown at the time, but which later proved to contain dioxin) to keep down dust. Many wild birds, cats, dogs, and horses died from subsequent contact with the dirt floor. Later dioxin was discovered in other parts of the town, which then had to be evacuated by all residents at a cost of millions of dollars.

Associated problems about dioxin are that it does not seem to leach out of soil once entrained, and although it is soluble to a limited extent in fats, it is insoluble in water. Thus, it can bioaccumulate in the food chain and is not readily destroyed by the many normal routes of decomposition that prove effective for other contaminants. However, photo-decomposition is effective when it is exposed to direct sunlight. Then, dioxin can be rapidly destroyed because it can have a half-life as short as one hour. Thus, when sprayed on crops, much of the dioxin is eliminated in a short period. The half-life in soils ranges up to 10 years or more.

In an attempt to address the growing problem of hazardous waste, in 1980, Congress enacted the Comprehensive Environmental Response, Compensation, and Liability Act (known as "Superfund"), which authorizes government responses to hazardous substance releases as well as cleanup of inactive hazardous waste-disposal sites. The Act defines "environment" as including groundwater and drinking water supplies, and established a $1.6 billion fund to be used by the government in the cleanup process. Unfortunately, the funds to carry through this mandate were not used for more than 2.5 years. Finally, in the summer of 1983, further guidelines for the program were set, the 400 most hazardous sites were announced, and cleanup operations began at more than 100 localities (Fig. 12.9).

## Case Histories

The list of examples of hazardous waste contamination of the environment is growing at an alarming rate. The Love Canal near-disaster in 1976 served as a doom warning, and did more to galvanize interest and awareness

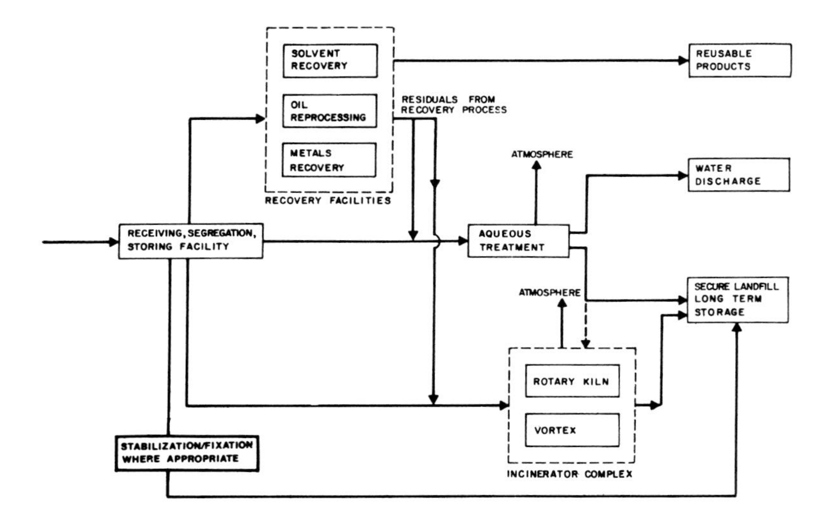


**FIGURE 12.9** Concept flow diagram for a hazardous waste treatment facility. Courtesy of N.Y. State Department of Environmental Conservation.

of the problem than any other case. The potential problem was first noted by homeowners who discovered chemical odors seeping from their basements. Sufficient evidence of abnormal contamination in this housing development led to an investigation of the site. It was discovered that a chemical company had buried waste in an unused canal, covered it, and later the site became a housing tract. Further study showed the inhabitants had above-normal rates of birth deformities, miscarriages, mental retardation, hearing loss, skin disorders, liver problems, and blood upsets. Although the cancer rate was also higher, the entire region suffers from air pollution, which could also be a contributor to the disease. In August 1978, the N.Y. State Department of Health declared an emergency, and began the evacuation and relocation of 240 families. By 1981, more than 1000 had left.

The United States does not have a monopoly on hazardous waste problems. For example, on July 10, 1976, chemical apparatus at a plant in Seveso, Italy, ruptured sending a plume of toxic chemicals 50 m into the sky. Minutes later the cloud cooled and settled in an area 2 km long and up to 700 m wide. In the following months, animals and vegetation began to die. People developed symptoms of illness with skin burns, stomach pains, internal bleeding, etc. The area of contamination was still a wasteland by 1984, isolated by barbed wire and armed guards.

## NUCLEAR WASTE

There are 166 nuclear power plants in 21 nations, but the problem of what to do with the radioactive waste (radwaste) is still unresolved. Radwastes result from a variety of activities including weapons manufacture, and from a variety of places such as hospitals, research laboratories, and nuclear power plants. Not only is danger posed by uranium and thorium, which are the most commonly used elements, but also by the entire series of associated radioisotope by-products that are released during their use, such as strontium, cesium, americium, and barium. Exposure to such radioactive wastes causes two types of biologic damages: somatic damage that affects tissue and genetic damage that alters hereditary materials. For example, the fission products of strontium-90 are so deadly because they can be absorbed like calcium into the body, where they cause leukemia.

Radwastes can be divided into three groups: (1) low-level wastes are those that can be diluted and released to the atmosphere after removal of the long-lived radioisotopes. These are commonly buried after solidification or occasionally placed in barrels and dumped at sea. (2) Intermediate-level wastes can be placed in concrete trenches in arid regions with deep water tables, where it is hoped the material will lose its destructive potency before it encounters the groundwater. (3) High-level wastes are the real problem. They are now contained on site, being placed in temporary storage in stainless steel and concrete underground containers. They need constant monitoring to determine whether leaks occur, and thus place in jeopardy waters of the surrounding area.

The moderator of a quiz show might ask the question, "What do Durango, Colo.; Grand Junction, Colo.; Edgemont, S. Dak.; Salt Lake City, Utah; and Shiprock, N. Mex., have in common?" The answer is mill tailings from now defunct plants, which formerly produced high-grade uranium, known as "yellow cake." In their wake, more than 27 million tons of radioactive residues were left at 20 locations. At Edgemont, all of the 2000 inhabitants have been exposed to the tailings, dust from which blows into homes, schools, and businesses. Abnormally high rates of birth defects and cancer have been reported at the nearby Pine Ridge Hospital. At the time of the operating plants, the spoil pile method of disposal had not been a concern. In Grand Junction, the material was used as aggregate in building blocks for homes. In Salt Lake City, it was used as fill, and buildings were constructed on top. For example, Fire Station No. 1 was built more than 20 years ago, and the radioactive count of the fill is still seven times greater than that allowed for uranium miners. In Grand Junction, millions of dollars were spent in decontaminating buildings and in renovating the more than 700 properties that required removal of radwaste tailings.

Of course, the real problem, or $64 question, is how to permanently dispose of the spent fuel from reactors, and other high-level wastes from industry and the military. Currently, the 84 U.S. reactors must temporarily store the nearly 6000 tons of spent fuel they have produced. The other radioactive materials are stored in solid or liquid form in tanks at Idaho Falls, Idaho; Savannah River, S. C.; and Hanford, Wash. Storage at such sites is not 100 percent foolproof. For example, at Hanford there have been 16 leaks within a 20-year period, totaling 1.35 million liters. The most recent, 1973, leaked 450,000 liters of high-level waste.

## Disposal Alternatives

*Shallow storage.* Mention has already been made of the temporary storage now employed throughout the country. A modification of this is the placement of waste into a borosilicate glass mixture such as done at Oak Ridge, Tenn. West Valley, N.Y., was the site for a commercial fuel reprocessing plant by the Getty Corporation, but it was not a financial success and was abandoned. The conversion of waste to a corrosion resistant glass is an attractive idea, but still needs further testing to determine if it would be universally safe. Surficial disposal has the advantage of being easily monitored and retrievable. However, surface materials are more vulnerable to geological processes and are difficult to isolate completely.

*Deep storage.* These methods are aimed at placing nuclear wastes into rocks that are considerably below the land surface, 1000 m or more. The idea being — the deeper the better, up to a point. If predictions on stability of the geological rock and hydrologic environment were wrong, and if leakage occurred, it would take many years, hundreds and more, to contaminate the earth's surface. Such questions arise whether deep wells or excavated tunnels would be safer. The problem of which type of host rock to use has not been settled either. Each rock type has its advocates. Some urge basalt, others crystalline rocks, but the majority of "experts" believe that the massive salt formations may be the most suitable. Whichever environment finally wins, the radwaste will need to be bound into some matrix and sealed with impermeable and non-corrosive materials. Furthermore, measures must be taken that permit monitoring, and the capability for retrieval if the unexpected should happen.

*Other methods.* Several other schemes have been advanced for storage, but none have received wide support. For example, some people urge sending the waste into outer space via rockets. Others suggest burial in deep sea trenches. More recently, it has been urged to study carefully the possibility for burial in subducting zones where lithospheric plates are colliding.

Some people even think burial under the 4000 m thick Antarctic ice sheet is the answer, but this has not received serious consideration.

## WASTEWATER DISPOSAL

Nearly all municipalities in the United States and the great majority of cities throughout the world have sewage treatment plants that provide for the disposal of fluid waste generated within the environs. Depending upon the needs of the community, and the financial structure of the government, such plants may possess one or more levels of treatment. **Primary treatment** is largely mechanical, whereby particles are settled or filtered out for collection and further disposal. The remainder is discharged to a nearby stream. **Secondary treatment** is mostly biological, whereby the waste undergoes conversion of the nutrient matter to biomass (sludge) and separation from the water. Thus, the biodegradable pollutants are largely removed. **Tertiary treatment** is somewhat rare, because it is so expensive and requires sophisticated equipment. It is largely chemical and consists of such operations as adsorption, ozonization, and demineralization techniques to remove the non-biodegradable materials and inorganic ions such as nitrogen and phosphorus.

Although sanitary engineers are in charge of sewage treatment plants, geologists can be involved in such matters as determining streamflow characteristics, which in turn regulate the amount and type of effluent that will be permitted to be released. For example, a river with continual lowflow properties would be a poor choice for a plant, because even a slight amount of poorly treated effluent would soon contaminate it. Disposal of sludge can pose a burial problem since it may contain toxic materials, such as trace elements. Sludge was disposed on a farm in the nearby Binghamton, N.Y., metropolitan area, but the practice had to be abandoned because of the abnormally high rates of cadmium found in the soil. Other landowners throughout the area refused to allow sludge to be placed on their property. It will now have to be hauled outside the entire area to a state-run facility.

Homeowners, farmers, and small businesses that are too remote from municipal sewage lines must depend upon some type of **septic tank–leach field operation** to dispose of their wastewater and sewage. Many millions are in use throughout the country, and when installed and used with care, this system provides a reasonable disposal entity. Waste enters an in-ground tank where the solids and other debris settle, and the fluids drain into an adjoining series of buried lateral conduits composed of sand and gravel or tiling. The trick is not to overload the system. The tank needs periodic pumping to remove accumulated debris and waste, and excessive amounts of fluids should not be allowed to enter the system so that the leach field effluent drains from the site to where off-site contamination could occur.

## RECYCLING

Of course, the best waste-disposal method is to produce less waste. As this seems to be a rarity, the next best method is to recycle materials. Not only can valuable resources be saved, but it places less of a burden of volume on the disposal facility. Furthermore, garbage can become part of a biomass energy system and provide additional savings in fuel consumption. Unfortunately, the United States has a very poor recycling record. Only 1 percent of all municipal waste is recycled. Contrast this with a report by the Worldwatch Institute of Washington, D.C., which reported that at least two-thirds of the material resources we currently waste could be reused without making important changes in our life-style. Indeed, the Japanese recycle nearly half their waste and import scrap, such as automobiles, from the United States. One joke making the circuit is that this year's Buick is next year's Toyota.

Not all waste lends itself to recycling; some is too low in value. Other waste may be too difficult to degrade, too contaminated to treat, or the wrong size to manipulate with the method being used. Societal factors may also interfere with an efficient operation. There may be a scarcity of recycling collection plans, inadequate market for recycled goods, inadequate funding for plants, consumer pressure against recycling (such as the massive lobbying against the "bottle bills," laws that prohibit throw-away-cans), tax incentives that favor mining and logging, and freight rates that favor movement of virgin materials. For example, railroad rate structures are designed so that it is two to three times cheaper to transport raw materials than recycled goods.

The advantages for recycling are nearly endless. Recycling of municipal waste could provide at least 2 percent of all U.S. energy requirements. For example, reprocessing aluminum saves 96 percent of the energy needed to manufacture new aluminum. Other metals also have large savings — copper, 87 percent, and lead and zinc, 63 percent. The conservation of natural resources would be greatly advanced by accelerated recycling emphasis, and a smaller burden would be placed on landfills, many of which are rapidly being overloaded. Even psychologically, recycling appeals to the human spirit of working with, not against environmental systems.

Bottles and cans provide many people's ideas of what recycling is about. However, that is only a start. Many other common substances are also drastically wasted. There are 100 million tires discarded each year in the United

States. They pose a monumental problem, but also an opportunity. Many landfills will not accept them, and when just dumped in the open, they are ugly and can become ideal breeding grounds for rodents and mosquitoes. New technology has shown, however, that they can be profitably processed with benefits providing more than a 15 percent return on investment.

Part of the recycling problem is public apathy and lack of knowledge of how to become involved. Education and dissemination of information by all media sources is the answer. It must be shown that the savings realized will actually affect the pursestrings of the consumer. The following procedures are used for recycling efforts:

1. Source separation. This is the segregation of selected waste by the consumer and accounts for about 90 percent of all current resource recovery.

2. Recycling centers. There are more than 3000 community recycling centers in operation across the country. They have developed from a sense of community cooperation, availability of a good core of participants, and usually involve only modest funding. The Santa Barbara, Calif., Community Environmental Council returns more than $100,000 annually to civic groups who cooperate.

3. Curbside collection. This is an attempt to make recycling convenient. In such programs, vehicles periodically (or on a preset schedule) collect recyclable and returnable containers. Without widespread cooperation, however, such programs may not be cost-effective.

4. Intermediate processing. The term is used to describe the processing of source-separated materials derived from large quantity-generating commercial and industrial operations. Ecocycle in Boulder, Colo., and Ecolohaul in Los Angeles are typical facilities that handle such processing.

5. End-use manufacturing. This is the final step along the road to recycling — the plant that reuses recycled materials. These include steel production in which scrap iron is mixed with raw ore, glass manufacturing, aluminum processing, paper mills, and a host of others.

Not only the United States, but the entire world, has a long way to go to conserve natural resources, eliminate waste, and manage a closed-loop environment whenever possible, so that multiple uses will be derived from societal materials. It would greatly help if more funding were available to provide for more technological breakthroughs on the benefit and cost factors involved. Until that time arrives, however, there are still multiple options — especially an informed public — that can produce enormous savings in money and energy by the currently available programs.

## POSTLUDE

The enormity of environmental contamination problems throughout the world is almost beyond comprehension. No nation has a monopoly on problems or their solutions. Although industrialized countries have the most costly problems, those in Third World nations may actually be greater in terms of percent of gross national product. Communist countries are also severely affected. In East Germany, only 16 percent of total river lengths are adequate for drinking purposes. Fifty percent of rivers in Poland are even unfit for industrial use, and sulfur dioxide levels in Prague are three times higher than levels rated as acceptable.

The United States receives mixed grades on the status of environmental contamination. A total of $70 billion was spent during a 12-year period after the landmark legislation of the Clean Air Act. In this time, 23 major cities instituting pollution controls cut the number of days when breathing outside air was unhealthful by 39 percent. In Chicago where air was considered dangerous to breathe during 240 days in 1974, the figure was cut to 48 by 1980. During the 12-year period 1968–80, nationwide particulate matter was reduced 55 percent, sulfur dioxide emissions were 24 percent lower than in 1974, and carbon monoxide was 90 percent less than in 1968. However, by 1982, EPA had allowed more than 12 states to raise their limits of sulfur dioxide emission by more than 1 million tons a year. It also permitted carbon dioxide levels to increase 33 percent. In 1981, the EPA had also exempted 160 of the 167 tall stack emitters from any additional pollution controls.

There are many reasons why environmental contamination problems will continue to be a critical problem in the decades to come. Many communities have not faced up to the problem — a type of out-of-sight, out-of-mind syndrome. Just as long as waste is not placed in their backyard, communities have postponed, or transported their problems elsewhere. Modern society continues to be geared to an endless cycle of consumption, throw-aways, and generation of mountains of waste of every description. Mining and extractive industries are given more favorable tax treatment than industries involved with recycling material. The full cost of environmental damage is not internalized but instead is exported to become a social burden on the consuming and affected populace. Thus, incentives for waste reduction or operation efficiency to counteract contamination is not rewarded.

CHAPTER 13

# ENVIRONMENTAL MANAGEMENT

Although environmental managers are usually social scientists, there are numerous requirements within the management matrix that need the input of geologists. Because environmental management cuts across nearly all segments of human activities, it is vital that most societal segments be represented on decision-making bodies. Environmental management consists of deliberate planning and administrating policies that determine the status of the biosphere. It involves all levels of human endeavors, agencies, companies, and governments — local, regional, and national. Furthermore, because of the extreme complexity of environmental matters, and the different constituencies, involvement of many different disciplines is required, especially those in the social and natural sciences. Also, guidelines for management necessitate flexibility, because the ultimate objectives and how to reach them depend upon a variety of factors. For example, the strategies to be adopted may vary depending upon what is being managed; is it a process, a resource, a land parcel, or some other type of entity?

Before any type of environmental management can be fulfilled, there must be prior recognition that something is in need of being managed, that a problem already exists or one is percevied to be in the making. Therefore, sensitivity on the part of managers or the public is required, and their awareness and perception are needed to motivate some sort of action. To be effective, the goals of the action must be realistic, not pie in the sky, but formulated with a proper design that is possible within the limits of available funding, and carrying the approval of most societal and political organizations. And in the final analysis, such environmental actions must be consistent within legal and legislative mandates and directives.

## THE FORMAT OF MANAGEMENT

Environmental management requires decision making with the objective of following some course or plan or action. Thus **planning** is one of the components of the management system, and in order for planning to occur, policies must be set. **Policy** consists of an operational mandate that provides guidelines to the planning and ultimate management of human affairs. There are certain prescribed procedures that help establish coherent objectives. The problem(s) must be identified. Data must be collected and interpreted. Plans and objectives must be formulated as solutions. Decisions must be implemented.

### Definition of the Problem

This is prerequisite to all environmental programs. The type and scope of the problem must be determined. Does it affect a few or many people? Is it expedient to attempt a solution immediately or can the remedy be postponed? What people are best trained to deal with the problem? What sort of funding allocations should be set aside or obtained to reach a decision or a solution? Which type of environmental ethic should be employed? These, and many others, pose a never-ending series of decisions that must be reached by those who plan and articulate policy. Linked to the problem are the goals that must be achieved and the values obtained under a certain course of action. It should also be determined whether there are worthy alternatives that should be considered. Often, it is common practice to launch into a series of feasibility and validity studies. At a minimum, it is necessary to discover the scope and magnitude of the problem and the financial limits. These operate as a system of constraints that harness the ultimate objectives of the "game plan."

### Conflicts, Constraints, and Priorities

It is inherent in the character of environmental management that many policies will bring conflicts (Fig. 13.1). The old adage, "You can't please all the people all the time," holds. The principal problem is that the environment belongs to all the people, so there are always multiple constituencies in such matters that deal with the governing of the earth's land–water ecosystem.

The environmental manager is faced with a variety of constraints that restrict the flexibility of his response to any particular problem.

1. Boundaries. What are the physical limits? The size of the area to be managed is of utmost importance. Political subdivisions are often not consistent with natural barriers. Thus, the assignment of a problem on jurisdictional bases may not be a perfect solution and can interfere with the complete integration of the natural system. For example, a watershed comprises a natural entity, but it may be impossible to administrate it as a coordinated

FIGURE 13.1 There are many conflicts to address in environmental management. Courtesy of Caterpillar Inc.

unit because the drainage ditch may cut across county and state lines. Thus, spatial constraints obviously can pose a severe limit on the types of planning and policies that are adopted.

2. Time. What is the length of time that the object being planned for is to be managed? If flood-control structures are to be designed, should they be built for the 50-year, 100-year, or 200-year flood event? The recurrence interval of natural processes also determines planning options in other high risk areas – volcanic terranes, and earthquake zones. For example, nuclear plants can only be built at sites where earthquake events are unlikely during the 40-year life of the plant for which they are designed. What are the rates of allowable resource depletion – the safe annual yield of water – the har-

vesting of timber — the tolerable rate of soil erosion? The time period of management, therefore, influences the character of the project, its size, and the costs that will be involved.

3. Money. Good intentions are generally not sufficient for prudent environmental management. In the vernacular, "money talks." To accomplish a perceived result, a budget and financial structure must be adopted to attain the desired objectives. Although there is no panacea concerning the magnitude of fiscal affairs, the most generally approved practice is to establish a type of benefit/cost analysis. An environmental program, on this basis, should only be designed when the benefits that flow from the action are greater than the costs to produce the perceived benefits. Obviously, the successs of such a venture is determined by the accuracy of predictions as to what the real benefits and real costs will be. When such assessments were first done in the United States, only tangible benefits were considered, but more recently secondary benefits have also been included in the final formula — the saving of aesthetic values, of human life and health. A strong data base is vital to success.

4. Ethics. The purpose of environmental management often becomes linked with the type of ethic system that is adopted. These viewpoints fall into three categories: (1) The business or utilitarian-developmental ethic. This is the belief that all of nature should be subservient to mankind and used to the fullest for human gratification. (2) The conservation ethic. This is the view that nature best serves mankind when resources are prorated through time with allowance for prudent continued usage. (3) The preservation ethic. Under this goal, mankind and nature are equals. Nature also has rights that need protection. Thus, wilderness is important not only for its own sake, but also for the spiritual fulfillment of the human soul. The assignment of management priorities becomes the clue in the final decision, but whatever strategy is finally adopted will never have 100 percent concurrence with the total populace.

## OBJECTIVES

The final equation for environmental management of human affairs rests with three fundamental ingredients — health, safety, and welfare. These factors determine life support systems and what the quality of that life will be (Fig. 13.2). This encompasses a comprehensive array of endeavors, including such wide-ranging involvements as the maintenance of soil productivity and a flow of minerals and fuel, to the location of dwellings, and industry at safe sites.

| *HEALTH* | *SAFETY* | *WELFARE* |
|---|---|---|
| Severity of harm, risk, and disease from: | Processes that threaten life and property | Factors that determine the quality of life |
| 1. Air contamination | 1. Volcanoes | 1. Mineral resources |
| 2. Water contamination | 2. Earthquakes | 2. Materials resources |
| 3. Earth contamination | 3. Landslides | 3. Energy resources |
| 4. Food contamination | 4. Floods | 4. Aesthetic resources |
| | 5. Coastal storms | 5. Living space |
| | 6. Wind storms | 6. The Human spirit |

**FIGURE 13.2** Societal Goals for Environmental Management.

The maintenance of human health is of vital concern to all. The environmental components that endanger it consist of the manner in which societal wastes are disposed. As indicated in Chapter 12, the full range of contamination of the environment with pollution, toxic waste, and acid rain all contribute their poisons and toxins to the land, water, and air. Although not yet recognized as a full-fledged subdiscipline, **geomedicine** is becoming a more important aspect of the interaction of geology and such problems as those that trace elements pose to humans.

Public safety also ranks high on the list of areas of concern to the environmental manager. Items of concern include establishment of a data base to help predict and warn of imminent danger. Whenever possible, geologic hazards need careful study and control if possible. Reports and maps and other forms of communication are needed to alert developers and people of building sites where their very lives may be in jeopardy.

The quality of life should not be an extraneous feature because human welfare determines the outlook of citizens and flavors their spirits, souls, happiness, and mental health. These amenities depend upon many avenues and include material resources that provide the raw materials for existence as well as for recreational opportunities in parks and playgrounds.

To obtain such objectives requires enlightened management of the environment and wide participation by government. Effectiveness of the planning, and programs that are adopted, become a measure of the communications systems and the ability to implement these goals.

## ROLE OF THE GEOLOGIST

As part of the total team that oversees the management of the environment, it is necessary for the geologist to become involved in policy decisions at their inception. The environmental manager may determine that a particular amount of iron or petroleum is needed to maintain a certain industrial capacity. However, if such materials are not available, or cannot be discovered by the geologist, the attainment of such requested goals becomes an exercise in futility.

The geologist should be involved in as much of the planning, managing, and implementing of policies as possible. He is the guardian and the surrogate of nature to assure that landforms, earth materials, and processes are not wantonly abused. It is his responsibility to devise appropriate strategies that enable the natural world to be kept in as much equilibrium and harmony as possible. Once a course of action has been determined, alternatives considered, and decisions justified, the geologist must continue to be concerned about the ultimate impacts of such actions as the following:

Will the proposed action cause deleterious feedback to other parts of the environmental system, such as is the case with construction of many dams?

Will the proposed action cause an acceleration of the degradation of the land–water ecosystem? Unfortunately, farming *does* increase erosion and sedimentation, so what are the steps that can be taken to minimize such deterioration of the soil resource?

What are the geologic constraints in the system? The resource base needs proper evaluation and projections calculated for their proper management. Short-term goals need to be balanced with long-term needs, as this can play a part in how the materials, processes, or lands are used.

Is it possible to institute a program of conservation, or even preservation, within the context of the project? To what level should an attempt be made to husband resources and lands?

To answer such questions, there can be no substitute for obtaining a strong data base, along with the communication of that information to the policy authorities. This requires careful study of the environmental setting that will be involved or impacted by the action. The collection and inventory of all pertinent physical factors is a prerequisite. Maps are of prime use because they show the extent, distribution, location, and magnitude of significant features that may be the crucial ingredients to the system being changed. The depiction of these features should be intelligible to planners, as well as the report that accompanies them. Unfortunately, the proper mix of these did not occur in Anchorage, Alaska. Although the U.S. Geological Survey had made known the fragility of the sedimentary materials in parts of the city, these sensitive sites were still the places where extensive development occurred. As a result of this lack in communication, the 1964 Alaskan earthquake produced great losses, both in terms of property and lives. On a smaller scale, the Blue Bird Canyon area, Calif., had been designated by geologists as unstable for building, yet a housing development was placed on the hillside. The landslide of 1978 demolished numerous homes in the area (Fig. 13.3). In a 1971 study of California, it was projected that unless there was appropriate geological input into the planning decision, the State was likely to suffer damages of $9.8 billion by the year 2000.

Finally, it should be realized that even natural science does not speak with a uniform and monolithic voice. For example, scientists are on opposite sides of the phreatophyte question. Hydrologists wish to destroy such plants because they consume prodigious amounts of groundwater that might otherwise be used by man. However, wildlife biologists wish to retain the plants

**FIGURE 13.3** Close-up view of Bluebird Canyon landslide of October 2, 1978. The houses that overhang in the background are situated immediately above the head scarp at Laguna Beach, Calif. Although heavy rainfall triggered the landslide, this location is hazardous because of low strength of the rocks in the substrate and downgradient channel entrenchment that helped undermine the foot of the slide. Courtesy of L. Beach Leighton.

because they offer a refuge for animals and birds. And then, scientists are in conflict with developers. In the National Seashore enclaves administered by the U.S. National Park Service, the battle is now raging whether to allow natural processes to take their course without interference by man, or should remedial measures be continually undertaken in attempts to protect property and public utilities? The searching question that will continue to haunt industrialized nations throughout the 1980s is: should energy procurement be hastened at all speeds without the introduction of providing appropriate safeguards to citizens? Or similarly, should costly antipollution devices and abatement be vigorously pursued and required of an industry if it is likely to bankrupt the company, make it economically uncompetitive, and idle hundreds of workers in the community?

## URBAN AREAS

The urban environment is one that requires an unusual amount of special planning and management because once the development has been determined, roads graded, pipes installed, buildings emplaced, there is no turning back. The site is fixed and permanent. Thus, there is no room for error.

The urban area is the place where mankind has most severely changed the environment. Here, a natural locale has been metamorphosed into an anthropogene (created by humans) landscape — the ultimate environmental change (Figs. 13.4, 13.5). Within the twentieth century, several new approaches have been made in the management of congested communities, the deliberate design of open space and of underground space.

### Open Space

The term **open space** has become associated with the deliberate maintenance of natural areas within an urban area. It is on the opposite side of the spectrum where urban sprawl resides — the rampant and unrestricted growth of developed areas. Instead, open space is characterized by such qualities as natural scenic beauty, or places necessary to help maintain natural systems. It can take several forms and is embraced under such related planning conditions as the **green belt**, **new town**, or **garden city** (Fig. 13.6). The principle dates back at least to the 13th century B.C. when the Levitical cities of the Near East were surrounded by pasture lands for use by city dwellers. Sir Thomas More also provided green belt ideas, where his imagined cities of Utopia were surrounded by agricultural belts. Open space provides communities an opportunity to manage resources, hydrologic processes, and

FIGURE 13.4 Plethora of highways required to transport people in urban areas because of the automobile revolution and style of living, Los Angeles, Calif.

FIGURE 13.5 Urban areas are often composed of nearly continuous concrete, brick, and blacktop. Parking lots comprise large portions of many cities. The channelized stream has highly erodible banks.

**FIGURE 13.6** Manmade lake in the urban setting of Radisson, a "new town" near Baldwinsville and Syracuse, N.Y. Courtesy of Gerald Olson.

reduce erosion and sedimentation, in addition to maintaining aesthetic and cultural qualities, for the preservation of land and historic sites. Thus, it serves a diversity of functions such as: (1) providing for resource production in forestry, agriculture, natural resources, and water supply; (2) preservation of natural and human resources; (3) provision for water and air quality and visual and recreational amenities; (4) protection of public safety in hazard-prone areas and fire zones; (5) filtering of air-born pollutants by vegetation; and (6) helping to compensate for the heat island effect and micro-climate disturbances caused by the city.

New towns and garden cities are planned communities where environmental factors are considered in the layout of the developments. Planning maintains the harmony of topographic and hydrographic features. It can occur on a large scale such as the complete design of cities like Reston, Va., and Columbia, Md., with 30,000 or more inhabitants, or on more isolated real estate development of several ten's of homes.

## Underground Space

Under the appropriate geological conditions, another option exists for cities burdened with ground space problems — the development of underground space. When surface land is at a premium, it may be possible to use abandoned mines or carve new rock-cut tunnels and rooms below the ground. This is an especially appealing strategem where abandoned mines exist, or where development can take place as new rock is being mined thereby providing for multiple-staged use of such areas. This has occurred at Kansas City, Mo., where limestone was extensively mined, and is still being mined. A total of one-seventh of all warehouse space for the city is now located in such underground space. Sweden has especially developed the technology of underground space where it is used for such a variety of endeavors as sewage facilities, air bases, and storage of many products including fish and petroleum (Figs. 13.7, 13.8). There are many advantages that flow from the use of underground space: (1) thermal dampening of daily and seasonal temperature fluctuations; (2) reduction of energy consumption; (3) protection from such hazards as tornadoes, hurricanes, and fire; (4) conservation of resources because fewer building materials are required; and (5) in some instances, lower building costs.

## NATURAL RESOURCES

One of the most active roles of geologists, in the entire field of environmental management, is the part they play in the exploration and procurement of non-renewable resources. They are the only ones trained to evaluate these materials and make judgments about their quality, quantity, and future potential.

### Mineral and Fossil-Fuel Resources

These geological resources are non-renewable because when mined out, there can be no "second crop." Thus, such materials as iron, copper, petroleum, and even coal need careful management strategy because their mining and financial status is so different from other resources. Here, it is vital that there be cooperation between the government and the corporations to assure a sustained and wise mining policy. This involves not only the withdrawal of the resources from the ground, but the reclamation of the area for those sites where great surface disruption has occurred.

The mining of geologic resources has often proceeded on the basis of mining the best ores first, because this brings maximum and quick profits to the investors. However, for the long run, such a policy is short-sighted because it makes additional mining more expensive, and may ultimately mean the shutdown of the mine because ores are too lean to reap a large profit. Such a practice is referred to as "highgrading." A management

**FIGURE 13.7** Construction for underground space in the bedrock of Sweden. Courtesy of Goran Wettlegren.

policy that can provide for the more uniform supply of the materials would be beneficial for the long run and help minimize ore depletion.

On the other hand, society may place impediments to mining by instituting repressive ordinances or policies that prohibit mining of certain areas. Such actions can be classed as **cultural nullification**. They lead to mining at less advantageous sites and cause higher commodity costs. Environmental

FIGURE 13.8 Oil storage caverns, Stockholm, Sweden. The roof of the cavern is 20 m below the water table. Height of cavern is 30 m and width is 20 m. Courtesy of Sentab, Sweden.

management should seek to abate noise, dust, and other nuisances created by mining, but to close down operations entirely would be counter-productive.

Whenever possible, multiple use of the land, or a sequentially phased program that permits different land uses should be another goal of management. Thus, an intelligent choice would be to permit mining of materials, and when depleted, to place the land into some other type of usage. Such was the case in Denver when first, sand and gravel was mined out, then the site was used as a landfill, and then when completed, the Denver Colosseum was built on the site. In California, the loss of mineral resources may be $17 billion during the 1970–2000 period unless communities plan properly for mining within their jurisdictions.

The management of strip-mined areas has been a hotly debated issue for decades. Although several states passed laws on the matter, federal legislation did not occur until 1977 with passage of the Surface Mining Control and Reclamation Act. The increased mining of coal finally prompted such an act. Along with the national policy, many mines have had programs for minimizing environmental degradation. This consists of such measures as: (1) cover being placed on spoil piles to reduce erosion, sediment production, and acid drainage, (2) the mining of as small an area as possible to minimize weathering on rock exposures, (3) development of catchment basins to trap sediment, and (4) revegetation and regrading of areas as soon as mined out (see also Chapter 3).

### Water Resources

Water, because it is nearly universal and is mobile, is a resource that has received more management exposure than all others. Since it is basic to life, all governments have needed to initiate management plans for its procurement. In the United States, the two largest water-management programs are those of New York City and the California Water Plan. Extensive litigation was required in the construction of dams and aqueducts and the apportionment of waters to the states of New York, Pennsylvania, and Delaware. For example, water moves more than 160 km to the City area from reservoirs in the Catskill Mountains. Even more dramatic is the California water diversion from the northern to the southern part of the State. It moves more than 1000 km through a system of 21 dams and reservoirs, 22 pumping stations, and extensive tunnels, pipelines, and canals. The Snowy Mountain Hydroelectric Project in Australia also required massive environmental management. The flow of rivers draining eastward was diverted by dams and tunnels, so water could move to the drylands of the interior part of the continent.

Groundwater has also been the subject of environmental management (see Chapter 14). Long Island, N.Y., has long been a leader in trying to conserve water resources by allowing for infiltration of surface water into the groundwater aquifers by the means of more than 2500 recharge pits. In the West, there are various schemes, such as the water-spreading operations in the San Fernando Valley, Calif., that aim to conserve water. Here, surface water is diverted to giant spreading grounds and allowed to percolate below the surface into the groundwater zone. Because of the scarcity of groundwater in many localities, more and more litigation and laws have resulted. To allow for its fair apportionment, such landmark cases as the Raymond Basin Adjudication in 1947 settled the lawsuit of Pasadena v. Alhambra. In such instances, it has been the courts who have determined the type and degree of environmental management for water resources.

### Land Resources

Along with water, soil is the most important resource for mankind. To assure its continuing productivity is a goal that all environmental managers must face. The U.S. Department of Agriculture's Soil Conservation Service became the government's way of helping farmers manage their soil resources. Soil is the basic building block for the production of crops, timber, and grazing animals, all of which utilize the nutrients that flow from the soil and the photosynthesis process. Management of forestlands has seen a nearly constant confrontation over the character of their usage. On one side, Gifford Pinchot and his followers of the school of conservation-utilitarian ethics have been arrayed against John Muir and his followers of the preservation ethic. The passage of the Multiple Use-Sustained Yield Act of 1960 arranged a compromise with its five provisions. Thus, forests were to be managed for the functions of (1) timber resources, (2) recreation, (3) wildlife, (4) forage for grazing, and (5) protection of rivers. The geology side of forest management emerges when improper techniques are used that cause accelerated soil erosion and increased sedimentation rates, and that produce imbalances in hydrologic systems.

Land-use and landscape policies comprise another aspect of land resources. Land and land use comprise elements of the landscape. Thus, land is now recognized as constituting both a commodity and a resource. Land-use policies have become more of a perogative of government due to the extension of its powers. These are exercised in such forms as: (1) eminent domain or condemnation of private land, (2) taxation power, (3) police power as in zoning and restriction codes, and (4) funding power of the government to supply subsidies, service, and protection. To establish a proper management base, land-classification systems and land-capability maps have been developed. These are necessary to catalog the essential ingredients as well as to show their spatial arrangements. Studies with such a goal seek to identify the type of land use for which the land is best suited. The various parcels may be given numerical designations that reflect their ratings for possible different uses.

Land use federal legislation dates back to the establishment of Yellowstone National Park in 1872. This established the principle that some lands could be set aside for purely recreational use. From this policy has emerged the entire range of land-use legislation that sets aside lands as national parks, national monuments, national seashores, wilderness and primitive areas, wild rivers, and others. Thus, the government has acted as a massive entrepreneur in deciding that lands must be held in trust for the people for certain specific and prescribed uses. These actions have at times caused societal polarization. Business and commercial groups have argued for increased development of such areas, whereas private interests on the environmental side have advocated increased extraction of lands from any type of enterprise. These include the Sierra Club, Environmental Defense Fund, Wilderness Society, Nature Conservancy, Izaac Walton League, Audubon Society, National Wildlife Federation, and others.

Starting in the 1960s, various states began initiatives to become environmental managers of the land. Hawaii passed the Land Use Law in 1961, which gave State agencies a degree of control over the use of land resources. From 1969 to 1971, the State of Maine passed a series of legislative acts, including the Site Location Law, whereby approval had to be obtained from the State before certain types of new development could take place. The purpose of such acts was to ensure that new developments would be located in a manner that would have minimal adverse impacts on the environment. Many other states have passed legislation that covers a variety of special environments. Delaware and California have significant coastal acts. Perhaps wetlands have received the greatest amount of protection within the past 10 years. The fragility of these environments has become recognized, and their importance in the food chain is of great concern. Most coastal states have now passed legislation for their protection, and many states, in addition, have laws for the safe keeping of freshwater wetlands.

### Federal Lands

Land-management policies of the United States started with acquisition of huge territories in the West from the Louisiana Purchase of 1803, and essentially ending the procurement stage was the buying of Alaska from Russia in 1867. The major part of these vast regions became part of the public domain encompassing about 2 billion acres. Eventually, federal ownership dwindled to 740 million acres when the other lands were sold or given away to new states as they joined the country, to railroads as they developed new routes, and to homesteaders settling the frontier. Responsibility for governing these lands and the resources they contain is divided among a host of different federal agencies. The National Park Service supervises 10 percent of public lands through their park, wilderness, and wild and scenic rivers programs. The Fish and Wildlife Service has 410 refuges on 12 percent of federal lands to protect habitats of waterfowl, endangered species, big game, and other animals. The largest overseer is the Bureau of Land Management with jurisdiction over 43 percent of the lands. Such lands are used to maintain livestock grazing, forests, and mineral extraction. The

Department of Agriculture largely through the U.S. Forest Service controls 25 percent of the land. Woodlands are used for lumbering, public recreation, watershed management, grazing, and even oil and mineral leasing. The Bureau of Indian Affairs is trustee for 7 percent of the land, acting in a capacity as steward for Indian-owned properties and fostering social and education services for 700,000 residents. Additional land is administered by a variety of other agencies. The Department of Defense has military bases and facilities on 3 percent of the land. The U.S. Army Corps of Engineers manages 12 million acres for flood control and navigation purposes.

A severe challenge to the type of management policy that should be followed on federal lands began in 1980 in a movement that has been called "the sagebrush rebellion." With the blessing of President Reagan, the Secretary of the Interior, James G. Watt, reversed public land management from the preservation ethic of predecessors to one of resource development. This was in direct response to the business community that felt in today's world it was inconsistent for the government to hold so much land with rather minimal use. For example, federal ownership in western states amounts to 86 percent in Nevada, 81 percent in Alaska, 64 percent in Utah and Idaho, and more than 40 percent in Arizona, California, Oregon, and Wyoming. It is still too early to tell what the final outcome will be for this new conflict. Although Watt resigned after three years, his legacy still endures in many parts of government and in the states.

## Adirondack State Park

The idea for establishment of parks and commons-type areas is centuries old. Central Park in New York City was the first major park in America in an urban setting, established in 1857 under the driving force of Frederick Law Olmsted. Although Yellowstone was the first federal park, the complete wilderness concept was not defined until 1885 when New York State founded a "Forest Preserve" for the purpose of maintaining it as "wild forest lands." This idea became known as the "forever wild" park, so that in 1892, the area was expanded to include more than 3 million acres (1.2 million ha) and renamed the "Adirondack State Park." This precursor subsequently led to the development of the Wilderness Act of 1964, which by 1977, included 163 wilderness areas throughout the country. The Wild and Scenic Rivers Act of 1968 further expanded this policy for the preservation of selected rivers that have outstanding scenic, recreational, geologic, and other important qualities.

Administration of the Adirondack State Park is vested with the Adirondack Park Agency. The Park now contains 2.4 million acres of Forest Preserve lands that are State owned, and 3.5 million acres that are privately owned containing about 100,000 residents. The Agency is charged with the purpose of preserving open space and the ecological integrity and beauty of the region. More than half the private land is classed as "resource management," which limits any development to 15 buildings per square mile. A total of 976,000 acres of the State land is classed as "wilderness" in which there can be no permanent dwellings and which must remain free from any type of development or emplacement of utilities that would impair its natural condition.

## GEOLOGIC HAZARDS

The public generally views the government as responsible for protecting them from natural disasters. It looks for guidance, protection, salvation, and beneficence whenever trouble strikes (Fig. 13.9). The earth forces of volcanic activity, earthquakes, landslides, and floods are seen as acts of a whimsical god and thereby as an enemy of the people. Such alien forces can only be combatted by governmental power, so it becomes the duty and obligation of environmental managers to provide safety for citizens. Therefore, to obtain public trust, the managers must provide effective policies and plans that mitigate the risks and costs to citizens of threats to their property or existence. Before anything can be done, there must be clear recognition that an actual hazard has high potential of happening. An evaluation must be made of the severity and frequency of occurrence (Fig. 13.10). Then, the risk must be calculated in terms of public benefits and costs. The planning options that are available whenever the possibility for disaster exists include avoiding or restricting habitation at the site, developing a program for financial remuneration if disaster strikes, undertaking some type of prevention or control of the hazard-producing force, and providing an effective monitoring and warning program that will permit safe and orderly evacuation. Of course, in some instances, the only alternative is to sustain losses and take the chance that nothing will happen (Fig. 13.11). A problem with predictions and forced evacuation arises if the disaster fails to strike when foretold. If scientists cry wolf too often, governments and the public will eventually fail to heed such warnings. This places a great burden on scientists wherein a loss of credibility, or a miscalculated precursor, could eventually lead to ruin if they are not sufficiently cautious in future danger signals.

Figure 13.10 presents an overview of factors involved with hazards, the risk assessments that are made, and the character of societal responses, which lead to management guidelines and programs. However, it must be added

**FIGURE 13.9** A view of North Ogden, Utah, with subdivisions nestled against the Wasatch Mountains, and the fault that marks the foot of the mountains and the flatlands of the basin. The fault throughout this region has been activated by earthquakes during historic times. Courtesy of Earl Olson.

that different societal groups would have different responses to the same event. Thus, a new range of factors also becomes involved in the final decision or management procedure. These would include education and historical background, experience with similar prior events, age, and confidence in government and authority. As indicated in the chapters dealing with hazards (and in Chapter 14), laws and ordinances are in effect in some states and on the federal level that treat the hazard menace and provide the governmental attitude and management criteria for preventing and mitigating geologic hazards.

## ENVIRONMENTAL COSTS

One of the least researched and most controversial issues facing environmental managers are budgetary considerations involved with policies that

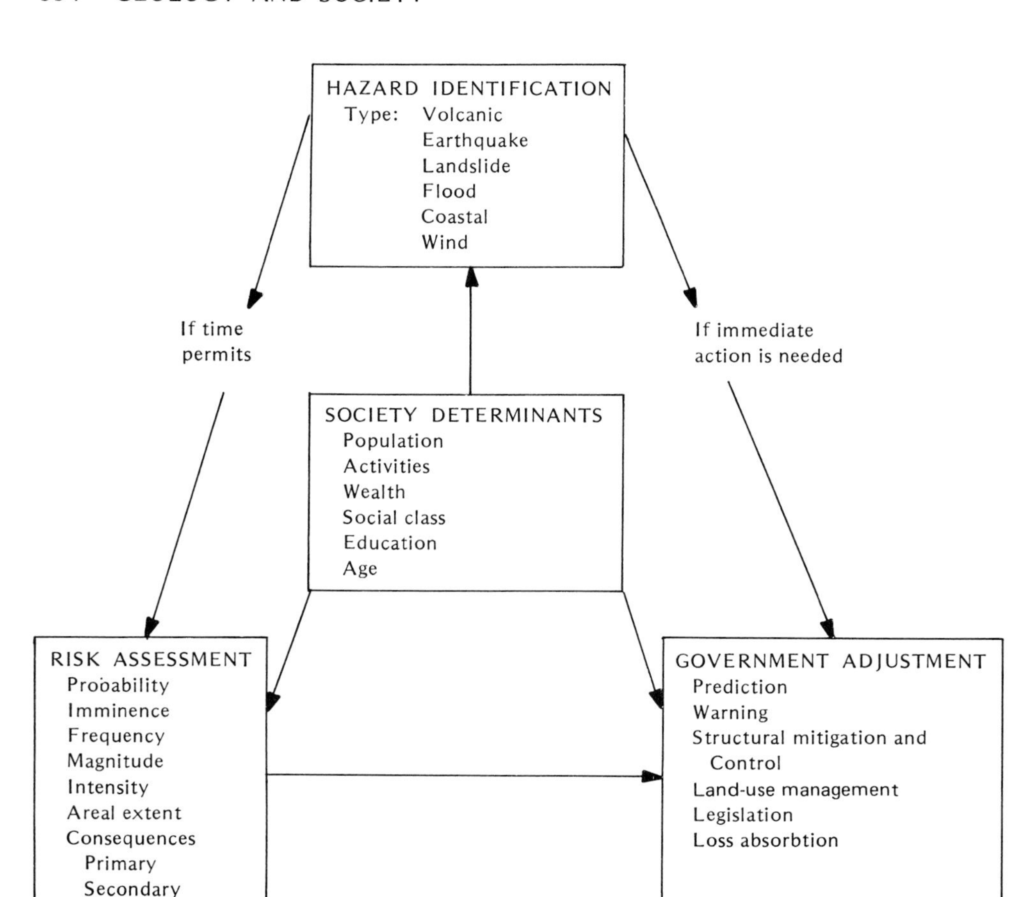


**FIGURE 13.10** Flow chart showing the decision-making process for hazard policies. Society determinants enter all phases of the process. The degree of involvement and attitude are a function of many factors. When there has been hazard identification, the reaction by those who are likely to be affected is governed by the type of people and their background, and the time available to make whatever adjustment is possible. If there is sufficient time, a careful risk assessment program can be launched, and thereafter, governmental adjustment will be undertaken. If the hazard is about to occur, time may not allow for assessment other than to take immediate government adjustment to the problem.

**FIGURE 13.11** San Francisco, Calif. This area was devastated by the disastrous 1906 earthquake. Today, residents apparently take a "fatalistic" attitude that they are not in severe jeopardy from a recurrence of such an event. Courtesy of U.S. Department of Housing and Urban Development.

affect the environment and society. As mentioned in the previous section, hazards are a vital part of environmental planning. However, the cost of hazards that turn into disasters is generally not considered as a cost item in budgetary matters. This is wrong according to such specialists as Mark and Stuart-Alexander (1977) who urge that costs of potential disasters should be a necessary component in any benefit/cost analysis for determining whether a particular engineering structure, such as a dam, should be built. Ruckelshaus (1983), a former director of the Environmental Protection Agency, also urged that the scientific process of assessing risk must be coupled with economic aspects of projects and statutory decisions.

A particularly vexing problem of environmental management deals with legal, practical, and economic ways to internalize damages to others that are perpetrated by a known pollutant source. Thus, how can those who have been harmed by acid rain obtain funds to cover the cost of damages they have suffered?

In the early days of the environmental movement, there was great inertia by many who made the argument that "environmental projects and statutes" would be the ruination of business and industry because it would cost so much in terms of lost revenue and jobs. For example, in 1971, the U.S. Department of Commerce warned that environmental regulations would cause severe economic dislocation and unemployment and that new major relief programs would have to be created to take their place. However, by 1976, EPA had spent $3.34 billion in wastewater facilities alone and reported there were 46,000 workers on construction projects and another 46,000 on off-site manufacturing, transportation, mining, and services. A report by the National Research Council also said that by 1974, 84,000 were employed in private industry as a result of mandated air and water pollution control systems. The Arthur D. Little Company estimated that by 1983, this number had increased 23 percent. When all environmental regulations are considered, a recent determination by the Council on Environmental Quality showed that a total of 2.2 million new jobs had been created. The study also showed that since 1970, the annual benefits of such statutes and programs have been $21.4 billion. The 1985 projected figure for improved water quality alone was calculated to be $12.4 billion.

### The 1980–83 Environmental Hiatus

Policies that deal with environmental management are subject to the perceptions of those who are empowered to oversee the decision-making process. With the American election of President Reagan in 1980, a new administrative posture was installed in the purview of environmental affairs. Many projects and programs were "put on hold," abandoned, or cut back. The reasons given for these actions included admonitions that the regulations were bad for business, that the statutes had gone too far, and that it was necessary to obtain new revenues by the sale of many government lands. Directors were placed in charge of many environment-oriented agencies that would be sympathetic for such policy reversals. Budget cutting was one avenue used to stop environmental objectives. For example, the EPA budget was consistently cut from $1.8 in 1981 to $1.2 billion in 1982, and to $916 million in 1983. The Council on Environmental Quality was cut during this period from an annual budget of $3 million to $700,000. When inflation is added to these budgets, the loss for environmental programs becomes even more drastic. This meant that funding for toxic waste programs was cut one-third, clean water programs were cut one-half, and safe drinking water projects were sliced 15 percent.

## ENVIRONMENTAL POLICIES IN OTHER COUNTRIES

In 1972, the United Nations took stock of its own interests and organized a special conference in Stockholm, Sweden, which resulted in the declaration, "Man has the fundamental right to freedom, equality and adequate conditions of life, in an environment of a quality that permits a life of dignity and well-being, and he bears a solemn responsibility to protect and improve the environment for present and future generations." The conference established the United Nations Environmental Programme (UNEP) to help improve and protect the world's human and natural environment. Six broad areas were identified as needing special attention and projects: (1) human settlements and human health, (2) terrestrial ecosystems, (3) environment and development, (4) oceans, (5) energy, and (6) natural disasters. Other positive steps and outgrowths from policies developed at the conference included:

1. Establishment of Global Environmental Monitoring System (GEMS) to coordinate monitoring of all participating countries.
2. International Register of Potentially Toxic Chemicals (IRPTC) was formed in 1975 and now has information on more than 40,000 chemicals, with precise data on those 400 deemed most harmful.
3. Inauguration of the Regional Seas Programme with early concentration on the Mediterranean region. A network of 84 laboratories in 16 countries was financed, and principal attention has been paid to problems of deforestation and to desertification. For example, it fostered the first international conference on desertification in 1977. The Law of the Sea conferences are discussed in Chapter 14.

Although China has had a long history of making many accommodations for Nature, it was not until 1978 that the first journal on environmental science was published there. During the 1970s, there was growing support to coordinate environmental policies and research efforts in order to devise regulatory approaches commensurate with a developing economy. Such efforts were prompted by awareness that air and water pollution had become serious and that environmental mismanagement had led to the degradation of soil and water resources, thereby endangering agricultural stability and production. The Environmental Protection Office was formed in 1974, but it only had a small staff and little influence over environmental research and development. However, by 1979, environmental policy had become more entrenched in the government, and a Forestry Act was adopted along with the Aquatic Products Act, which now provides protection for waterways, fisheries, and aquaculture. The major focus of Chinese policy has been toward harnessing and controlling water power and improvement of water quality. Problems of air pollution have not yet been seriously addressed.

Environmental policies in such countries as Hungary and Poland are a mixture of attempts at some protection and a do-nothing approach. Hungary passed the Environmental Protection Act in 1976, which established a National Council for Environment and Nature Protection. Forest conservation programs date from 1872, but the present stimulus stemmed from the professional community instead of the general public or government. Budapest suffers from pollution by cement dust, chemical and metallurgical discharges, and particulate matter from the burning of soft coal. The Danube and Tiszka rivers are badly polluted from industrial waste and human sewage. A Five Year Plan (1976–81) was instituted, which created a Central Environmental Protection Fund to be used to develop low-waste technologies, and a Water Quality Fund was established with authority to fine the most flagrant polluters.

Poland's principal goal has been economic and the development of industrialization. Industrial pollution is severe in southern Silesia and has not been corrected. Also the numerous small and privately owned farms make the use of advanced large-farm technology impractical. Environmental management is aimed at adapting the environment to industry rather than providing safeguards for the environment. The major emphases have been on developing non-waste and low-waste technologies, plus the better utilization and recycling of waste.

## POSTLUDE

Managing environmental affairs is certainly one of today's most difficult assignments. Because people are so diverse and their goals in life so variable, environmental managers know that all of the people cannot be pleased all the time. In a manner of speaking, conflicting interests almost invariably become "the name of the game." Thus, it is heartening to know that new approaches are now being taken in attempts to reduce lawsuits and courtroom battles. The technique is known as "mediation," a style of interaction for conflict resolution that has been used for centuries in international relations, labor disputes, and a variety of conflict-intervention processes. Environmental mediation has now evolved from a novel concept to one that is more accepted but still often misunderstood. The first explicit effort to mediate an environmental dispute in the United States began in the fall of 1973 over a flood control–land use planning conflict in the state of Washington. It culminated in December 1974, with a unanimous written agreement among the 12 parties that were involved. Under mediation, the mediator does not have the authority to impose a settlement, but instead serves to help the parties reconcile their differences. Another notable success for this method is known as the Homestake Pitch Mine Dispute, Colo. Here, the Homestake Corporation and seven environmental coalitions reached an agreement that settled their differences over the operation and reclamation of a uranium mine in Gunnison National Forest. Both state and federal agencies concurred in the agreement.

Governments, and indeed other societal groups, need all the wisdom they can muster to adopt environmental policies that deal effectively with the problems of today and with a prudent legacy for the future. The conflicts that emerged with attempts to dismantle federal land ownership in the United States emphasized the growing problems of management. As mineral, oil, and water resources become even more precious, their significance on federal lands will increase. Even as urban and suburban congestion continue, the need for many to find tranquility in peaceful settings will endure (Fig. 13.11). Thus, the environmental management of the total land-water ecosystem will remain a top priority of the human spirit as well as tangible assets of the body.

CHAPTER 14

# ENVIRONMENTAL LAW

Environmental law as a subdiscipline of the legal profession became popular starting in the late 1960s. Although prior to that time, geologists and lawyers were involved with legal matters that were environmentally oriented, there was no formalization with such rubrics in university courses or in law societies. Of course, governing bodies have passed many statutes and laws dealing with the environment throughout recorded history, and the courts have dealt with environmental decisions for millennia.

Law consists of societal rules established by governing authorities. Regulations that are imposed or mandated have the force of the general public behind them and noncompliance can result in some type of penalty. Laws may be written or unwritten, such as English Common Law. They consist of a wide range of legislative actions, ordinances, agency regulations, and codified directives of recognized public officials. Environmental law is now being interpreted to cover an extremely broad spectrum of human endeavors, but this chapter will emphasize the geological aspects of the subject.

Geologists can be involved in a wide range of legislation and litigation through a great variety of environmental laws. These laws run the gamut from the general to the specific, and their impact on nature and society has been steadily increasing in the past 30 years, with exceptional growth in the last 15 years. A 1982 study of lawyers in the United States by William Futrell, president of the Environmental Law Institute, showed that 8000 lawyers devoted at least half time to environmental law. About 3000 environmental lawyers are employed by corporations, 2000 by law firms, 1500 by federal agencies, 750 by state and local governments and 750 by public interest firms and universities.

## GENERAL LAWS

These are laws that are applicable to many areas of societal needs and justice systems, but which have been adopted for application into the environmental arena. In the United States, the basis for these legalities is founded in the Constitution, especially the Bill of Rights and amendments, and in the English Common Law. The Fifth and Fourteenth amendments provide that no person can be deprived of life, liberty, or property without due process of law and that just compensation must be made when property is taken for public use. Such statutes provide the basis for the doctrine of eminent domain – governments may acquire private land when in the interests of the public good. The Ninth Amendment is being increasingly used by environmental groups when specific issues arise that do not seem to have been specified in the Constitution. English Common Law has also provided a powerful heritage, and its many aspects have been repeatedly stated in courtroom trials. The combination of the Constitution and English Common Law produced the prevailing theories for environmental litigation until the 1960s, which witnessed passage of the new type of environmental laws. Thus, the usual environmental grievance being debated in the courts involved a plaintiff bringing suit under the claim of damage caused by acts of nuisance, negligence, trespass, or public trust.

## Nuisance

The common law of nuisance concerns the right to be free from unreasonable interference that disrupts the enjoyment of land. It has provided the most important legal basis for dealing with pollution problems prior to passage of federal statutes in the early 1970s. Nuisance is viewed as the common law foundation for modern environmental law. Before 1972, this doctrine consisted of little more than a theoretical means to remedy serious instances of interstate pollution. In that year, a landmark U.S. Supreme Court ruling in City of Milwaukee v. State of Illinois allowed there should be proper redress for problems associated with interstate pollution. The case involved water pollution by Milwaukee from its combined storm runoff and sanitary sewage system. The Court cited as precedent the Georgia v. Tennessee Copper case of 1907, which stated, "It is a fair and reasonable demand on the part of a sovereign that the air over its territory should not be polluted on a great scale by sulphurous acid gas, that the forests on its mountains should not be further destroyed or threatened by the act of persons beyond its control, that the crops and orchards on its hills should not be endangered by the same source."

Although in the 1972 decision the Supreme Court held that damages caused by interstate water pollution could support a federal common law nuisance lawsuit by the victim, a new ruling in 1981 reversed the case. Instead, in an opinion by the Supreme Court, it held that passage of the Clean Water Act of 1972 so occupied the field of water pollution that Congress had preempted the federal common law of nuisance. As a result,

Illinois was not permitted to seek additional federal court action against Milwaukee. Thus, the primary effect of the decision shifted to Congress the burden of resolving such environmental disputes. Since then, the Supreme Court has repeatedly said it will not become an activist in favor of either the environment or industry when the problem deals with federal statutes.

### Public Trust

Much of the new land-use legislation stems from the governmental ethic of public trust – certain lands and waters are retained in the public domain for the use and enjoyment by everyone. In the United States, establishment of Yellowstone National Park in 1872 paved the way for such precedent. The earliest major controversy that erupted in this connection was the Hetch Hetchy Case (Nash, 1967). Preservation advocates such as John Muir argued that to use this valley, which had previously been declared a wilderness preserve, to build a dam and store water for San Francisco, was a breach of public faith. After several years, the federal government ruled it could be used for a water reservoir.

The most recent landmark case has just recently been decided in National Audubon Society v. Superior Court of Alpine County California (1983). Los Angeles had been granted permits in 1940 to appropriate water from streams feeding Mono Lake on the eastern slope of the Sierra Nevada. However, in recent years, the lake level has been steadily dropping with a reduction in surface water area. Birds, which previously nested on an island that is now a peninsula, are subject to predation by marauding animals. In addition, scenic values and other ecological merits have been degraded. In their ruling, the California Supreme Court cited the public trust doctrine and extended the idea that the State had a continuing duty to protect inland navigable lakes. They quoted Roman Law as an important historical base, "By the law of nature these things are common to mankind – the air, running water, the sea, and consequently the shores of the sea."

## SPECIFIC ENVIRONMENTAL LAWS

Although specific environmental legislation largely dates from the 1960s, two types of environmental laws were enacted through many of the preceding decades. The Rivers and Harbors Act of 1899 was revolutionary in its purpose and represented a model to be followed more than 60 years later with clean air, clean water, and clean land acts. The Act prohibits the dumping or discharging of industrial "refuse" into navigable waters without the permission of the U.S. Army Corps of Engineers. Polluters can be fined up to $2500 per day as well as jailed for a period up to one year. Not until the 1960s was this Act enforced, however, at which time many individuals sought to take advantage of the clause whereby a person who provided information that leads to a conviction is entitled to one-half the fine as a bounty for discovery. The authority of the Corps to grant such permits was challenged in Zabel and Russell v. U.S. Army Corps of Engineers (1970). The plaintiffs brought suit in 1967 after the Corps had refused them a permit to dredge and fill wetlands in Tampa Bay, Fla., for construction of a trailer park. The decision of the appellate court upheld the authority of the Corps, which had based its denial on adverse ecological effects that would be caused by the construction. Also, on February 22, 1971, the U.S. Supreme Court sustained the decision.

During the past quarter century a great variety of specific environmental laws have been passed, both in the United States and throughout the world. Mining laws and land-use laws have become much more prevalent, and also coastal laws (see Chapter 9).

Specific acts that relate to the environment can be placed into two broad categories: (1) those that mandate user standards and (2) those that set management policy. The Clean Air Act of 1967 is typical of legislation that provides numerical and quantitative stipulations as to the limits that air can be polluted. By their very nature, such acts are highly prone to disagreements in interpretation and thereby result in many court cases. In United States v. Bishop Processing Company (1968), the government brought action against a rendering plant in Maryland because its odors crossed the state line into Delaware. The injunction was upheld by the U.S. Supreme Court to prohibit continuation of such pollutants. The National Environmental Policy Act of 1970 provides another approach for setting standards by environment users. This act can be considered a Bill of Rights for the environment because it stipulates a set of guidelines that must be followed by governmental agencies whenever construction projects are planned that might affect the environment. The environmental impact statement clause in the act paved the way for an entirely different type of appraisal system that must be initiated by the user. It requires that alternative plans must be considered, and that the type and scope of impact on the environment must be analyzed and predicted before construction can commence.

Numerous lawsuits have resulted from various interpretations given EIS (Environmental Impact Statements) (Coates, 1976). A more recent case occurred in a residential section of Ashville, NC, where a group of residents were affected by interstate highway construction that had received approval under NEPA (National Environmental Policy Act). In Mountainbrook Homeowners v. Adams (Secretary of Transportation), the 1979 lawsuit claimed property values had been diminished by a type of construction that was not consistent with the EIS. The case was dismissed by a federal district court by ruling a federal court has no jurisdiction to order a remedy on behalf of citizens. Instead, real estate use is the province of state and local law and cannot be resolved by federal action.

Many states have also passed their versions of NEPA, and in New York State, such legislation was passed in 1976 known as SEQR (State Environmental Quality Act). Its purpose is to encourage, "productive and enjoyable harmony between man and his environment; promote efforts which will prevent or eliminate damage to the environment and enhance human and community resources; and enrich the understanding of the ecological systems, natural, human and community resources important to the people of the state." The Act provides for a broad-based review of all actions funded or approved by local, regional, and state agencies. Projects are reviewed in their entirety rather than in isolated parts, and the primary tool for evaluation is the mandated EIS. Since the law was implemented in 1978, there has been an average of 250 projects each year that fall within the guidelines. A study of 996 positive declarations showed that 670 were from local agencies, and 326 were filed by the State. Of the 848 negative declarations, 556 were local and 292, state.

There is an unusually broad spectrum of laws that provides perspectives about the environment and controls management policies by which it is to be governed. Each sector of the environment is covered by a variety of laws and jurisprudence as formulated in court actions in the United States. Many of these have similar stature in other countries of the world. For example, in Brazil, the Forest Law of 1959 prohibits construction work above specified slope levels in certain parts of the country to reduce landsliding. In an early recognition of the deleterious aspects of urban sprawl, England passed the New Towns Act of 1946, which gave the government the authority to designate which land areas were most appropriate for the development of new urban sites. Japan had laws about parks as early as 1873, a forest preserve law in 1907, and a National Park Law in 1931.

## LAWS FOR GEOLOGISTS

Although geologists may be involved in many types of environmental law, some of this law is aimed entirely at the geologist and requires his participation. In the Raymond Basin Adjudication of 1949 (Pasadena v. Alhambra), the California Supreme Court established that hydrologists were the ones to determine the allocation of water in the region. The grading ordinances in southern California, such as the City of Los Angeles Grading Ordinances of 1952, 1963, and 1969, require that engineering geologists must make the decisions for developments taking place on hillslopes of the city. The Pennsylvania Solid Waste Management Act of 1969 requires geological inspection, analysis, and approval of sites that are to be used as disposal sites for solid refuse. In addition, governmental agencies may extend their powers as administrative overseers of environmental affairs into codified regulations for adherence by users. For example, in 1972, the New York State Department of Transportation issued guidelines entitled "Requirements for Geologic Source Resports." Before a mining company can qualify to use products for use on State highways, it must file a special report, which can only be undertaken by a certified geologist.

## MINING LAW

Laws regarding mining in the western world had their first formulation in the Napoleanic Code of 1804, which held that each landowner possesses everything on and below his property. Nineteenth century monarchies, however, excluded gold and silver, which was owned by the crown as a regalian right. In early mining law cases, the word "mineral" was interpreted by the courts to mean metalliferous deposits, but by 1854, the Hartwell v. Camman case in New Jersey established that minerals need not contain metal. Additional cases extended minerals to include granite (Armstrong v. Granite Company, 147 N.Y., 495) and limestone (Brady v. Smith, 181 N.Y., 178).

The California Gold Rush of 1849 helped establish new local mining laws that later formed the basis for common law, which became incorporated into federal statutes in 1866 and then into the Mining Act of 1872. Such rules held that (1) discovery and working gave the right to a mineral, and (2) the discoverer had the right to continue mining a vein even when it extended under the surface claim of another. Such law was appropriate only for deposits in place, and guidelines were set and gradually enlarged that circumscribed the type of assessment work that was necessary each year to sustain a claim.

The Federal Mineral Leasing Acts of 1920, and 1971 removed coal, petroleum, oil shale, potash, phosphate, and sodium salts from mining controls. The multiple Mineral Development Law of 1954 attempted to resolve problems that had arisen between mining laws and leasing laws, and set the stage for the Multiple Surface Use Act of 1955, which stipulated that underground mining was subordinate to surface extraction of most marketable mineral products.

Because of their mobility, oil and natural gas laws have certain differences from the fixed-mineral laws. Although the courts have generally held them to be minerals, they cannot be claimed as such until extracted by a landowner and put in his possession, as in Funk v. Holdeman (53 Pa. St., 229, 1866).

The Mining Act of 1872 is an anachronism, but is still the foundation for the national policy toward development of hard-rock minerals on public lands. On governmental lands, oil companies pay royalties for resources they extract, cattlemen and sheepmen pay a grazing fee, and people who use parks are charged admission, but hard-rock minerals on about 500 million acres are open to mining at no charge. The Lode Law of 1866 was the precursor, and it determined that mining was the best, and often the only, use for most western lands. The sand plains, alkaline deserts, and dreary mountains of rock were said to be worthless, and it was deemed that mineral supplies were inexhaustible.

In 1974, the U.S. Forest Service issued regulations aimed at reducing harm to the environment caused by mining and cited their authority as stemming from the Organic Administration Act of 1897. The 1976 Federal Lands Policy and Management Act further declares that it is the policy of the United States to retain and manage public lands to protect their scenic, scientific, and ecological values. On November 24, 1980, the Bureau of Land Management issued its own regulations whereby for large mining operations, they may designate a bond to cover the costs of reclamation. Mining companies now argue that the spate of laws and regulations is too restrictive and prohibits some of their development. Indeed, have the laws gone too far (Fig. 14.1)? The reclamation side of mining is discussed in Chapter 2.

The largest land acquisition of recent times in the United States was finalized by law on December 2, 1980, when the Alaska National Interest Lands Conservation Act was passed. This single statute doubles the area of the National Parks and Wildlife Refuge System and expands the Wilderness Preservation System to three times its former size. The area so declared is the size of California, which is now safeguarded, with two-thirds of it receiving full protection as a wilderness area. On overall balance, the Act does provide a compromise with developmental interests. For example, 95 percent of the proved oil, and 66 percent of the minerals are unaffected by the bill. The legislation also transfers 105 million acres to Alaska, along with 44 million acres to native inhabitants.

## WATER LAW

Modern water law comprises a mixed blend of statutes, practices, and court decisions from many countries. Roman law, Moslem law, Spanish law,

## Major Federal Actions Limiting Mineral Development on Public Lands

These laws restrict mineral exploration and development on certain public lands. Together, they comprise a procession of direct and indirect "public land withdrawals." The cumulative effect of these federal actions, corresponding regulations, and related management decisions may in application delay or prevent the development of valuable minerals, thereby increasing U.S. dependence on other nations for critically needed metals and minerals.

1. **Forest Service Organic Act of 1897, 16 U.S.C. §§475-482.**
2. **Forest & Rangeland Renewable Resources Planning Act of 1974, 16 U.S.C. §§1600-1614.**
3. **National Forest Management Act of 1976, 16 U.S. §§1600-1676.**
4. **The Federal Land Policy and Management Act, 43 U.S.C. §§1701, *et. seq.***
5. **Reclamation Act of 1902, 32 Stat. 388; 43 U.S.C. §416.**
6. **Wilderness Act, 16 U.S.C. §§1131-1136.**
7. **The Wild & Scenic Rivers Act, 16 U.S.C. §§1271, *et. seq.***
8. **National Trails System Act, 16 U.S.C. §§1241, *et. seq.***
9. **Coastal Zone Management Act, 16 U.S.C. §§1451, *et. seq.***
10. **Protection of Wetlands, Executive Order 119900.**
11. **National Wildlife Refuge System Administration Act of 1966, 16 U.S.C. §§668 (dd) and (ee).**
12. **Endangered Species Act of 1973, 16 U.S.C. §1531, *et. seq.***
13. **Fish and Wildlife Coordination Act, 16 U.S.C. §§661-666c.**
14. **The Antiquities Act of 1906, 16 U.S.C. §§431, *et. seq.***
15. **Archaeological and Historic Preservation Act, 16 U.S.C. §§469 (a-1), *et. seq.***
16. **National Historic Preservation Act of 1966, 16 U.S.C. §470.**
17. **Alaska Native Claim Settlement Act of 1971, 43 U.S.C. §1601.**

## Major Federal Environmental Laws

These laws limit and control mineral exploration and mining, irrespective of whether or not it occurs on public lands. They comprise a series of largely uncoordinated special purpose laws to protect and improve the air, water, land and aesthetic values of the overall environment. As with the land withdrawal laws, the environmental laws are implemented and administered through the use of thousands and thousands of pages of detailed regulations. Differences in interpretation between government regulators and industry operators lead to frequent and drawn-out court proceedings, further increasing the cost of doing business.

1. **National Environmental Policy Act of 1969 (NEPA), 42 U.S.C. §4321, *et. seq.***
2. **Clean Air Act, 42 U.S.C. §1857, *et. seq.***
3. **Clean Air Act Amendments of 1977, 42 U.S.C. §4701, *et. seq.***
4. **Federal Water Pollution Control Act (FWPCA), 33 U.S.C. §466, *et. seq.***
5. **Clean Water Act Amendments of 1977, 33 U.S.C. §1751, *et. seq.***
6. **Safe Drinking Water Act, 42 U.S.C. §§300 (F) to 300 (j-9).**
7. **Federal Resource Conservation & Recovery Act (RCRA), 42 U.S.C. §6901, *et. seq.***
8. **Noise Control Act of 1972, 42 U.S.C. §§4901-4918.**
9. **Toxic Substances Control Act, 15 U.S.C. §§2601-2629.**
10. **Uranium Mill Tailings Radiation Control Act of 1978, 42 U.S.C. §§2021, *et. seq.***

FIGURE 14.1 Graphic summation of Federal laws that affect mining operations. Courtesy American Institute of Professional Geologists.

the Code of Napoleon, and English Common Law have all provided elements in the matrix of surface and subsurface legal rights.

### Surface-Water Law

In the United States, the two principal theories that govern ownership rights of rivers are the riparian doctrine and the prior appropriation doctrine. Under special circumstances, the principle of prescriptive rights is used.

The riparian doctrine holds that water use accompanies ownership of the land adjacent to the stream. Such a right exists even if the water is not physically used by the landowner. However, when the water is used, it must be for reasonable and beneficial use by the owner on his property. Typical lawsuits establishing these principles were in Colburn v. Richards (Mass., 1816), Cook v. Hull (Mass., 1820), and Merritt v. Brinkerhoff (N.Y., 1820). Strobel v. Kerr Salt Co., (N.Y., 1900) set an important precedent whereby the relative economic merits and investments of property owners should not be a consideration in determining rights. The riparian system of water law is especially adaptable for climates where water is abundant, as in the eastern United States. However, it is not practical in drier climates where water is scarce and rivers may not flow throughout the year, as in much of the western United States.

In the more arid parts of the West, a water system of rights evolved that was termed the "prior appropriation doctrine." Land ownership is irrelevant to water rights, and instead, such rights are acquired by those who first use the water for some beneficial purpose. Thus, priority in time is the principal determinant when there is an insufficient amount of water to satisfy all who wish to use it. Lawsuits that established such legality include the California court decision in Tartar v. Spring Creek Water and Mining Company in 1855. The Desert Land Act of 1877 affirmed such rights to the various states for all non-navigable waters of the public domain.

### Groundwater Law

Because less is known of the movement and occurrence of subsurface waters when compared to surface water, laws are in greater conflict when dealing with groundwater. The usual rule established in most courts is to regard percolating waters as part of the property, thus subject to ownership by the landholder, according to the Napoleanic Code, as in Acton v. Blundell in 1843 (Fig. 14.2). However, if water was judged to flow in "underground

**FIGURE 14.2** Subdivision on outskirts of Phoenix, Ariz. Residents of such sites always take it for granted they will be supplied with plentiful groundwater resources without any legal entanglements.

streams,'' the courts ruled it was governed by riparian doctrine, just as surface streams.

With increasing demands for subsurface water in arid regions, other approaches to water law have been made. In the classic case of Pasadena v. Alhambra (Calif., 1949), the court formulated a new doctrine of mutual prescription whereby the original users were given correlative rights, and the total usage of each was to be determined by scientists in the state government.

## Other Water-related Laws

In the 1970s, the biggest push in water law were those specific statutes that addressed the growing problem of water contamination. The Clean Water Act was passed in 1972, and most of the effort was concentrated on the more than 60,000 point sources (industries, and municipal treatment facilities) that directly discharge pollutants into the nation's waters. The heart of this effort is the National Pollutant Discharge Elimination System (NPDES) permit program under which cleanup requirements are applied to individual sources.

The Corps of Engineers was given most responsibility for major policy of the Clean Water Act, along with the EPA. However, the Corps was reluctant to enlarge their idea of what constitutes navigable water, and contended they should only be involved with waters that are "subject to the ebb and flow of the tide and/or...susceptible for use in interstate or foreign commerce." Instead, environmental groups and the EPA argued for a broader mandate and contended their jurisdiction should include "waters of the United States including the territorial seas." In the lawsuit Natural Resources Defense Council v. Callaway (1975), the District Court agreed the Corps must play an expanded role. Thus, they now have jurisdiction of 25,000 mi of waterways, 3 million mi of streams, 124,000 mi of tidal shoreline, 4.7 million mi of lake shoreline, 30,000 mi of canals, and 148 million acres of wetlands.

The first federal legislation with authority to control groundwater pollution was the Water Pollution Control Act Amendments of 1972. Incorporated into the Clean Water Act of 1977, this granted jurisdiction to the EPA over both surface water and groundwater. In addition, the Safe Drinking Water Act of 1974 established a federal regulatory mechanism to ensure the quality of publicly supplied drinking water and to control groundwater pollution. The Act especially embodied a preventative approach to protect public health against harm from toxic substances in drinking water. Further support was obtained by passage of the Toxic Substances Control Act of 1976, which authorized EPA to regulate pollutants throughout their manufacture, transportation, use, and disposal cycles. Additional emphasis in the disposal of waste that can contaminate water occurred from the Resource Conservation and Recovery Act of 1976. This enabled state governments and the EPA to regulate disposal of municipal solid and hazardous waste. The Act broadened federal supervision over the most dangerous and widespread sources of groundwater contamination. It established a mechanism for controlling the generation, transportation, treatment, storage, and disposal of hazardous waste.

The Surface Mining Control and Reclamation Act of 1977 was the first federal attempt at land-use policy wherein regulation and rehabilitation of damaged mining environments is mandated. The Act gives the Office of Surface Mining (Department of Interior) authority to issue permits for mining and reclamation plans and to establish a bonding system to assure that mined-out areas will be reclaimed. The Act contains strict provisions for the prevention of chemical contamination and hydrologic disruption of groundwater.

## MARITIME LAW

In 1609, a treatise entitled *Mare Liberum* was written by a young Dutch jurist named Hugo Brotium, and it became recognized as the basic principle of international law. This Magna Carta of maritime law advanced the notion that the seas were to remain free and open to all nations except for a narrow strip of coastal waters. The concept made sense in a world of warring nations each seeking naval supremacy and sovereignty of the high seas. Not until President Harry Truman claimed the Outer Continental Shelf and its underlying oil and gas as U.S. territory was the principle seriously challenged. This triggered a series of other claims, and Chile, Peru, and Ecuador were the first to extend their rights to a 200 nautical mile distance from the shore. In many instances, these and other claims have prompted a series of disputes over fishing rights such as those between Iceland and England.

Negotiations concerning marine science and other issues began at the United Nations Conference on the Law of the Sea in 1974. On April 30, 1983, a Law of the Sea Treaty was approved by a vote of 130 to 4. Israel, Turkey, United States, and Venezuela voted against the treaty, and there were 17 abstentions. When finally ratified by 60 nations, the treaty will be consummated. It establishes boundaries between jurisdictional regions and provides for rules of conduct and for methods by which disputes can be settled. It imposes a 12 mile limit of territorial sovereignty from the coastline, and a 200 mile economic zone where the coastal nation has exclusive rights to fishing and mineral exploitation.

Most abstainers were from the Soviet group, and those nations that voted against the treaty called it a "Something for Nothing Treaty." The major stumbling block for the United States was the assignment of an International Sea-Bed Authority that would control the technology, development, and mineral resources of the oceans. In the Pacific Ocean, for example, it is predicted there are 16 billion tons of mineral nodules. Their recovery will require massive infusions of capital for research and equipment. However, the 36 member council that directs the Sea-Bed Authority would control everything – the sites, permissions, technology, profits, volume, fees, taxes, and royalties. It would act as a supranational mining corporation with power to compete with private companies and to share their technology. This is viewed as patently unfair with insufficient protection for the American firms that have pioneered costly underwater mining technology. Furthermore, the elaborate administrative structure that will be required to oversee the treaty and all of its manifestations will derive funding from the general United Nations budget – a financial enterprise in which the United States is expected to continue to pay the largest share.

### Coastal Law

One of the earliest legal cases involving the coastal zone occurred in England in the 1820s. Storm waves had eroded a beach and spilled over into the coastal lands at the community of Pagham. A group of property owners convinced the Pagham Commissioner of Sewers to construct a rock groin as protection from further destruction. Unfortunately, the groin induced increased erosion on the downdrift beach. Owners of these properties complained in a court case but to no avail because in the ruling, Lord Tentreden declared, "Each landowner may erect such defenses for the land under their care as the necessity of the case requires, leaving it to others in like manner, to protect themselves against the common enemy," (Marx, 1967). This "common enemy" doctrine has been the stimulus for many coastal engineers to erect whatever fortifications they can devise in attempts to prevent the destruction of shorelines by coastal processes.

In the United States, the U.S. Army Corps of Engineers has the primary statutory responsibility for administration and protection of coastal areas. The enabling legislation is provided by the Rivers and Harbors Act of 1899. It stipulates that a permit must be obtained before any type of construction can be done in navigable waters. The law also requires a permit for dredging, filling, excavation, and the discharge of refuse in such waters. Navigable waters have been defined as nearly all waterways including tidal waters, ocean waters, and other waters that affect interstate commerce and adjacent wetlands.

In the 1960s, the coastal states started to enact legislation that would help preserve the coastal zone from uncontrolled development. In 1963, Massachusetts passed a Coastal Wetlands Protection Act, which prohibited alteration of wetlands without authorization. The Rhode Island Salt Marsh Act has gone through several stages and was updated with important amendments in 1967 and 1969. Other early wetland legislation was also enacted by Maine in 1967, North Carolina in 1969, and by Maryland and Georgia in 1970. All coastal states in the United States now have wetland protection legislation. However, several lawsuits have challenged the legality of such legislation. For example, in Carton et al. v. State of New Jersey (A-636-73, 1978), the claimants argued that the Wetland Act was vague, unreasonable, and unconstitutional. They also contended that the Act amounted to taking of private property without just compensation. However, the court upheld the Act and ruled it was a valid exercise of governmental power and did not constitute a taking.

The coastal zone has also been the site where theories of the public trust doctrine were tested. Two lawsuits between the borough of Neptune

City and the borough of Avon-by-the-Sea in New Jersey resulted in the New Jersey Supreme Court ruling in Avon v. Neptune in 1976. Avon had charged non-residents higher fees than local people for use of beaches. The court ruled that charging a discriminatory fee to users was analogous to the erection of physical barriers. It held that the beach area was subject to the public trust doctrine.

The fragility of the coastal environment has been the focus of many lawsuits that would change or destroy the sand composition of the coastal zone. In Potomac Sand and Gravel v. Governor (293 A 241, 266 Md., 1972), the company argued that laws which prohibited the removal of sand and gravel from the tidal wetlands were unconstitutional. However, the court ruled such laws were valid and specifically recognized the need to preserve natural resources.

Since 1964, the Town of Islip, N.Y., has had ordinances regarding the use of off-road vehicles (ORVs) on the beaches of Fire Island. The laws provide for restriction of use and require permits for travel:

> The preservation of... Fire Island is of paramount importance to the citizens of the Town of Islip...The proliferation of motor vehicles on the barrier beach runs counter to the public policy...and is detrimental to the preservation of the barrier beach. The limitation of the use of motor vehicles on the island will promote the health, safety and general welfare of the community, including the protection and preservation of the town.

The ordinance has been repeatedly challenged by recreationists, dune buggy clubs, and some residents. For example, in Sak v. Fire Island Advisory Commission (July 22, 1977), the plaintiff argued the permits were discriminatory and unconstitutional. The Suffolk County Supreme Court upheld the administrative procedures of the ordinance and stated, "It is now beyond dispute, that vehicular traffic on Fire Island is injurious and that efforts to curtail such traffic by a scheme which differentiates among applicants for driving permits on a reasonable basis is constitutional."

## LAND-USE LAW

There are many laws that govern land utilization by man that have been passed on all government levels. Such legislation is aimed at protection of man or is for the necessity of providing assurance of his health, safety, welfare, and sustenance.

### Water and Soils

Prior to 1849, flood control had been considered only a local problem, but the Swamp Land Acts of 1849 and 1850 recognized the severity of flooding. These acts set the stage for states to adopt planning of flood-prone areas and led to the establishment of the Mississippi River Commission in 1879 whose mandated duty was to improve navigation and "prevent destructive floods." A Flood Control Act of 1928 authorized expenditure of $60 million for flood-control work on the lower Mississippi but it took the disastrous floods of 1935 and 1936 for Congress to pass the much more inclusive Flood Control Act of 1936. This latter statute, along with the Soil Conservation Act of 1935, put the federal government in the business of large-scale land managers. The Department of Agriculture (through the Soil Conservation Service) was to assume responsibility for upstream flood-control measures, and the Army Corps of Engineers was authorized to undertake projects on navigable rivers throughout the United States. The scope of the Department of Agriculture was enlarged in 1954 with passage of the Watershed Protection and Flood Prevention Act, which allowed this agency to construct flood-prevention structures that would impound up to 4000 acre-feet of water if costs did not exceed $250,000.

Another type of breakthrough in environmental land law occurred in 1902 with the Reclamation Act, which established irrigation in the West as a national policy for implementation. Although lawsuits were brought against the government challenging such powers, the Supreme Court affirmed the principle of governmental promotion for the general welfare through "...large-scale projects of reclamation, irrigation, or other internal improvement" (United States v. Gerlach Livestock Co., 339 U.S., 723, 738, 1950). An unusual expansion of governmental perogatives for the general welfare occurred in 1933 with enactment of the Tennessee Valley Authority Act, which charged the new agency with a land development and rehabilitation program that would involve land use, water control, and soil conservation throughout the region.

### Range and Forest Lands

The first federal environmental legislation of forested lands was passed in 1891 with the Forest Reserve Act, and followed in 1897 with the Forest Management Act. These acts were aimed at the West, and their purpose was to ensure a continuous timber supply and provide favorable conditions for streamflow. The Weeks Act of 1911 laid the groundwork for federal programs on a country-wide basis, and the Clarke-McNary Act of 1924, the McSweeney-McNary Act of 1928, and the Sustained Yield Forest Management Act of 1944 combined to provide a comprehensive environmental package of the legal stewardship of the government for forest lands. The 1944 Act summarized this posture by stating their purpose was "...in the maintenance of water supply, regulation of stream flow, prevention of soil erosion, amelioration of climate, and preservation of wildlife." Again in 1960, the Multiple Use-Sustained Yield Act emphasized the necessity of maintaining forests in perpetuity for the purpose of assuring: (1) timber from trees, (2) forage from grazing lands, (3) recreation, (4) wildlife management, and (5) protection of rivers and consideration for their constant flow.

The Taylor Grazing Act of 1934 was the first federal legislation that directed attention to a public policy for establishment of grazing districts and raising of forage crops. It provided the Secretary of Interior with powers to undertake erosion and flood-control measures that will protect and rehabilitate grazing lands.

### Parks and Wilderness

Governmental attitudes for the establishment of park-like areas were greatly influenced by Roman law and the English "commons," which held that certain lands be set aside for the people as a public trust. Passage of the Yellowstone Act of 1872 by Congress made Yellowstone National Park the first of its kind in the world, and put the federal government into the business of preserving terrain for recreational purposes. The New York legislature enacted a bill in 1885 to establish a forest area that was to remain permanently "...as wild forest lands" and led to the Adirondack State Park. This "forever wild" concept culminated in the Wilderness Act of 1964, which set aside millions of acres throughout the United States for the purpose of benefitting "...the American people...an enduring resource of Wilderness."

Additional environmental legislation was implemented by such federal laws as the Land and Water Conservation Act of 1965 and the Wild and Scenic Rivers Act of 1968. These bills amplified the government's resolve to plan and preserve lands for outdoor recreation and "...enjoyment of present and future generations."

## LEGAL RIGHTS FOR NATURAL FEATURES

It is becoming increasingly common that people, the courts, and legislative bodies are recognizing the importance of maintaining balance in natural systems. Since ecosystems cannot speak or act for themselves in legal proceedings, various surrogate organizations are becoming their advocates and are championing the view that natural features and objects have certain legal rights. The large number of lawsuits, as well as their significance, that challenged unnecessary and environmentally harmful developments were instrumental in passage of NEPA and establishment of the Environmental Protection Agency. Furthermore, the states, such as Hawaii, Vermont, and Maine, were passing land-use laws whereby the land was to be treated as a natural resource. The dual rationale for these trends was that man no longer was to be considered as omnipotent and that destruction of nature ultimately created feedback mechanisms that proved harmful to man as well.

Perhaps tampering with rivers and floodway corridors received the earliest attention and prohibitions of construction. More recently, coastal areas have been selected as sites for preservation. Typical of these laws is the Delaware Coastal Zone Act of 1971 to "...protect the natural environment of its bay and coastal areas..." The Coastal Zone Management Act of 1972 was the first land-control legislation ever passed by Congress, and it has led to much additional legislation by states that border coastal areas. Wetlands are another environmental feature that are being increasingly protected by law. Massachusetts and Connecticut were some of the earliest states to pass such legislation, and the Freshwater Wetlands Act of 1975 in New York is one of the most comprehensive of its type for the protection of these features.

## EMINENT DOMAIN

Governments have always used their power to acquire private lands in the interest of the public good. Such lands in the United States cannot be condemned for whimsical or non-beneficial uses, and the rights of property owners to obtain just compensation are assured by the Constitution. However, numerous environmental lawsuits result from condemnation practices when:

1. Plaintiffs believe abnormal environmental damage will occur. For examples, see Coates (1973, p. 383–387) in discussion of such cases as Hetch Hetchy in California, Boundary Waters Canoe Area in Minnesota, Mineral King in California, and Florida jetport hearings.

2. Plaintiffs feel aggrieved because of the lowness of the monetary assessment of their property (see Coates 1971, 1976).

## HAZARDS AND DISASTERS

Geologists have played important roles in developing legislation pertaining to hazards and disasters, and in implementing their regulations. Furthermore, lawsuits that have resulted from such events have often called upon geologists for testimony. There is a strong relationship between disasters and legislation that is thereafter passed in an attempt to prevent a recurrence of the event, or to minimize losses. The St. Francis Dam failure in 1928 that claimed more than 500 lives resulted in an amendment to the California State Water Resources Code in 1929 that thereafter required all private dams to be inspected. A geological evaluation of the dam became one of the inspection requirements. When five lives were lost with the failure at Baldwin Hills Reservoir, Calif., again, the Water Resources Code was amended to include all reservoirs, storage areas, and dams under State jurisdiction.

A 5.5 M earthquake occurred in 1933 at Long Beach, Calif., that caused much damage and toppled many brick school houses. If the earthquake had occurred when classes were in session, the loss of life would have been enormous. This reality prompted the California State Legislature to pass the Field Act a few months later. It mandated that all new public school buildings must be constructed of earthquake-resistant materials with a design approved by the State architect. The 7.7 M Kern County earthquake in 1952 showed that schools built in conformity with the Act withstood damage better than others, and such results prompted increasing compliance and renovations with the Act guidelines. The destructive San Fernando earthquake of 1971 prompted passage of the California Hospital Safety Act of 1972. This law requires a report by a certified geologist when new hospital structures are designed. Additional California geological involvement stems from passage of the Alquist Priolo Geologic Hazard Zones Ace of 1973. Fault zone areas must be delineated, and cities and counties are required to approve projects only after a geological report has been filed that provides data and maps that have information about potential hazards and surface fault ruptures.

Federal involvement with statutes pertaining to geologic hazards started with the various flood-control acts. The most far-reaching was the Flood Control Act of 1936, passed after the disastrous Mississippi River basin floods in 1935 and 1936 (see also Chapter 8). The National Flood Insurance Act was part of the Housing and Urban Development (HUD) Act of 1968 and provided low-cost insurance to homeowners in the 100- and 500-year floodplains once the local government adopted and enforced floodplain land-use measures. Later amendments extended the coverage to include mudslides. The Flood Disaster Protection Act of 1973 upgraded the early legislation. This was prompted by the devastating floods and landsliding connected with Hurricanes Agnes and Camille. Lending institutions must require homeowners to purchase Flood Insurance if they reside within the 100-year floodplain. The Federal Disaster Relief Act of 1974 provides guidelines for providing loans to communities following a disaster. The Act requires that any state or local government that receives property loans must agree to study the hazard and take measures to mitigate a recurrence. Perhaps the most comprehensive federal legislation concerning geological hazards is the Earthquake Hazards Reduction Act of 1977. The President is required to establish and maintain a coordinated program that aims to predict seismic hazard, design resistance structures, develop models for land-use decisions, and provide research on mitigation of hazards.

### Lawsuits

In Washington, a motorist was injured when a landslide from a steep hill adjacent to the highway struck his car. A lower court ruled the motorist was entitled to recover damages because the government had failed to post warning signs. However, in Boskovich v. King County (1936), the case was appealed and the verdict overturned. The appeal court said the county was not the insurer of road safety and was not guilty of negligence. Just the reverse was held by the majority of the court in Brown v. McPhersons Inc. (1975). The plaintiff suffered losses because of a landslide. The court record showed that a geologist had warned state officials and McPhersons, a realty firm, of the possibility of landsliding in the area. However, the State failed to warn local residents, and thus the court ruled, they had failed in their duty and so were culpable for the action.

## POSTLUDE

In the United States, the final authority for the interpretation of environmental law with all of its ramifications is the U.S. Supreme Court. Its opinions and rulings dominate the behavior of other courts and provide guidelines concerning the direction legislation is allowed to pursue. During environmental lawsuits, the Supreme Court has shown three important trends: (1) it has strictly enforced choices made by the Congress in passage of their environmental statutes, (2) it has sharply limited judicial expansion of substantive duties of procedural rights in environmental lawsuits, and (3) it has refused to allow lower federal courts to become involved in significant environmental policy. Through their rulings, the Supreme Court has held that the legislative branch of government must settle environmental and public health issues. This has led to sustaining environmentally protective measures authorized by congressional statute, even when under severe attack by mining and industry. Thus, these three trends show a decidedly conservative Court but one that is not completely adverse to environmental claims. However, the primary effect of the decisions has been to shift to Congress the burden of resolving most environmental disputes. Such trends have left plaintiffs, who have suffered important environmental injury, without the full spectrum of judicial remedies. For example, the Court has limited their involvement with NEPA, thereby leaving agency decisions virtually irreversible by courts so long as the minimum number of formal procedures are followed. The following are some of the typical decisions that set these precedents:

1. In Kleppe v. Sierra Club (1976), the Court ruled that an agency could not be compelled to prepare an EIS at any time other than that specified under NEPA. It based its decision on strict interpretation of the constitutional separation of powers, on the statutory history of Administrative Procedure acts, and on a literal reading of NEPA.

2. In Vermont Yankee Nuclear Power Corp. v. Natural Resources Defense Council (1978), the Court reversed a lower court's findings that would have expanded the procedural duties under NEPA. Instead, the Supreme Court held that only those procedural duties actually written were required and that no additional demand could be made to change them.

3. In EPA v. National Crushed Stone Association (1980), industry had attempted to interject economic considerations into environmental decisions where Congress had not so stipulated. The Court upheld EPA's decision under the Clean Water Act that in setting "best practicable control technology currently available," the agency was not required to consider economic variances from its nationally uniform standards.

APPENDIX A

# TABLE OF CONVERSIONS

1 in = 2.54 cm
1 ft = 0.305 m
1 m = 1.609 km
1 $mi^2$ = 2.589 $km^2$
1 acre = 0.405 hectare
1 acre-ft = 325,851 gallons
1 gallon = 3.785 liters
= 0.003785 $m^3$
1 ton = 1.016 metric tons
1 $ft^3$ = 0.0283 $m^3$
1 cm = 0.03937 in
1 m = 3.28 ft
1 km = 0.621 mi
1 hectare = 2.47 acres
1 $m^3$ = 35.314 $ft^3$
1 metric ton = 2.204.62 lb
1 kg = 2.2046 lb
1 barrel (oil) = 42 gal

APPENDIX B

# VOLCANIC DISASTERS

| | | |
|---|---|---|
| 1500 B.C. | Thera (Santorini volcano) Aegean Sea | Destroyed cities on Thera and tsunami affected coastal communities throughout region. |
| A.D. 260 | Ilopango (in what is now El Salvador) | Affected a region of 1000s of kilometers and caused displacement of more than 30,000 Mayans. |
| A.D. 79 | Vesuvius, Italy | Destroyed Pompeii, Herculaneum, and other villages, killed 2000. |
| 1669 | Mt. Etna, Sicily | Earthquakes and associated volcanism killed 20,000 and ruined 14 villages. |
| 1783 | Mt. Skaptar, Iceland | Deadly gases from eruption killed 10,000 along with thousands of cattle; other deaths from starvation. |
| 1815 | Tamboro, Indonesia | Direct effects from explosion and resulting tsunami killed 12,000; an additional 92,000 died from disease and starvation. |
| 1883 | Krakatau, Indonesia | Island exploded and resulting tsunami drowned 36,000. |
| 1902 | Mt. Pelée, Martinique | 30,000 people in St. Pierre killed by deadly gases and suffocating ash. |
| 1902 | La Soufriere, St. Vincent | 2000 killed; Carib Indian population largely eliminated. |
| 1911 | Taal, Philippines | Killed 1300. |
| 1909 | Kelut, Indonesia | Killed 5500. |
| 1951 | Mt. Lamington, New Guinea | Killed 6000. |
| 1963 | Mt. Irazu, Costa Rica | 2000 cattle killed, 40,000 ha of farmland destroyed. |
| 1963 | Mt. Agung, Bali | 161 killed, 78,000 homeless. |
| 1965 | Taal, Philippines | 180 killed. |
| 1968 | Arenal, Costa Rica | Fiery gas cloud killed 80; ash destroyed surrounding farmland and buildings. |
| 1977 | Nyragongo, Zaire | Killed 70. |
| 1979 | Mt. Sinila, Java | 175 killed, 17,000 evacuated. |
| 1980 | Mount St. Helens, U.S.A. | 50 killed; nearly $3 billion in property, timber, wildlife, and development damages. |

APPENDIX C

# MAJOR EARTHQUAKE DISASTERS THROUGHOUT THE WORLD

| Year | Locality | Deaths |
|---|---|---|
| 856 | Greece, Corinth | 45,000 |
| 1038 | China, Shensi | 23,000 |
| 1057 | China, Chihli | 25,000 |
| 1268 | Asia Minor, Silicia | 60,000 |
| 1290 | China, Chihli | 100,000 |
| 1293 | Japan, Kamakura | 30,000 |
| 1531 | Portugal, Lisbon | 30,000 |
| 1556 | China, Shensi | 830,000 |
| 1667 | Caucasia, Shemaka | 80,000 |
| 1693 | Italy, Catania | 60,000 |
| 1737 | India, Calcutta | 300,000 |
| 1755 | Northern Persia | 40,000 |
| 1755 | Portugal, Lisbon | 60,000 |
| 1759 | Lebanon, Baalbek | 30,000 |
| 1783 | Italy, Calabria | 50,000 |
| 1797 | Ecuador, Quito | 41,000 |
| 1811 | U.S., New Madrid, Mo. | Several |
| 1819 | India, Cutch | 1,543 |
| 1822 | Asia Minor, Aleppo | 22,000 |
| 1828 | Japan, Echigo (Honshu) | 30,000 |
| 1868 | Peru and Ecuador | 25,000 |
| 1875 | Venezuela and Colombia | 16,000 |
| 1886 | U.S., Charleston, S.C. | 60 |
| 1896 | Japan, Sea Wave, Sanriku Coast | 22,000 |
| 1897 | India, Assam | 1,542 |
| 1905 | India, Kangra | 20,000 |
| 1906 | U.S., San Francisco, Calif. | 700 |
| 1906 | Chile, Valparaiso | 1,500 |
| 1908 | Italy, Messina | 75,000 |
| 1915 | Italy, Avezzano | 29,970 |
| 1920 | China, Kansu | 200,000 |
| 1923 | Japan, Tokyo-Yokohama | 143,000 |
| 1932 | China, Kansu | 70,000 |

| Year | Locality | Deaths |
|---|---|---|
| 1935 | Pakistan, Quetta | 60,000 |
| 1939 | Chile, Chillan | 30,000 |
| 1939 | Turkey, Erzincan | 23,000 |
| 1946 | Eastern Turkey | 1,300 |
| 1946 | Japan, Honshu | 2,000 |
| 1948 | Japan, Fukui | 5,131 |
| 1949 | Ecuador, Pelileo | 6,000 |
| 1950 | India, Assam | 1,500 |
| 1953 | Northwestern Turkey | 1,200 |
| 1954 | Algeria, Orleansville | 1,657 |
| 1956 | Northern Afghanistan | 2,000 |
| 1957 | Northern Iran | 2,500 |
| 1957 | Outer Mongolia | 1,200 |
| 1957 | Western Iran | 2,000 |
| 1960 | Morocco, Agadir | 12,000 |
| 1960 | Southern Chile | 5,700 |
| 1962 | Northwestern Iran | 10,000 |
| 1963 | Yugoslavia, Skopje | 1,100 |
| 1964 | Southern Alaska | 131 |
| 1965 | Chile, El Cobre | 400 |
| 1966 | Eastern Turkey | 2,529 |
| 1967 | Venezuela, Caracas | 236 |
| 1968 | Northeastern Iran | 11,588 |
| 1970 | Western Turkey | 1,086 |
| 1970 | Northern Peru | 66,794 |
| 1971 | U.S., San Fernando, Calif. | 65 |
| 1971 | Chile | 90 |
| 1972 | Southern Iran | 5,000 |
| 1972 | Nicaragua, Managua | 10,000 |
| 1973 | Central Mexico | 500 |
| 1974 | Pakistan | 5,300 |
| 1975 | Turkey | 2,400 |
| 1976 | Guatemala | 23,000 |
| 1976 | Italy | 900 |
| 1976 | Indonesia, West Irian | 5,000 |
| 1976 | Indonesia, Bali | 400 |
| 1976 | China, Tangshan | 655,000 |
| 1976 | Turkey | 4,000 |
| 1977 | Romania | 1,300 |
| 1977 | Iran | 1,200 |
| 1977 | Indonesia, Sumbawa | 187 |
| 1978 | Iran | 25,000 |
| 1979 | Yugoslavia | 235 |
| 1979 | Iran | 1,500 |

| Year | Locality | Deaths |
|---|---|---|
| 1979 | Colombia-Ecuador border | 400 |
| 1979 | Indonesia | 122 |
| 1980 | Algeria | 3,000 |
| 1980 | Italy | 3,000 |
| 1980 | Azores | 300 |
| 1981 | Iran | 4,000 |
| 1981 | China | 150 |
| 1981 | Pakistan | 212 |
| 1981 | Indonesia | 250 |
| 1982 | North Yemen | 1,340 |
| 1983 | Turkey | 2,000 |
| 1983 | Colombia | 240 |
| 1983 | Guinea | 200 |

*Source:* Office of Emergency Preparedness, 1972, Disaster preparedness: U.S. Government Printing Office, and various news accounts.

APPENDIX D

# LANDSLIDE DISASTERS

| | Date | People killed | Remarks |
|---|---|---|---|
| Brenno Valley, Switzerland | 1512 | 600 | Rockslide dammed valley; dam broke in 2 yr causing destruction |
| Tour d'Ai, Switzerland | 1584 | 300 | Landslide devastated village of Yvorne in Rhone valley |
| Mount Conto, Switzerland | 1618 | 2,430 | Rockslide |
| Goldau, Switzerland | 1806 | 457 | Landslide destroyed village |
| Mt. Ida, Troy, N.Y. | 1843 | 15 | Sediment slump and flow |
| Elm, Switzerland | 1881 | 115 | Rock avalanche also demolished 83 houses |
| Trondheim, Norway | 1893 | 111 | Liquefaction flow in marine clays |
| Frank, Canada | 1903 | 70 | Rock avalanche destroyed most of town |
| Kansu Province, China | 1920 | 100,000–200,000 | Earthquake caused loess flows |
| Nordfjord, Norway | 1936 | 73 | Rockfall created 74 m wave |
| Kobe, Japan | 1938 | 461 | Rocky mudflows |
| Kure, Japan | 1945 | 1,154 | Rocky mudflows |
| Yokahama, Japan | 1958 | 61 | Rocky mudflows |
| Madison, Montana | 1959 | 28 | Rock avalanche buried campers |
| Vaiont, Italy | 1963 | 2,000 | Rockslide into reservoir created wave that flooded below dam |
| Anchorage, Alaska | 1964 | 114 | Combined toll from landslides and earthquake |
| Aberfan, Wales | 1966 | 144 | Manmade mining spoiled hill; landsliding buried mostly children |
| Brazil | 1966–67 | 2,700 | Combined toll from landslides and floods |
| Nelson County, Va. | 1969 | 150 | Combined total from debris avalanches and floods |
| Huascaran area, Peru | 1970 | 21,000 | Combined rock avalanche and debris flow buried two cities |
| St. Jean Vianney, Canada | 1971 | 31 | Slab flows buried people and houses |

*Note:* See Coates (1976) for sources.

APPENDIX E

# MAJOR LAND SUBSIDENCE AREAS

| Name of Locality | Maximum subsidence (meters) | Area affected (kilometers) | Cause | Damages (millions of dollars) |
|---|---|---|---|---|
| San Joaquin Valley, Calif. | 9 | 13,500 | a | 100 |
| Houston-Galveston, Tex. | 2.75 | 12,170 | a | > 1000 |
| Eloy-Picacho, Ariz., and adjacent area | 3.6 | 8,700 | a | Several million |
| Tokyo area, Japan | 4.6 | 2,400 | a | 225 from 1957–70 |
| Nobi Plain, Japan | 1.5 | 800 | a | |
| Po Valley, Italy | 3 | 780 | d | |
| Santa Clara Valley, Calif. | 3.9 | 650 | a | > 25 |
| Baton Rouge, La. | 0.5 | 650 exceed 5 cm subsidence | a | > 1 |
| Sacramento Valley, Calif. | 0.7 | 500 | a | |
| Osaka, Japan | 3 | 500 | a | Tens of millions |
| San Joaquin Valley (southwest), Calif. | 5 | > 500 | b | |
| Lake Maracaibo, Venezuela | 3.9 | 450 | c | 35 up to 1976 |
| London, England | 0.35 | 450 | a | |
| Niigata, Japan | 2.6 | 430 | d | Tens of millions |
| Saga Plain, Japan | 1.2 | 400 | a | 523 from 1960–79 |
| Venice area, Italy | 0.14 | 400 | a | |
| Debrecan, Hungary | 0.42 | 390 | a | |
| Savannah, Ga. | 0.15 | 330 exceed 2 cm | a | |
| Las Vegas, Nev. | 1.7 | 300 | a | Several million |
| Raft Valley, Idaho | 2.8 | 260 | a | |
| Taipei, Taiwan | 1.9 | 230 | a | |
| Mexico City, Mexico | 8.7 | 225 | a | > 500 |
| New Orleans, La. | 0.8 | 150 | a | |
| Victoria-Gippsland, Australia | 1.6 | 102 | f | |
| Saxet oil field, Tex. | 0.93 | 92 | c | |
| Wilmington, Long Beach, Calif. | 8.8 | 78 | c | 200 |

| Name of Locality | Maximum subsidence (meters) | Area affected (kilometers) | Cause | Damages (millions of dollars) |
|---|---|---|---|---|
| Chocolate Bayou oil field, Tex. | 0.53 | 40 | c | |
| Visonta, Hungary | 0.5 | 40 | a | |
| Haranomachi City, Japan | 2 | 25 | a | |
| Goose Creek oil field, Tex. | 1 | 10 | c | |
| Baldwin Hills, Calif. | 3 | 5 | c | 25 |
| Wairakei, New Zealand | 4.8 | 1.3 | e | |

Explanation:
a = groundwater wells
b = hydrocompaction from surface water
c = oil and gas wells
d = methane in water wells
e = geothermal wells
f = dewatering coal mines

# SEVERE FLOOD DISASTERS IN THE UNITED STATES

| Year | Place | Lives Lost | Million dollars lost | Cause |
|---|---|---|---|---|
| 1831 | Barataria Isle, La. | 150 | | |
| 1844 | Upper Mississippi river | | | Rainfall-river flood |
| 1889 | Johnstown, Pa. | 2,200 | 20 | Dam failure |
| 1903 | Passaic & Delaware rivers | 100 | 25 | Rainfall and dam failure |
| 1903 | Missouri River basin | | 50 | Rainfall-river flood |
| 1903 | Heppner, Oregon | 247 | | Rainfall-river flood |
| 1913 | Ohio River basin | 467 | 150 | Rainfall-river flood |
| 1913 | Brazos & Colorado rivers, Tex. | 177 | 128 | Hurricane rainfall-river flood |
| 1921 | Arkansas River | 120 | 13 | Rainfall-river flood |
| 1926 | Miami & Clewiston, Fla. | 350 | 70 | Hurricane tidal & river flood |
| 1926 | Illinois River | | 25 | Rainfall-river flood |
| 1927 | New England | | 50 | Rainfall-river flood |
| 1927 | Lower Mississippi | 100 | 284 | Rainfall-river flood |
| 1927 | Vermont | 120 | | Rainfall-river flood |
| 1928 | Lake Okeechobee, Fla. | 2,400 | | Hurricane tide & flood |
| 1935 | Susquehanna & Delaware rivers | 52 | 36 | Rainfall-river flood |
| 1935 | Republican River, Kan.-Neb. | 110 | | Rainfall-river flood |
| 1936 | Northeastern U.S. | 107 | 221 | Rainfall-river flood |
| 1936 | Ohio River basin | 137 | 150 | Rainfall-snowmelt flood |
| 1937 | Ohio River basin | | 418 | Rainfall-river flood |
| 1938 | New England streams | 200 | 125 | Hurricane tidal & river flood |
| 1938 | California streams | 79 | 100 | Rainfall-river flood |
| 1939 | Licking & Kentucky rivers | 78 | 1.7 | Rainfall |
| 1940 | Southern Virginia, Carolinas, E. Tennessee | 40 | 12 | Rainfall |
| 1942 | Mid-Atlantic coastal streams | | 28 | Rainfall-river flood |
| 1943 | Central States | 60 | 172 | Rainfall-river flood |
| 1944 | South Florida | | 63 | Hurricane tidal & river flood |
| 1944 | Missouri River basin | | 52 | Rainfall-river flood |
| 1945 | Hudson River basin | | 24 | Rainfall-river flood |
| 1945 | South Florida | | 54 | Hurricane tidal & river flood |

| Year | Place | Lives Lost | Million dollars lost | Cause |
|---|---|---|---|---|
| 1945 | Ohio River basin | | 34 | Rainfall-river flood |
| 1947 | South Florida | | 60 | Hurricane tidal & river flood |
| 1947 | Missouri River basin | 29 | 178 | Rainfall-river flood |
| 1948 | Columbia River basin | 35 | 102 | Rainfall-river flood |
| 1950 | San Joaquin River, Calif. | | 32 | Rainfall-river flood |
| 1950 | Central West Virginia | 31 | 4 | Rainfall |
| 1951 | Kansas River basin | 28 | 883 | Rainfall-river flood |
| 1952 | Missouri River basin | | 180 | Snowmelt floods |
| 1952 | Upper Mississippi River | | 198 | Rainfall-river flood |
| 1954 | New England streams | | 180 | Hurricane tidal floods |
| 1955 | Northeastern U.S. | 185 | 684 | Hurricane tidal & river flood |
| 1955 | California & Oregon streams | 61 | 271 | Rainfall-river flood |
| 1957 | Ohio River basin | | 65 | Rainfall-river flood |
| 1957 | Texas rivers | | 144 | Rainfall-river flood |
| 1959 | Ohio River basin | | 114 | Rainfall-river flood |
| 1960 | South Florida | | 78 | Hurricane tidal & river flood |
| 1960 | Puerto Rico | 107 | | Hurricane tidal & river flood |
| 1963 | Ohio River basin | 26 | 97.6 | Rainfall |
| 1964 | Montana | 31 | 54.3 | Rainfall |
| 1964 | Florida | | 325 | Hurricane tidal & river floods |
| 1964 | Ohio River basin | | 106 | Rainfall-river floods |
| 1964 | California streams | 40 | 173 | Rainfall-river floods |
| 1964 | Columbia River-N. Pacific | | 289 | Rainfall-river floods |
| 1965 | South Florida | | 139 | Hurricane tidal & river flood |
| 1965 | Upper Mississippi River | 15 | 158 | Rainfall-snowmelt river flood |
| 1965 | Platte River, Colo.-Neb. | | 191 | Rainfall-river flood |
| 1965 | Arkansas River, Colo.-Kan. | | 61 | Rainfall-river flood |
| 1965 | New Orleans & vicinity | | 322 | Hurricane tidal flood |
| 1965 | Sanderson, Tex. | 26 | 2.7 | Flash flood |
| 1967 | South California | 12 | | Rainfall-river flood |
| 1969 | California | 103 | 399.2 | Rainfall |
| 1969 | Northern Ohio | 30 | 87.9 | Rainfall |
| 1969 | James River, Va. | 154 | 116 | Rainfall |
| 1971 | Chester, Pa. | 10 | | Rainfall-river flood |
| 1972 | Rapid City, S. Dak. | 245 | 200 | Rainfall-river flood |
| 1972 | E. United States | 118 | 3,500 | Hurricane Agnes |
| 1973 | Mississippi basin | 11 | 1,500 | Rainfall |
| 1974 | U.S. Midwest | 28 | | Flash floods |
| 1976 | Big Thompson Canyon, Colo. | 140 | 16.5 | Flash floods |
| 1977 | Johnstown, Pa. | 49 | 117 | Flash floods |
| 1977 | Texas | 31 | | Rainfall |
| 1977 | Kansas City, Mo. | 23 | 50 | Rainfall |

| Year | Place | Lives Lost | Million dollars lost | Cause |
|---|---|---|---|---|
| 1978 | Central Texas | 52 | | |
| 1979 | Pearl River, Miss. | | 600 | Largest flood damage on record at Jackson |
| 1979 | Florida & East coast | | 50 | Hurricane David |
| 1979 | Gulf Coast | 13 | 1,000 | Hurricane Frederic |
| 1980 | Texas | 24 | 600 | Hurricane Allen |
| 1980 | California-Arizona | 31 | 425 | Rainfall-river flood |
| 1980 | Pennsylvania—West Virginia | 6 | 42 | Flash flooding |
| 1981 | Texas | 17 | 30 | Flash floods |
| 1982 | California | 31 | 500 | Rainfall, floods, mudslides |
| 1982 | Ft. Wayne, Ind. | | 100 | Rainfall |
| 1982 | Kansas City, Mo. | 4 | 30 | Rainfall |
| 1982 | Lower Mississippi valley | | 50 | Rainfall, snowmelt |
| 1983 | Lower Mississippi valley | 7 | 250 | Rainfall, snowmelt |
| 1983 | Colorado River | 11 | 100 | Rainfall, snowmelt |
| 1983 | Arizona | 13 | 300 | Rainfall; worst floods on record |
| 1983 | Utah | | 200 | Rainfall, snowmelt |
| 1983 | Texas | | 200 | Hurricane Alicia |
| 1983 | Missouri & Tennessee | 7 | 50 | Rainfall |

Sources: Office of Emergency Preparedness, 1972, Disaster Preparedness: U.S. Government Printing Office, and various news accounts.

# COASTAL AREA DISASTERS FROM TROPICAL STORMS AND TSUNAMIS

| Year | Locality | Deaths | Remarks |
|---|---|---|---|
| 1500 B.C. | Thera Islands, Mediterranean | | Devastated coastal area. May have contributed to decline of Minoan civilization. Tsunami after volcanic eruption. |
| 1755 | Lisbon, Portugal | 60,000 | Tsunami caused by earthquake |
| 1815 | Tamboro, Indonesia | 12,000 | Tsunami caused by volcanic eruption |
| 1856 | Isle Derniere, La. | 320 | Hurricane |
| 1875 | Indianola, Tex. | 176 | Hurricane |
| 1883 | Krakatau, Indonesia | 36,000 | Tsunami caused by volcanic eruption |
| 1886 | Sabine, Tex. | 150 | Hurricane |
| 1896 | Japan | 27,000 | Tsunami, 10,000 houses lost |
| 1900 | Galveston, Tex. | 6,000 | Hurricane |
| 1906 | Gulf Coast, U.S.A. | 151 | Hurricane |
| 1909 | Gulf Coast-New Orleans | 700 | Hurricane |
| 1915 | Louisiana & Texas | 550 | Hurricane |
| 1919 | Louisiana & Texas | 284 | Hurricane |
| 1926 | Miami area, Fla. | 350 | Hurricane; $70 million losses. |
| 1928 | Lake Okeechobee, Fla. | 2,400 | Hurricane |
| 1928 | Puerto Rico | 300 | Hurricane; $50 million damages |
| 1932 | Puerto Rico | 225 | Hurricane |
| 1935 | Florida Keys | 400 | Hurricane |
| 1938 | New England | 200 | Hurricane & stream flooding; $125 million damages |
| 1946 | Hawaii | 173 | Tsunami from earthquake in Aluetians |
| 1955 | Northeastern U.S. | 185 | Hurricane Diane with stream flooding; $684 million damages |
| 1957 | Louisiana | 556 | Hurricane |
| 1960 | Hilo, Hawaii | 61 | Tsunami from earthquake near Chile |
| 1964 | Florida | | Hurricane & stream floods; $325 million damages |
| 1965 | New Orleans region | | Hurricane; $322 million damages |
| 1969 | India | 600 | Cyclone |
| 1969 | Taiwan | 177 | Typhoon |

| Year | Locality | Deaths | Remarks |
|---|---|---|---|
| 1970 | Philippines | 2,011 | Typhoons: George, Kate, Sening, Titang & Patsy |
| 1970 | Bangladesh | 300,000 | Cyclone. Second worst disaster of 20th century |
| 1970 | Texas | 9 | Hurricane Celia; $500 million damages |
| 1971 | South Korea | 178 | Typhoons Olive & Hester |
| 1974 | Honduras | 8,000 | Hurricane Fifi |
| 1975 | West Indies & U.S. | 86 | Hurricane Eloise |
| 1976 | Baja, California, Mexico | 500 | Hurricane Liza |
| 1976 | Philippines | 215 | Typhoon; 60,000 homeless |
| 1976 | La Paz, Mexico | 500 | Hurricane; 14,000 injured |
| 1976 | Philippines | 3,000 | Tsunami from an earthquake |
| 1977 | India | 8,400 | Cyclone |
| 1978 | Sri Lanka | 500 | Cyclone; 70,000 homeless, much damage |
| 1978 | Acajutla, Ecuador | 100 | Tsunami from earthquake |
| 1979 | Southeast India | 600 | Cyclone; 1 million homeless |
| 1979 | Caribbean & southeastern U.S. coastal area | 1,300 | Hurricane David; much damage |
| 1979 | U.S. Gulf Coast | 13 | Hurricane Frederic; 4700 injured. Greatest coastal property damage on record: $2.3 billion |
| 1979 | French Riviera | 13 | Tsunami apparently induced by dredging operations |
| 1980 | Central America & southeastern U.S. | 273 | Hurricane Allen; great damage throughout area with $600 million in Texas alone |
| 1980 | Southwest Japan | 3 | Typhoon Orphid; much damage in region |
| 1981 | Philippines | 120 | Tropical Storm Kelly |
| 1981 | North-central Japan | 14 | Typhoon Thad |
| 1981 | India & Arabian Sea | 470 | Cyclone |
| 1981 | Philippines | 50 | Typhoon Irma; battered northern Luzon |
| 1981 | Eastern India & Bangladesh | 93 | Cyclone; 2 million homeless |
| 1981 | Philippines | 50 | Typhoon Lee; 200,000 homeless |
| 1982 | Tonga | 24 | Cyclone Isaac; worst on record |
| 1982 | Philippines | 38 | Typhoon Nelson; $7.4 million damage, 83,000 homeless |
| 1982 | Orissa State, India | 200 | Cyclone; great damage |
| 1982 | Nicaragua & Honduras | 226 | Tropical Storm Aleta; 80,000 homeless |
| 1982 | Honshu Island, Japan | 80 | Typhoon Bess |
| 1982 | Central Japan | 22 | Typhoon Judy; 60,000 houses flooded |
| 1982 | Gugarat State, India | 300 | Cyclone; slammed west coast |

| Year | Locality | Deaths | Remarks |
|---|---|---|---|
| 1983 | Texas coast | | Hurricane Alicia; more than $200 million damage, 11,410 families received $158 million in flood insurance claims |

# GLOSSARY

**Acid precipitation** Includes any rain, snow, or atmospheric moisture with water whose pH is acidic (less than 7). Some analysts would like to restrict the term to only those acid increments caused by manmade, airborne contaminants such as sulfur and nitrogen compounds.

**A-horizon** The top part of the soil profile. May contain humus or organic-rich matter.

**Anticline** A fold that is an upward bend or arch in rocks. The rock strata dip away and down from the crest, with oldest rocks in the center of the structure.

**Aquifer** A body of permeable earth material through which groundwater moves readily, thereby making possible its use as a water supply.

**Area (landfill) method** That technique used to bury solid waste inside the earth when the land is flat.

**Artificial recharge** Techniques used to increase groundwater storage; includes injection wells, infiltration pits, and manmade ponds.

**Ash (volcanic)** Rock particles less than 4 mm in diameter ejected by volcanic activity.

**Barrier island** A low sandy offshore island. Sometimes called "offshore bar" or "barrier beach." Barrier islands usually have multiple beach ridges and are roughly parallel with the mainland separated by a salt wetland or lagoon.

**Basalt** An igneous extrusive rock formed by volcanic surface action. It is fine-grained, dark, and consists of ferromagnesian minerals. The most common igneous rock.

**Beach** A body of wave-washed sediment on the coast between the ocean or lake and the mainland.

**Beach nourishment** A method used to artificially install sediment at the sea–land interface. The material is usually dredged from another site and pumped onto the beach through pipes.

**Berm** A nearly level part of the beach formed by the deposition of materials by wave action. It may resemble a small terrace, and in some instances, there may be a higher winter berm, and a lower summer berm.

**Benefit/cost ratio** A number that expresses all benefits that would result from a construction project in relation to those costs that would be required for its completion. A ratio greater than 1.0 is generally necessary before the project can be authorized.

**B-horizon** The soil zone of accumulation below the A-horizon. Contains leached materials derived from the soil above.

**Breaker** A wave form that topples over. There are three types of breakers – depending on their geometry – *spilling*, *plunging*, and *surging*.

**Breakwater** A manmade structure installed offshore to act as a barrier to wave action and thereby protect the shore and property.

**Bulkhead** A manmade structure installed at the sea–land interface. Bulkheads may consist of any material such as wood, metal, or concrete and usually positioned vertically.

**Buttress** Earth material emplaced at the toe of a slope to provide mass so as to prevent gravity movement of the hillside.

**Carbonates** Those earth materials that are rich in the chemical radical $CO_3$. Limestone and dolomite are the most common carbonates. Such materials weather rapidly under the action of water and can form caves and sinkholes.

**Channelization** The deliberate modification of a stream by straightening, deepening, narrowing or widening, and changing the streambed.

**C-horizon** The lowest soil zone that contains partially decomposed or disintegrated parent material within the soil profile.

**Clearcutting** A technique of timbering whereby all trees are cut and removed during the same time span. The environmental impact of such action is highly controversial.

**Coal** A sedimentary rock composed chiefly of partly decomposed plant matter that contains less than 40 percent inorganic matter. *Anthracite* and *bituminous* are types of coal.

**Collapse** A gravity phenomenon that produces a sudden, mostly vertical, displacement of earth material without a free face.

**Common law** A type of legal system of understandings that have evolved through time without statutory assignment. English Common Law is typical of that body of governmental order that has an historical base.

**Commons** Original usage dates to "the commons," a place where English citizens had equal use of the land by royal decree.

**Conservation** The ethic for management of earth, soil, crops, and timber resources that will provide for long-term utilization. The programs used to accomplish conservation may involve measures to reduce soil, water, or other resource loss, such as soil conservation measures of terracing.

**Contamination** The introduction of harmful impurities into the environment, usually from man-derived actions. Some would restrict the term entirely to societal activity, whereas *pollution* would embrace both man-derived materials, such as hazardous chemicals, and natural processes, such as volcanic gases.

**Contour farming** The techniques of planting crops in rows that conform to similar hillside elevations.

**Contour (strip) mine** The mining method that extracts earth materials along cuts that are made parallel to similar hillside elevations.

**Creep** The imperceptible downslope movement of surface materials by gravity processes.

**Crust** The outer layer of the planet Earth. Generally considered to be composed of rocks that have brittle characteristics. Also called the "lithosphere."

**Cultural nullification** Terms used to describe potentially valuable rock and mineral resources that cannot be mined or extracted because of societal prohibitions.

**Cyclone** In this book, the term is used to describe large tropical disturbances in the Indian Ocean. It is the equivalent of hurricanes in the Atlantic and typhoons in the Pacific.

**Deforestation** The cutting down and removal of an excessive number of trees. It generally implies the absence of conservational measures and the lack of replanting provisions.

**Degradation** In this book, the term is used to describe human activity that causes the accelerated deterioration of earth materials and land surfaces.

**Desertification** The process of changing terrain from non-desert conditions to one where plant growth is severely inhibited or impossible. The term is usually associated with land use or atmospheric changes produced as a result of human intervention.

**Desiccation** The process of dehydration of earth materials.

**Dilatancy** The inelastic increase in volume of earth materials.

**Dimension stone** Bedrock that is quarried and cut into specific sizes. Marble and granite are typical rocks that are fashioned for buildings and other structures.

**Dioxin** An extremely toxic group of chemicals that are unintentionally created as byproducts in the manufacturing of such materials as herbicides, detergents, pesticides, disinfectants, and wood preservatives.

**Dredging** The artificial excavation of earth materials below water. The purpose of such action is for some reason other than use of the material for manufacturing enterprises.

**Drought** A prolonged dryness caused by the absence of rain.

**Drumlin** A hill consisting of unconsolidated sediment and caused by glacial processes. Most drumlins are asymmetrical and streamlined, which reflect the direction of ice movement.

**Drylands** Regions that have a moisture deficiency.

**Dust Bowl** Term used to describe conditions throughout the Great Plains of the United States during the 1930s. This period was exceptionally dry, and abnormally large amounts of soil were deflated from farmlands causing many dust storms and clouds.

**Earthquake** The sudden shaking of the earth in response to rock movement, as along a fault, or related to volcanic activity.

**Ecosystem** A grouping of animals and plants that live together and interact in a particular environment.

**EIS (Environmental Impact Statement)** A required report under the National Environmental Policy Act of 1970 that discusses effects of a construction project on the environment.

**El Nino effect** Water temperature changes in the southeastern Pacific that produce upsets in anchovy and tuna patterns, and which may lead to unusual weather conditions.

**Eminent domain** A governmental power that allows land to be confiscated for use in the public interest. Such condemnation and appropriation require just compensation to the original owner.

**Environment** The total combination of the surroundings and circumstances that denote a particular set of conditions for a locality.

**Environmental geology** The practical applications of geologic principles in the solution of environmental problems. It involves geologists acting on societal matters.

**Environmental management** The deliberate planning, policies, administration, and implementation of actions that relate to and govern the status of the biosphere.

**Environmental mediation** The settlement of environmental controversies without resort to legal action. It involves a structured dialogue between opposing sides with the aid of an impartial conciliatory convenor.

**Epicenter** The position on the earth's surface vertically above the earthquake focus.

**Erosion** A general term that describes the physical or chemical removal of material from a site. This is an active process of streams, glaciers, wind, gravity forces, groundwater, and oceans that displace rock and sediment and transport the material to a new location.

**Estuary** A drowned river mouth with tidal action that moves marine waters inland.

**Expansive soil** A general term for those earth materials that undergo swelling with addition of moisture. Other terms include expansive sediments, expansive clays, and others. Clay minerals such as montmorillonite produce maximum volumetric changes in the material.

**Fault** A fracture in earth materials along which the opposite sides have been relatively displaced.

**Feedback process** A resultant change caused by actions that have occurred in another part of the interrelated environmental system.

**Flood** High streamflow that overtops natural banks or manmade constraining structures wherein the water covers adjacent land.

**Flood fringe** Land located between the *floodway* and the maximum elevation subject to the 100-year regulatory flood.

**Floodplain** That part of the river valley capable of being flooded. The terrain is usually relatively flat and contains underlying alluvial deposits.

**Floodproofing** Structural changes made to a building for the purpose of making it stronger and more resistant to possible flood damages.

**Floodway** That part of the *floodplain* nearest the river that is subjected to most floods. For planning purposes, this is often designated as the 100-year flood zone.

**Focal depth** The vertical distance from an earthquake focus, or point of origin, to its *epicenter*.

**Fossil fuel** Those fuels that formed geologically by changes in animal and plant life. The resulting hydrocarbons generally date in past geologic time, and their development is measured in many thousands to many millions of years.

**Gas** The vapor phase of fossil fuel, usually considered to be *natural gas.* Differs from the liquid form, which upon refinement, is called *gasoline.*

**Geochemistry** The use and application of chemical principles to the study of geology.

**Geoengineering** The practical use of geology principles in the location and design of engineering structures. It is the function of the *geoengineer*, or *engineering geologist*, to assure that geological factors have been considered in the construction process.

**Geologic hazards** Those earth processes of the lithosphere and hydrosphere that are capable of producing disasters. These include earthquakes, floods, landslides, and volcanic activity. Such events as hurricanes and tsunamis are by-products from the interaction of several forces.

**Geology** The science of the earth. It includes the study of earth materials, processes, and time relations of events.

**Geomorphology** The study of the earth's surface features, and the *exogenic*, or surface processes, that sculptured them.

**Geophysics** The use and application of physics to the study of geology.

**Geopressured sand** Sands at considerable depths that exist in an environment of unusually high pressure. Typical locales for such material occur in deltas with very high rates of sedimentation.

**Geoscience** Those disciplines that are directly related to geology.

**Geothermal energy** The utilization of heat and gases produced by natural heat processes within the earth's crust.

**Greenhouse effect** The process of warming Earth's atmosphere by absorption of solar infrared radiation by gases such as carbon dioxide and water vapor. The term is unfortunate because manmade greenhouses stay warm mostly by heat retention from the conservation of warm air.

**Groin** A shore-protection structure built from the beach into the water to trap sediment in the littoral zone and prevent coastal erosion.

**Groundwater** Water in the *zone of saturation* below the water table.

**Groundwater mining** The excessive withdrawal of groundwater at rates that are substantially greater than natural recharge to the groundwater reservoir.

**Igneous rocks** Rocks that form as a result of molten activity. When the rock solidifies within the earth, it is called *intrusive* or *plutonic*. Molten material that hardens on land is *lava* or if ejected is *pyroclastic*. Such rocks are called *extrusive* or *volcanic.*

**Irrigation** The act of emplacing additional water on crops to augment natural rainfall. Irrigation water may have either groundwater or surface-water sources.

**Jetty** A manmade structure at coastal inlets designed to prohibit erosion and protect navigational routes. The structure extends from the land into the water.

**Karst** An assemblage of topographic depressions, such as sinkholes, caused by the solution of carbonate rocks. Named from the Karst region of Yugoslavia.

**Landslide** The rapid movement end of the gravity movement spectrum. Specifically, the motion and the landforms created as a result of rapid movement of rock and/or *regolith* on the surface. Such movement can occur by falling, sliding, or flowing.

**Land-use planning** The entire scope and strategy involved with systems analysis related to issues involving land utilization.

**Lateral accretion** Those alluvial deposits formed by the action of a migrating stream. Such materials commonly form along the bed of a stream on the inside bend of a meander.

**Lava** Molten silicate materials formed by the flowing action of a volcano on the Earth's surface.

**LDC** The initials are for those nations that can be classified as Less Developed Countries.

**Leachate** The contaminated waste fluids that result from water reacting with materials in a landfill. It may consist of many trace metals and hazardous and toxic substances.

**Leaching** The continual removal by water of soluble matter in the soil, regolith, or bedrock, generally by downward percolation.

**Limestone** A sedimentary rock composed mostly of calcite ($CaCO_3$).

**Magma** Molten silicate materials created by igneous activity within the earth.

**Mantle** The 2900 km thick zone between the crust or lithosphere and the Earth's core.

**Mercalli Scale (Modified)** A scale of earthquake intensity with divisions from I to XII that represent the degree of earth shaking and relative damage or changes at a site.

**Metal** A natural element with properties of losing electrons in chemical reactions, and generally is a good conductor of electricity and heat.

**Metamorphic rock** A rock formed in the solid state from a preexisting rock by changes in temperature, pressure, and chemistry. Such transformation occurs in deeper Earth zones than those of weathering. *Slate* and *marble* are representative metamorphic rocks.

**Mineral** A naturally occurring substance with characteristic physical properties and chemical composition with crystalline formation.

**Montmorillonite** A clay-mineral family of hydrous aluminum silicates. This mineral has the maximum shrink–swell potential and can produce costly structural damages.

**Moraine** A general term used to denote a glacial deposit, usually consisting of hilly terrain.

**Muck** A dark soil, commonly formed in wetlands, with high percentage of finely disaggregated organic matter.

**Natural gas** A mixture of gaseous hydrocarbons that forms in nature and commonly is associated with liquid petroleum deposits.

**Natural levee** Deposits formed at the side of a stream during flood stage that eventually build up a barrier separating the streambed from the adjacent flat floodplain alluvium.

**Nature** Inherent Earth processes and materials not related or created by mankind.

**Negligence (law)** The legal doctrine related to the action of carelessness by people who do not exercise standard and recognized codes of ethics.

**NEPA** Initials for the National Environmental Policy Act of 1970.

**Non-metallic minerals** Those minerals that do not possess metal properties.

**Non-renewable resources** Those geologic resources that are not being sufficiently replenished by natural processes for mankind's use in the foreseeable future.

**No-till agriculture** That type of farming that employs the minimum of cultivation methods during crop production. It commonly involves retention of stubble on the land.

**NRC** Initials for the Nuclear Regulatory Commission.

**Nuclear energy** That type of energy produced by an atomic power plant.

**Nuclear waste** Non-usable and contaminated materials and by-products that result from production of manufactured nuclear energy and associated industrial and medicinal uses.

**Nuisance (law)** The legal doctrine that relates to the unreasonable interference with a person whereby his property cannot be fully used or enjoyed.

**Oil** Liquid hydrocarbons of natural origin.

**Oil shale** A body of fine-grained sedimentary rocks rich in hydrocarbon derivatives. Such deposits are capable of yielding petroleum through rock cracking and distillation.

**Oligotrophication** The process that involves insufficient nutrient production in a lake.

**OPEC** Initials for the Organization of Petroleum Exporting Countries. This group of nations represents a cartel for the purpose of fixing oil prices sold on the world market.

**Open space** The planning concept that sets aside unused areas and disallows their development for commercial purpose. Usually situated in or contiguous to urbanized areas.

**Ore** Metallic minerals in sufficient concentration that can be profitably mined.

**ORV** Initials for off-road vehicles such as motorbikes.

**Overbank deposits** Alluvial sediment deposited by a stream during flood stage outside the pre-flood streambed.

**Overgrazing** The process of land deterioration caused by excessively large numbers of grazing animals that exceed the capacity of vegetation to regenerate.

**Permafrost** Permanently frozen ground, mostly found in the Arctic.

**Permeability.** The ability of earth materials to transmit fluids.

**Petroleum** Gaseous, liquid, or solid substances occurring naturally and consisting of chemical compounds of carbon and hydrogen.

**Phreatophytes** That type of vegetation capable of extending roots into the groundwater zone and using the water for growth.

**Piping** The formation of small passages and conduits by surface water percolating down through unconsolidated sediments above the water table.

**Placer deposit** An ore deposit formed by surface processes that have winnowed out less dense or softer materials and left behind a lag of heavy or resistant minerals, such as gold, diamonds, and tin-bearing minerals. Streams produce most placers.

**Plates** Those major segmented parts of the Earth's crust that move independently of other crustal masses. The two broad categories consist of continental plates, such as the North American Plate, and oceanic plates. The concepts of sea-floor spreading and global tectonics are based on large-scale motion of these lithospheric plates.

**Pollution** The introduction of foreign matter into a new environment. Common usage generally equates pollution to manmade products that intrude into the lands and waters thereby impairing the environment. When used in this manner, the term pollution becomes synonymous with *contamination.*

**Porphyry copper** Those copper or copper ore deposits that occur as disseminated minerals in a large body of igneous rocks composed of minerals of large crystalline size. Such deposits are the most important source of copper.

**Precursor** A warning signal that a hazardous process is about to occur, such as those phenomena associated with the impending onslaught of an earthquake or volcanic activity.

**Preservation** The ethic that believes mankind should not disturb natural features but retain them in as pure form as possible, such as wilderness areas.

**Public trust** The doctrine that holds it is the duty of government to provide materials and amenities for society and to safeguard their preservation.

**Radwaste** Shorthand for radioactive waste. See nuclear waste.

**Reclamation** The act of restoring previously used lands to new productive purposes, or the change in natural systems to accommodate societal utilization, such as draining wetlands.

**Recycling** Reprocessing waste debris to make useful materials.

**Regolith** Unconsolidated rock fragments that cover bedrock.

**Renewable resources** Those resources that can be regenerated naturally, such as timber and water.

**Reserves** Potential resources that have been determined to be useful and can be profitably mined under existing economic and technological conditions.

**Resources** Earth materials useful to and needed by society. In the broadest sense, they include minerals, rocks, soil, organic matter, water, and air.

**Revetment** A sloping surface lined with stone or with resistant manmade materials designed to prevent water erosion.

**Richter Scale** A logarithmic scale of numbers from 1 to 10 that represents the magnitude or quantity of total energy released by an earthquake.

**Ring of fire** The arcuate zone peripheral to the Pacific Ocean where much of the Earth's volcanic and seismic activity occurs.

**Rock** In geology, an aggregate of minerals. In engineering, any earth material that must be blasted in order to excavate the site.

**Rock avalanche** The largest and fastest moving of all landslide types (except for rock fall).

**Riparian rights** The privileges of waterfront owners to use stream and lake water. Such use must be reasonable. The water must be returned to the stream, but the owner has rights to the water, which is supposed to be maintained in its normal quantity and quality.

**Salinization** The process of increasing the salt content of earth materials and water. This is often produced by irrigation in dryland environments and freshwater withdrawal in coastal aquifers.

**Sand** Sediment particles with dimensions larger than 1/16 mm and smaller than 2 mm.

**Sand boil** Small-scale sand eruptions created by excessive hydraulic pressure in sensitive sediments along river levees.

**Sand bypassing** Dredging of sand in coastal areas and pumping the material to another site so that littoral currents must rework the sand again.

**Sand dune** A geomorphic feature consisting of a hill or ridge of windblown sand.

**Sandstone** A fragmental sedimentary rock consisting primarily of sand-size particles.

**Sanitary landfill** A solid waste-disposal method in which rubbish, garbage, and waste products of all descriptions are dumped on the ground and covered daily with a layer of fine-grained earth material that is compacted in place.

**Satellite imagery** The production of visual records by the use of Earth-orbiting, manmade satellites.

**Seawall** A coastal embankment emplaced along the shore to prevent water erosion.

**Sedimentary rock** A rock formed by lithification of a sediment at or near the Earth's surface.

**Seismic** Pertaining to shock waves of earthquakes.

**Seismic gap theory** The concept that future earthquake activity can be determined from a study of those sites along an earthquake-prone zone that has not experienced recent major earthquakes.

**Seismogram** The record made by an earthquake-recording instrument.

**Seismograph** An instrument that records seismic waves.

**Septic tank** A buried container designed to hold fluid and solid-waste products from a single residence or small facility. The holding tank permits the breakdown of anerobic bacteria, and fluids drain from the tank into adjacent structures or leaching field.

**Settlement** The sinking of manmade structures into the underlying earth materials.

**Siltation** The process of accumulation of silt and closely associated sediments.

**Sinkhole** A solution cavity open to the sky that is generally formed by subsurface erosion in carbonate rocks and the caving in of overlying rock. The term has been expanded to include surface depressions caused when there is subsurface failure in shallow mines.

**Slash and burn** A method of clearing land used by shifting cultivators to obtain new land for farming. Trees and brush are burned and crops are planted. Unless precautions are used, such lands generally become infertile in a few years.

**Slump** The downward and outward movement of a coherent body of earth materials. This is one type of *landslide.*

**Soil** In pedology, that part of the regolith that can support rooted plants. In engineering, any earth material that can be excavated without blasting.

**Solid waste** The full range of non-fluid waste produced by society.

**Strip farming** A land cultivation method of plowing along the terrain contours and planting alternate sections of land with different crops.

**Strip mining** The method of *surface mining* that progressively removes new rock layers row after row.

**Sturzstrom** A Swiss term applied to rock avalanches composed of finely disaggregated debris.

**Subduction zone** The location of a lithospheric plate where it descends under another plate during shifting of the crust.

**Subsidence** Differential sinking of the earth's surface in respect to the surrounding terrain.

**Surface mining** The extraction of mineral and rock resources by using methods that proceed from the top of the ground. *Strip mining* is a form of surface mining.

**Syncline** A downward bend in rock strata. This type of fold has the youngest rocks in the center.

**Synfuels** Products produced as result of phase changes of organic matter by manufacturing techniques, for example, the gasification of coal, and the production of fluid hydrocarbons from tar sands.

**Tailings** The waste residue from mining operations.

**TAPS** Initials of the Trans-Alaska Pipeline System.

**Tar sand** A sand or sandstone whose pores contain hydrocarbons in the solid state.

**Tide** Changes in sea-level position caused by the gravitational attraction of the moon and sun.

**Toxic** Substances that are especially injurious or lethal to humans and animals.

**Tragedy of the Commons** The theory that a locality becomes severely deteriorated when everyone has equal access to it and overloads the system.

**Tsunami** Ocean waves generated by displacement of the sea floor. They can be caused by earthquakes, volcanic activity, or massive submarine slides. The waves reach abnormally large size when they encounter land. Incorrectly, also called "tidal waves."

**Typhoon** The name applied to tropical cyclones in the Pacific Ocean region.

**Underground space** The use of underground openings by society. Such openings could originally have been mines, or new openings excavated deliberately for new utilization.

**Urban sprawl** The housing and commercial condition that results when land is developed in areas adjacent to cities without planning and management of the land-water ecosystem.

**Utilitarian** The ethic that the land and water should have maximum use for the benefit of society. Generally interpreted to mean only short-range goals are considered without long-range planning for future generations.

**Water augmentation** The act of increasing water supply. A variety of methods can be employed such as *water spreading, recharge pits,* and *injection well recharge.*

**Waterlogging** That condition of saturation of surface materials associated with a rise in the water table caused by excessive irrigation.

**Watershed** A drainage basin. A region wherein all precipitation occurs within the same area and runoff slopes toward a common elevation.

**Water table** The top surface of the groundwater zone. It separates saturated and non-saturated earth materials.

**Weathering** The disintegration and decomposition of rock materials during surface exposure to air, moisture, organisms, and chemicals. It does not involve transportation of material from the site.

**Wetland** A general term that applies to all water-saturated ground. The term replaces such previous words as bog and swamp. There are both freshwater and saltwater wetlands.

**Wilderness** A region where mankind has not changed natural systems.

**Zero demand distance** The distance beyond which the cost of transportation for minerals and rocks is greater than the value of the product if sold.

**Zone of aeration** The area above the water table where openings between earth materials are not completely occupied by water. Also called the *vadose zone.*

**Zone of saturation** The area below the water table where all openings between earth materials are completely filled with water. Also called the *groundwater zone* or the *phreatic zone.*

**Zoning ordinance** A statutory mandate that determines the specific usage for a land parcel.

# REFERENCES AND SOURCE MATERIALS


Abelson, P. H., and Hammond, A. L., 1976, Materials: Renewable and nonrenewable resources: American Association for the Advancement of Science: Washington, D.C., 196 p.

Acton v. Blundell, 12 M. & W. 324 (1843).

Aghassy, J., 1973, Man-induced badlands, *in* Coates, D. R., ed., Environmental geomorphology and landscape conservation: Volume 2: Non-Urban Regions: Stroudsburg, Pa., Dowden, Hutchinson & Ross, p. 124-136.

Alexander, D. E., 1982, Creeping disaster in Italy: Geotimes, v. 27, n. 11, p. 17-20.

Alfors, J. T., Burnett, J. L., and Gay, T. E., Jr., 1973, Urban geology: master plan for California: California Division of Mines and Geology Bulletin 198, 112 p.

Allen, A. S., 1978, Basic questions concerning coal mine subsidence in the United States: Association of Engineering Geologists, v. 15, n. 2, p. 147-162.

American Institute of Professional Geologists, 1981, Metals, minerals, mining: Golden, Colo., American Institute of Professional Geologists, 36 p.

Anderson, D. L., 1971, The San Andreas fault: Scientific American, v. 225, n. 5, p. 52-67.

Armstrong v. Granite Company, 147 N.Y., 495, 1854.

Arthur D. Little, Inc., 1972, Channel modifications: an environmental, economic, and financial assessment: Report to the Council on Environmental Quality: Environmental Protection Agency, var. pages.

Aspinall, W. P., Sigurdsson, H., and Shepherd, J. B., 1973, Eruption of Soufriére Volcano on St. Vincent Island, 1971-1972: Science, v. 181, p. 117-124.

Avon v. Neptune (N.J., 1976).

Bailey, R. G., 1971, Landslide hazards related to land use planning in Teton National Forest, northwest Wyoming: U.S. Department of Agriculture Forest Service, 131 p.

Barnes, J., 1972, Geothermal power: Scientific American, v. 226, n. 1, p. 70-77.

Barr, T. N., 1981, The world food situation and global grain prospects: Science, v. 214, p. 1087-1095.

Bell, B., 1971, The Dark Ages in ancient history: American Journal of Archeology, v. 77, p. 1-26.

Bell, F. G., 1983, Engineering properties of soils and rocks (2d ed.): London, Butterworths, 149 p.

Bell, F. G., 1983, Fundamentals of engineering geology: London, Butterworths, 656 p.

Belt, C. B., Jr., 1975, The 1973 flood and man's constriction of the Mississippi River: Science, v. 189, p. 681-684.

Bennett, H. H. and Chapline, W. R., 1928, Soil erosion a national menace: U.S. Department of Agriculture Circular 33, 83 p.

Bezuidenhout, C. A., and Enslin, J. F., 1970, Surface subsidences in the dolomitic areas of the Far West Rand, Transvaal, Republic of South Africa, *in* Land subsidence, v. 2, International Association of Scientific Hydrology Publication 89, p. 482-495.

Blanchard, N. C., ed., 1908, Proceedings of a conference of governors in the White House: U.S. House of Representatives Document 1425, 60th Congr. 2nd Sess.

Blumb, R. R., 1980, Trace chemistries of fire: A source of chlorinated dioxins: Science, v. 210, p. 385-390.

Bolt, B. A., 1978, Earthquakes, a primer: San Francisco, W. H. Freeman & Co., 241 p.

Bolt, B. A., et al., 1977, Geological hazards (2d ed.): New York, Springer-Verlag, 330 p.

Boskovich v. King County, 188 Wash. 63 (1963).

Boyle, R. H., and Boyle, R. A., 1983, Acid rain: Amicus Journal, v. 4, n. 3, p. 22-37.

Brabb, E. E., ed., 1979, Progress on seismic zonation in the San Francisco Bay Region: U.S. Geological Survey Circular 807, 91 p.

Brabb, E. E., and Fleming, R. W., 1983, Team effort met landslide emergency: Geotimes, v. 28, n. 12, p. 18-19.

Brady v. Smith, 181 N.Y., 178, 1854.

Brink, R. A., Densmore, J. W., and Hill, G. A., 1977, Soil deterioration and the growing world demand for food: Science, v. 197, p. 625-630.

Brobst, D. A., and Pratt, W. P., eds., 1973, United States mineral resources: U.S. Geological Survey Professional Paper 820, 722 p.

Brown, H. S., 1954, Challenge of man's future: New York, Viking Press, 290 p.

Brown, J., and Berg, R. L., eds., 1980, Environmental engineering and ecological baseline investigations along the Yukon River-Prudhoe Bay haul road: Cold Regions Research Engineering Laboratory Report 80-19, 187 p.

Brown, J., and Grave, N. A., 1979, Physical and thermal disturbance and protection of permafrost: Cold Regions Research Engineering Laboratory Special Report 79-5, 42 p.

Brown, L. R., 1978, The worldwide loss of cropland: Worldwatch Institute, Worldwatch Paper 24, 48 p.

Brown, L. R., 1981, World population growth, soil erosion, and food security: Science, v. 214, p. 995-1002.

Brown v. McPhersons Inc., 86 Wash. 2d 293, 545 P2d 13 (1975).

Bryson, R. A., and Baerreis, D. A., 1967, Possibilities of major climatic modifications and their implications: Northwest India, a case for study: American Meteorological Society Bulletin, v. 48, n. 3, p. 136-142.

Bullard, F. M., 1962, Volcanoes in history, theory, in eruption: Austin, University of Texas Press, 441 p.

Butcher, R. D., 1982, Utah's national parks: State of siege: National Parks, v. 56, n. 9-10, p. 16-27.

Butler, E. W., and Pick, J. B., 1982, Geothermal energy development: Problems and prospects in the Imperial Valley of California: New York, Plenum Press, 361 p.

California Legislature, 1972, The San Fernando earthquake of February 9, 1971, and public policy: Special Subcommittee of the Joint Committee on Seismic Safety, California Legislature, 127 p.

Campbell, A. N., et al., 1982, Recognition of a hidden mineral deposit by an artificial intelligence program: Science, v. 217, p. 927-928.

Canby, T. Y., 1984, El Nino's ill wind: National Geographic, v. 165, n. 2, p. 144-183.

Carolina Environmental Study Group v. United States, 510 Fed. Rep., 2nd ser. 796 (D.C. Cir., 1975).

Carson R., 1962, Silent Spring: Boston, Houghton Mifflin, 368 p.

Carton et al. v. State of New Jersey, A-638-73 (1978).

Caufield, C., 1983, Dam the Amazon, full steam ahead: Natural History, v. 92, n. 7, p. 60-67.

Chandler, W. U., 1983, Materials recycling: The virtue of necessity: Worldwatch Institute, Worldwatch Paper 56, 52 p.

Citizens for Safe Power v. Nuclear Regulatory Commission (NRC), 524 Fed. Rep., 2nd ser. 1291, 1300 (D.C. Cir., 1975).

City of Milwaukee v. State of Illinois, 49 U.S.L.W. 4445 (April 28, 1981), vacating 599 F. 2d 151 (7th Cir. 1979).

Clark, B. R., 1981, Stress anomaly accompanying the 1979 Lytle Creek earthquake: Implications for earthquake prediction: Science, v. 211, p. 51-43.

Clawson, M., 1983, Reassessing public lands policy: Environment, v. 25, n. 8, p. 6, 8-17.

Cleaves, A. B., 1961, Landslide investigations: A field handbook for use in highway location and design: Washington, D.C., U.S. Department of Commerce, Bureau of Public Roads, 67 p.

Coates, D. R., 1971, Legal and environmental case studies in applied geomorphology, *in* Coates, D. R., ed., Environmental geomorphology: Binghamton, State University of New York, Publications in Geomorphology, p. 223-242.

Coates, D. R., ed., 1972, Environmental geomorphology and landscape conservation: Volume 1: Prior to 1900 (Benchmark Papers in Geology): Stroudsburg, Pa., Dowden, Hutchinson & Ross, Inc., 485 p.

Coates, D. R., ed., 1973, Environmental geomorphology and landscape conservation: Volume 3: Non-urban regions: Stroudsburg, Pa., Dowden, Hutchinson & Ross, Inc., 483 p.

Coates, D. R., ed., 1976, Geomorphology and engineering: Stroudsburg, Pa., Dowden, Hutchinson & Ross, Inc., 360 p.

Coates, D. R., 1976, Geomorphology in legal affairs of the Binghamton, New York, metropolitan areas, *in* Coates, D. R., ed., Urban geomorphology: Geological Society of America Special Paper 174, p. 111-148.

Coates, D. R., ed., 1977, Landslides: Geological Society of America Reviews in Engineering Geology, v. 3, 278 p.

Coates, D. R., 1980, Evidence for lawyers: Geomorphology in practice: Geographical Magazine, v. 53, n. 1, p. 48-49.

Coates, D. R., 1981, Effects of man and law in the coastal corridor, *in* Leonard, J. E., and Maurmeyer, E., eds., Coastal and nearshore processes of the western Atlantic: Northeastern Geology, v. 3, n. 3-4, p. 162-170.

Coates, D. R., 1981, Environmental geology: New York, John Wiley & Sons, 701 p.

Colburn v. Richards, 13 Mass 419 (1816).

Columella, L.I.M., 1785, Columella's Husbandry: Anonymous translation, London, 600 p.

Conservation Foundation, 1983, A global chain links the economy and environment: Washington, D.C., Conservation Foundation Letter, June, 8 p.
Conservation Foundation, 1983, Egypt: A case of ecological vulnerability: Washington, D.C., Conservation Foundation Letter, August, 8 p.
Cooke, R. U., and Doornkamp, J. C., 1974, Geomorphology in environmental management: Oxford, England, Clarendon Press, 413 p.
Cooke v. Hull (Mass., 1820).
Council on Environmental Quality, 1981, The global 2000 report to the president: Washington, D.C., U.S. Government Printing Office, 3 volumes.
Craig, R. G., and Craft, J. L., eds., 1982, Applied geomorphology: London, George Allen & Unwin, 253 p.
Crandell, D. R., Mullineaux, D. R., and Rubin, M., 1975, Mount St. Helens volcano: Recent and future behavior: Science, v. 187, p. 438-441.
Crippen, J. R., and Bue, C. D., 1977, Maximum floodflows in the conterminous United States: U.S. Geological Survey Water Supply Paper 1887, 52 p.
Cross, R. F., 1981, Water, water, everywhere?: Conservationist, v. 36, n. 2, p. 2-7.
Crossette, G., ed., 1970, Selected prose of John Wesley Powell: Boston, David R. Godine, 122 p.
Crowe, B. L., 1968, The tragedy of the commons revisited: Science, v. 111, p. 1103-1107.
Culler, R. C., 1970, Water consèrvation by removal of phreatophytes: American Geophysical Union Translation, v. 51, n. 10, p. 684-689.
Daniels, R. B., 1960, Entrenchment of the Willow drainage ditch, Harrison County, Iowa: American Journal of Science, v. 225, p. 161-176.
Dasmann, R. F., 1959, Environmental conservation: New York, John Wiley & Sons, 375 p.
Deasy, G. F., 1960, Terrain damages resulting from bituminous stripping in Pennsylvania: Pennsylvania Academy of Science, v. 134, p. 124-130.
Detwyler, T. R., ed., 1971, Man's impact on environment: New York, McGraw-Hill, 731 p.
Dolton, G. L., et al., 1981, Estimates of undiscovered recoverable conventional resources of oil and gas in the United States: U.S. Geological Survey Circular 860, 87 p.
Doornkamp, J. C., 1982, The physical basis for planning in the Third World: Third World Planning Review, v. 4, n. 1, p. 13-30.
Dunrud, C. R., and Osterwald, F. W., 1978, Coal mine subsidence near Sheridan, Wyoming: Association of Engineering Geologists Bulletin, v. 15, n. 2., p. 175-190.
Eckel, E. B., ed., 1958, Landslides and engineering practice: National Research Council, Highway Research Board Special Report 29, 323 p.
Eckholm, E. P., 1975, Salting the earth: Environment, v. 17, n. 7, p. 9-15.
Eckholm, E. P., 1976, Losing ground: New York, W. W. Norton & Co., Inc., 223 p.
Eckholm, E. P., 1982, Human wants and misused lands: Natural History, v. 91, n. 6, p. 33-48.
Eckholm, E. P., and Brown, L. R., 1977, Spreading deserts: The hand of man: Worldwatch Institute, Worldwatch Paper 13, 40 p.
Erhlich, P. R., 1968, The population bomb: New York, Ballantine Books, 223 p.
Emerson, J. W., 1971, Channelization: A case study: Science, v. 1733, p. 325-326.

Environmental Defense Fund v. EPA, 636 Fed. Rep., 2nd ser. 1267, 1283 (D.C. Cir., 1980).

EPA v. National Crushed Stone Association, 49 LW 4008 (December 2, 1980).

Epstein, S. S., Brown, L. O., and Pope, C., 1982, Hazardous waste in America: San Francisco, Sierra Club, 593 p.

Everett, A. G., 1979, Secondary permeability as a possible factor in the origin of debris avalanches associated with heavy rainfall: Journal of Hydrology, v. 43, p. 347-354.

Exxon Corporation, 1981, World energy outlook: Exxon Corporation, 40 p.

Faulkner, E. H., 1943, Plowman's folly: New York, Editions for the Armed Services, 192 p.

Ferber, A. E., 1969 Windbreaks for conservation: U.S. Department of Agriculture, Agriculture Information Bulletin 339, 30 p.

Finkl, C. W., 1981, Beach nourishment, a practical method of erosion control: Geo-Marine Letters, v. 1, n. 2, p. 155-161.

Flawn, P. T., 1970, Environmental geology: New York, Harper & Row, 313 p.

Fleming, R. W., and Taylor, F. A., 1980, Estimating the costs of landslide damage in the United States: U.S. Geological Survey Circular 832, 21 p.

Font, R. G., 1977, Engineering geology of the slope instability of two overconsolidated north-central Texas shales, *in* Coates, D. R., ed., Landslides: Geological Society of America Reviews in Engineering Geology, v. 3, p. 205-212.

Foose, R. M., 1967, Sinkhole formation by groundwater withdrawal: Far West Rand, South Africa: Science, v. 157, p. 1045-1048.

Foose, R. M., and Hess, P. W., 1976, Scientific and engineering parameters in planning and development of a landfill site, *in* Coates, D. R., ed., Geomorphology and Engineering: Stroudsburg, Pa., Dowden, Hutchinson & Ross, p. 289-312.

Fradkin, P. L., 1981, A river no more; The Colorado River and the West: New York, Alfred A. Knopf, 384 p.

Francis, P., 1976, Volcanoes: Middlesex, England, Penguin Books, 368 p.

Frazier, J. W., ed., 1982, Applied geography; Englewood Cliffs, New Jersey, Prentice-Hall, Inc., 333 p.

Freedman, J. L., ed., 1977, Lots of danger: Property buyer's guide to land hazards of southwestern Pennsylvania: Pittsburgh Geological Society, Inc., 85 p.

Funk v. Holdeman, 53 Pa. St., 229 (1866).

Gates, G. O., ed., 1972, The San Fernando Earthquake of February 9, 1971, and public policy: Joint Committee on Seismic Safety, California Legislature, 127 p.

General Accounting Office, 1977, To protect tomorrow's food supply, soil conservation needs priority attention: Report to Congress CED-77-30, Washington, D.C., Comptroller General of the United States, 59 p.

Gentry, A. H., and Lopez-Parodi, J., 1980, Deforestation and increased flooding of the Upper Amazon: Science, v. 210, p. 1354-1356.

Gibbs, 1982, Love canal: My story: Albany, State University New York Press, 174 p.

Gifford, C. A., Fisher, J. A., and Walton, T. L., 1977, Floating tire breakwaters: Florida Sea Grant Publication SUSF-SG-77-002, 13 p.

Gilbert, G. K., 1917, Hydraulic-mining debris in the Sierra Nevada: U.S. Geological Survey Professional Paper 105, 154 p.

Glantz, M. H., ed., 1977, Desertification: Boulder, Colo., Westview Press, 346 p.

Glenn, L. C., 1911, Denudation and erosion in the Southern Appalachian Region and Monongahela Basin: U.S. Geological Survey Professional Paper 72, 137 p.

Godfrey, P. J., and Godfrey, M. M., 1973, Comparison of ecological and geomorphic interactions between altered and unaltered barrier island systems in North Carolina, *in* Coates, D. R., ed., Coastal Geomorphology: Binghamton, State University of New York, Publications in Geomorphology, p. 263-278.

Graham, F., Jr., 1978, The Adirondack Park, a political history: New York, Alfred A. Knopf, 314 p.

Gray, D. H., 1969, Effects of forest clearcutting on the stability of natural slopes: Association of Engineering Geologists Bulletin, v. 7 (1 and 2), p. 45-60.

Gregory, K. J., and Walling, D. E., eds., 1979, Man and environmental process: Kent, England, Wm. Dawson Westview Press, 276 p.

Gupta, H. K., 1976, Dams and earthquakes: Amsterdam, Elsevier, 229 p.

Haicheng Earthquake Study Delegation, 1977, Prediction of the Haicheng earthquake: EOS, American Geophysical Union Transactions, v. 58, n. 5, p. 236-272.

Hansen, J. E., Wang, W., and Lacis, A. A., 1978, Mount Agung eruption provides test of global climatic perturbation: Science, v. 199, p. 1065-1067.

Hardin, G., 1968, The tragedy of the Commons: Science, v. 162, p. 1243-1248.

Harte, J., and Socolow, R. H., 1971, The Everglades: Wilderness versus rampant land development in south Florida, *in* Patient Earth, Harte, J., and Socolow, R. H., eds., New York, Holt, Rinehart & Winston, p. 181-202.

Hartwell v. Camman, N.J. (1854)

Hasan, S. E., and West, T. R., 1982, Development of an environmental geology data base for land use planning: Association of Engineering Geologists Bulletin, v. 19, n. 2, p. 117-132.

Hastings, J. R., and Turner, R. M., 1972, The changing mile: Tucson, Arizona University Press, 317 p.

Hayes, D., 1976, Energy: The case for conservation: Worldwatch Institute, Worldwatch Paper 4, 77 p.

Hays, S. P., 1969, Conservation and the gospel of efficiency: New York, Athenium, 297 p.

Helfrich, H. W., ed., 1970, The environmental crises: New Haven, Yale University Press, 187 p.

Hey, R. D., 1974, Prediction and effects of flooding in alluvial systems, *in* Funnell, B. M., ed., Prediction of geological hazards: Geo. Soc. Misc. Paper 3, p. 42-56.

Hoffman, J. S., and Barth, M. C., 1983, Carbon dioxide: Are we ignoring a vital environmental issue?: Amicus Journal, summer, p. 24-28.

Holzer, T. L., 1981, Preconsolidation stress of aquifer systems in areas of induced land subsidence; Water Resources Research, v. 17, n. 3, p. 693-704.

Holzer, T. L., Davis, S. N., and Lofgren, B. E., 1979, Faulting caused by groundwater extraction in south-central Arizona: Journal Geophysical Research, v. 84, p. 603-612.

Hoyt, W. G., and Langbein, W. B., 1955, Floods: Princeton, Princeton University Press, 469 p.

Hsu, K. J., 1975, Catastrophic debris streams (Sturzstroms) generated by rockfalls: Geological Society of America Bulletin, v. 86, p. 129-140.

Hughes, G. M., Landon, R. A., and Farvolden, R. N., 1971, Hydrogeology of solid waste disposal sites in northeastern Illinois: U.S. Environmental Protection Agency, 154 p.

Hughes, J. D., 1973, Ecology, in ancient civilization: Albuquerque, N. Mex., University New Mexico Press, 181 p.

Hultberg, H., and Johansson, S., 1981, Acid groundwater: Nordic Hydrology, v. 12, p. 51-64.

Hunt, C. A., and Garrels, R. W., 1972, Water, the web of life: New York, W. W. Norton & Sons, 308 p.

Hurley v. Kinkaid, 285 U.S. 95 (1932).

Hutchinson, C. S., 1983, Economic deposits and their tectonic setting: New York, John Wiley & Sons, Inc., 365 p.

Illinois v. Costle, 9 ELR 20243, D.D.C. (1979).

Iltis, H. H., 1983, Tropical forests: What will be their fate?: Environnment, v. 25, n. 10, p. 55-60.

Jacks, G. V., and Whyte, R. O., 1939, Vanishing lands: New York, Doubleday, 332 p.

Jacobsen, T., and Adams, R. M., 1958, Salt and silt in ancient Mesopotamian agriculture: Science, v. 128, p. 1251-1258.

Johnson, T. M., and Cartwright, K., 1980, Monitoring leachate migration in the unsaturated zone in the vicinity of sanitary landfills: Illinois State Geological Survey Circular 514, 82 p.

Jones, D. E., and Holtz, W. G., 1973, Expansive soils: The hidden disaster: American Society Civil Engineers, v. 43, n. 8, p. 49-51.

Jones, F. O., Embody, D. R., and Peterson, W. L., 1961, Landslides along the Columbia River Valley, northeastern Washington: U.S. Geological Survey Professional Paper, 367, 98 p.

Kadomura, H., 1983, Some aspects of large-scale land transformation due to urbanization and agricultural development in recent Japan: Advances in Space Research, v. 2, n. 8, p. 169-168.

Kahn, H., 1982, The coming boom: New York, Simon & Schuster, 237 p.

Kaye, C. A., 1976, Beacon Hill end moraines, Boston: New explanation of an important urban feature, *in* Coates, D. R., ed., Urban geomorphology: Geological Society of America Special Paper 174, p. 7-20.

Kazis, R., and Grossman, R. L., 1982, Environmental protection: Job-taker or job-maker: Environment, v. 24, n. 9, p. 13-20, 43.

Kedar, Y., 1957, Water and soil from the desert: Some ancient agricultural achievements in the central Negev: The Geographical Journal, v. 123, p. 179-187.

Keller, E. A., 1976, Channelization: Environmental, geomorphic, and engineering aspects, *in* Coates, D. R., ed., Geomorphology and engineering: Stroudsburg, Pa., Dowden, Hutchinson & Ross, p. 115-140.

Kerr, R. A., 1983, New signs of Long Valley magma intrusion: Science, v. 220, p. 1138-1139.

Kiersch, G. A., and Cleaves, A. B., eds., 1969, Legal aspects of geology in engineering practice: Geological Society of America Engineering Geology Case Histories No. 7, 112 p.

Kimball, L., 1983, The law of the sea: On the shoals: Environment, v. 25, n. 9, p. 14-20.

King, C. A. M., 1972, Beaches and coasts (2nd ed.): London, Arnold, 570 p.

Kleppe v. Sierra Club, 427 U.S. 390 (1976).

Knox, G. B., Parker, O. E., and Milfred, C. J., 1972, Development of new techniques for delineation of flood plain hazard zones: Water Resources Center, University of Wisconsin, 15 p.

Kolb, C. R., 1976, Geologic control of sand boils along Mississippi River levees, *in* Coates, D. R., ed., Geomorphology and engineering: Stroudsburg, Pa., Dowden, Hutchinson & Ross, p. 99-113.

Kreig, R. A., and Reger, R. D., 1976, Preconstruction terrain evaluation for the Trans-Alaska Pipeline Project, *in* Coates, D. R., ed., Geomorphology and engineering: Stroudsburg, Pa., Dowden, Hutchinson & Ross, p. 55-76.

Kriebel, D., 1981, The dioxins: Toxic and still troublesome; Environment, v. 23, n. 1, p. 7-13.

Krug, E. C., and Frink, C. R., 1983, Acid rain on acid soil: A new perspective; Science, v. 221, p. 520-525.

Lamoreaux, P. E., 1979, Remote-sensing techniques and the detection of karst: Association of Engineering Geologists Bulletin, v. 16, n. 3, p. 383-392.

Landau, N. J., and Rheingold, P. D., 1971, The environmental law handbook: New York, Friends of the Earth/Ballantine, 496 p.

Larson, W. E., Pierce, F. J., and Dowdy, R. H., 1983, The threat of soil erosion to long-term crop production: Science, v. 219, p. 458-465.

Lave, L. B., and Seskin, E. P., 1970, Air pollution and human health: Science, v. 169, p. 723-733.

Laycock, G., 1970, The diligent destroyers: New York, Doubleday, 223 p.

Leatherman, S. P., 1979, Barrier dune systems: A reassessment: Sedimentary Geology, v. 24, p. 1-16.

Leatherman, S. P., and Godfrey, P. J., 1980, The impact of off-road vehicles on coastal ecosystems in Cape Code National Seashore: An overview: University of Massachusetts NPSCRC Report 34, 34 p.

Legget, R. F., 1974, Glacial landforms and civil engineering, *in* Coates, D. R., ed., Glacial geomorphology: Binghamton, State University of New York, Publications in Geomorphology, p. 351-374.

Legget, R. F., and Karrow, P. F., 1983, Handbook of geology in civil engineering: New York, McGraw-Hill, 1340 p.

Leighton, F. B., 1980, Bluebird Canyon landslide, Laguna Beach: A geomorphic threshold event, *in* Coates, D. R., and Viteck, J. D., eds., Thresholds in geomorphology: London, George Allen & Unwin, p. 387-400.

Lennett, D. J., 1980, Handling hazardous waste: Environment, v. 22, n. 8., p. 6-15.

Leopold, L. B., 1968, Hydrology for urban land planning: A guidebook on the hydrologic effects of urban land use: U.S. Geological Survey Circular 554, 18 p.

Leopold, L. B., and Maddock, T. Jr., 1954, The flood control controversy: New York, Ronald Press Co., 278 p.

Levine, A. G., 1982, Love canal: Science, politics and people: San Diego, Lexington Books, 288 p.

Liroff, R. A., and Davis, G. G., 1981, Protecting open space: Land use controls in the Adirondacks: New York, Ballinger Publishing Co., 302 p.

Lizarraga-Arciniega, J. R., and Komar, P. D., 1975, Shoreline changes due to jetty construction on the Oregon Coast: Oregon State University Sea Grant College Program Publication ORESU-T-75-004, 85 p.

Lofgren, B. E., 1969, Land subsidence due to the application of water: Geological Society of America Reviews in Engineering Geology, v. 2, p. 271-303.

Lowrance, W. W., 1983, The agenda for risk decision making: Environment, v. 25, n. 10, p. 4-8.

MacDonald, G. A., 1972, Volcanoes: Englewood Cliffs, N.J., Prentice-Hall, 510 p.

Mallio, W. J., and Peck, J. H., 1981, Practical aspects of geological prediction: Association of Engineering Geologists Bulletin, v. 18, n. 4, 369-374.

Marine, G., 1969, America the raped: New York, Discuss Books, 331 p.

Mark, R. K., and Stuart-Alexander, D. E., 1977, Disasters as a necessary part of benefit-cost analyses: Science, v. 197, p. 1160-1162.

Marsh, G. P., 1864, Man and nature: or Physical geography as modified by human action: New York, Charles Scribners & Sons, 560 p.

Mathewson, C. C., 1981, Engineering geology: Columbus, Charles E. Merrill Publishing Co., 410 p.

Mathewson, C. C., Castleberry, J. P., II, and Lytton, R. L., 1975, Analysis and modeling of the performance of home foundations on expansive soils in central Texas: Association of Engineering Geologists Bulletin, v. 12, n. 4, p. 275-302.

Mathewson, C. C., and Clary, J. H., 1977, Engineering geology of multiple landsliding along I-45 road cut near Centerville, Texas, *in* Coates, D. R., ed., Landslides: Geological Society of America Reviews in Engineering Geology, v. 3, p. 213-223.

Martinez, J. D., et al., 1981, Catastrophic drawdown shapes floor: Geotimes, v. 26, n. 3, p. 14-14.

Marx, W., 1967, The frail ocean: New York, Ballantine Books, 274 p.

McHarg, I. L., 1969, Design with nature: Garden City, N.Y., Doubleday & Co., 197 p.

McKelvey, J., 1982, Implications of Law of the Sea: Geotimes, v. 27, n. 10, p. 21-22.

McMullen, W. B., McPhail, J. F., and Murfitt, A. W., 1975, Design and construction of roads on muskeg in arctic and subarctic regions: Muskeg Research Conference, 16th, Montreal, 51 p.

Meadows, D. H., and Meadows, D. L., 1974, The limits to growth (2d ed.): Washington, D.C., A Potomac Associates Book, 207 p.

Menard, H. W., 1974, Geology, resources, and society: San Francisco, W. H. Freeman, 621 p.

Merritt v. Brinkerhoff, 17 Johns. 306 (1820).

Mildner, W. F., 1982, Erosion and sediment: Association of Engineering Geologists Bulletin, v. 19, n. 2, p. 161-166.

Miller, G. T., 1980, Energy and environment: The four energy crises (2d ed.): Belmont, Calif., Wadsworth Publishing Co., 162 p.

Miller, H. C., 1981, The barrier islands: A gamble with time and nature: Environment, v. 23, n. 9, p. 6-11, 36-42.

Mirsky, A., 1982, Influence of geologic factors on Ancient Mesopotamian civilization; Journal of Geological Education, v. 30, p. 294-299.

Mogi, K., 1979, Dilatancy of rocks under general triaxial stress states with special reference to earthquake precursors, *in* Kissling, C., ed., Earthquake precursors: Tokyo, Cen. Acad. Publ., p. 203-217.

Morton, R. A., 1976, Effects of Hurricane Eloise on beach and coastal structures, Florida Panhandle: Geology, v. 4, p. 277-280.

Mountainbrook Homeowners Association v. Adams, 9, ELR 20686, W.D.N.C. (1979).

Murphy, E. F., 1971, Man and his environment: Law: New York, Harper & Row Publishers, 168 p.

Murray, C. R., and Reeves, E. B., 1977, Estimated use of water in the United States in 1975: U.S. Geological Survey Circular 765, 39 p.

Nash, R., 1967, Wilderness and the American mind: New Haven, Yale University Press, 256 p.

National Academy of Science, 1975, Earthquake prediction and public policy: Washington, D.C., National Academy of Science, 142 p.

National Academy of Science, 1975, Mineral resources and the environment: Washington, D.C., Committee on Mineral Resources and the Environment, National Academy of Science, 348 p.

National Audubon Society v. Superior Court of Alpine County, S.F. 24368 (Cal. February 17, 1983).

National Research Council, 1982, Carbon dioxide and climate; Washington, D.C., 72 p.

National Research Council, 1983, Acid deposition: Atmospheric process in eastern North America; Washington, D.C., Committee on Transport and Chemical Transformation in acid precipitation, Environmental Studies Board, 392 p.

National Science Foundation, 1980, A report on flood hazard mitigation: Washington, D.C., National Science Foundation, 253 p.

Natural Resources Defense Council v. Callaway 392 Supp. 685 (D.D.C. 1975).

Natural Resources Defense Council v. Watt, 520 F. Suppl. 1359 (C.D. Cal. 1981); (9th Cir. Aug. 12, 1982).

New England River Basin Commission, 1976, The ocean's reach: Boston, New England River Basin Commission, 91 p.

Niedoroda, A. W., and Limeburner, R., 1975, A preliminary report on the geomorphic effects of off-road vehicles on coastal systems: National Park Service C.P.S.U. Report 7, 48 p.

Niering, W. A., 1970, The dilemma of the coastal wetlands: Conflict of local, national and world priorites, *in* Helfrich, H. W., ed., The environmental crises: New Haven, Conn., Yale University Press, p. 142-156.

Nilson, T. H., Taylor, E. A., and Brabb, E. E., 1976, Recent landslides in Alameda County, California (1940-71): An estimate of economic losses and correlations with slope, rainfall, and ancient landslide deposits: U.S. Geological Survey Bulletin 1398, 21 p.

Noble, C. C., 1976, The Mississippi River flood of 1973, *in* Coates, D. R., ed., Geomorphology and engineering: Stroudsburg, Pa., Dowden, Hutchinson & Ross, p. 79-98.

Office of Emergency Preparedness, 1972, Disaster preparedness: Washington, D.C., U.S. Government Printing Office, var. pages.

Office of the Chief of Engineers, 1976, A perspective on flood plain regulations for flood plain management: Department of the Army, Washington, D.C., EP 1165-2-304, 156 p.

Olson, E. P., 1978, Landslide investigation Manti Canyon, in 2 parts: U.S. Department of Agriculture, Forest Service Intermountain Region, 51 p.

Olson, E. P., 1983, The day the earth shook: U.S. Department of Agriculture, Forest Service Intermountain Reporter, November, p. 1-5.

Olson, G. W., 1976, Landuse contributions of soil survey with geomorphology and engineering, *in* Coates, D. R., ed., Geomorphology and engineering: Stroudsburg, Pa., Dowden, Hutchinson & Ross, p. 23-42.

Olson, G. W., 1981, Soils and the environment: New York, Chapman & Hall, 178 p.

Osborn, F., 1948, Our plundered planet: Boston, Little, Brown, 217 p.

Palmer, J.E., 1981, Coal resources and reserves estimates; Some unresolved problems: Geological Society of America Bulletin, v. 92, n. 8, p. 519.

Palmer, L., and KOIN-TV Portland, 1980, Mount St. Helens the volcano explodes: Northwest Illustrated, 119 p.

Paone, J., Morning, J. L., and Giorgetti, L., 1974, Land utilization and reclamation in the mining industry, 1930-71: U.S. Bureau of Mines Information Circular 8642, 61 p.

Pasadena v. Alhambra, 33 Cal. 2d 908, 3, 207 P2D 17 (1949).

Perry, R. W., and Greene, M. R., 1983, Citizen response to volcanic eruptions: The case of Mount St. Helens: New York, Irvington Publishers, 145 p.

Peters, W. C., 1978, Exploration and mining geology: New York, John Wiley & Sons, Inc., 696 p.

Phillips, R. E., et al., 1980, No-tillage agriculture: Science, v. 208, p. 1108-1113.

Pihlainen, J. A., 1963, A review of muskeg and its associated engineering problems: Cold Regions Research Engineering Laboratory Tech Report 97, 56 p.

Poland, J. F., and Davis, G. H., 1969, Land subsidence due to withdrawal of fluids, *in* Varnes, D.J., and Kiersch, G., eds.: Geological Society of America Reviews in Engineering Geology, v. 2, p. 187-269.

Pomeroy, J. S., 1980, Storm-induced debris avalanching and related phenomena in the Johnstown Area, Pennsylvania, with references to other studies in the Appalachians: U.S. Geological Survey Professional Paper 1191, 24 p.

Pomeroy, J. S., 1982, Landslides in the Greater Pittsburgh Region, Pennsylvania: U.S. Geological Survey Professional Paper 1229, 48 p.

Potomac Sand and Gravel v. Governor, 293 A241, 266 Md. (1972).

Powell, J. W., 1878, Report on the lands of the arid region of the United States, with more detailed account of the lands of Utah: 45th Congress, 2nd Session, House of Representatives Executive Document 73, Washington, D.C., var. pages.

Press, F., 1975, Earthquake prediction: Scientific American v. 232, p. 14-23.

Price, H., and Johnson, A., 1978, Relationship between alluvial soils and flooding in the Piedmont lowland of southeastern Pennsylvania: Water Resource Research, v. 14, p. 1189-1194.

Pringle, L., 1982, Water: The next great resource battle: New York, Macmillan, 144 p.

Prokopovich, N. P., 1983, Neotectonic movement and subsidence caused by piezometric decline: Association of Engineering Geologists Bulletin, v. 20, n. 4, p. 393-404.

Prokopovich, N. P., and Marrioot, M. J., 1983, Cost of subsidence to the Central Valley Project, California: Association of Engineering Geologists Bulletin, v. 20, n. 3, p. 325-332.

Pryde, P. R., 1983, The "Decade of the Environment" in the U.S.S.R.: Science, v. 220, p. 274-279.

Raleigh, C. B., et al., 1982, Forecasting southern California earthquakes: Science, v. 217 p. 1097-1104.

Ramparts Editors, 1970, Eco-catastrophe: New York, Canfield Press, 158 p.

Rampino, M. R., et al., 1979, Can rapid climatic change cause volcanic eruptions?: Science, v. 206, p. 826-829.

Regenstein, L., 1983, The toxics boomerang: Environment, v. 25, n. 10, p. 36-43, 43-54.

Rex, R. W., 1971, Geothermal energy: The neglected energy option: Atomic Science Bulletin, v. 27, n. 8, p. 52-56.

Ricci, P. F., and Moloton, L. S., 1981, Risk and benefit in environmental law: Science, v. 214, p. 1096-1100.

Rienow, R., and Rienow, L. T., 1967, Moment in the sun: New York, Ballantine Books, 305 p.

Risser, J., 1978, Soil erosion creates a crisis down on the farm: Conservation Foundation Letter, December, 16 p.

Risser, J., 1981, A renewed threat of soil erosion: It's worse than the Dust Bowl: Smithsonian, v. 11, n. 12, p. 120-131.

Roth, R. A., 1983, Factors affecting landslide susceptibility in San Mateo County, California: Association of Engineering Geologists Bulletin, v. 20, n. 4, p. 353-372.

Royster, D. L., 1979, Landslide remedial measures: Association of Engineering Geologists Bulletin, v. 16, n. 2, p. 301-352.

Ruckelshaus, W. D., 1983, Science, risk, and public policy: Science, v. 211, p. 1026-1028.

Ruhe, R. V., 1971, Stream regimen and man's manipulation, *in* Coates, D. R., ed., Environmental geomorphology: Binghamton, State University of New York, Publications in Geomorphology, p. 9-23.

Sak v. Fire Island Advisory Commission, NY Suffolk Co. Supreme Court (July 22, 1977).

Sax, J. L., 1970, Defending the environment: New York, Vintage Books, 252 p.

Scholle, S. R., 1983, Update: Acid deposition and the materials damage question: Environment, v. 25, n. 8, p. 25-32.

Schneider, W. J., and Goddard, J. E., 1974, Extent and development of urban flood plains: U.S. Geological Survey Circular 601-J, 14 p.

Schuster, R. L., 1983, Engineering aspects of the 1980 Mount St. Helens eruptions: Association of Engineering Geologists Bulletin, v. 20, n. 2, p. 125-143.

Schuster, R. L., and Krizek, R. J., 1978, Landslides: Analysis and control: Washington, D.C., National Academy of Sciences, Special Report 176, 234 p.

Seaburn, G. E., and Aronson, D. A., 1973, Catalog of recharge basins on Long Island, New York: N.Y. State Department of Conservation Bulletin 70, 80 p.

Sears, P. B., 1947, Deserts on the march: Norman, Okla., University of Oklahoma Press, 178 p.

Secretary of the Interior, 1977, Mining and minerals policy, annual report: Washington, D.C., U.S. Government Printing Office, 154 p.

Seldman, N., and Huls, J., 1981, Waste management: Beyond the throwaway ethic: Environment, v. 23, n. 9, p. 25-36.

Sharpe, C. F. S., 1939, Landslides and related phenomena: New York, Cooper Square Publishing, Inc., 137 p.

Shaeffer, J. R., 1960, Flood proofing: An element in a flood damage reduction program: University of Chicago, Department of Geography Research Paper 65, 160 p.

Shepard, F. P., and Wanless, H. R., 1971, Our changing shorelines: New York, McGraw-Hill, 579 p.

Sheridan, D., 1981, Western rangelands: Overgrazed and undermanaged: Environment, v. 23, n. 4, p. 15-20, 37-39.

Sheridan, D., and Carroll, A., 1979, Off-road vehicles on public land: Washington, D.C., Council on Environmental Quality, U.S. Government Printing Office, 83 p.

Shulters, M. V., and Clifton, D. G., 1980, Mount St. Helens volcanic ash fall in the Bull Run watershed, Oregon, March-June, 1980: U.S. Geological Survey Circular 850-A, 15 p.

Sierra Club v. Morton, 405 U.S. 727,738 (1972).

Silvester, R., 1974, Coastal engineering: New York, Elsevier Scientific Publishing Co., 450 p.

Simkhovitch, V. G., 1916, Rome's fall reconsidered: Political Science Quarterly Journal, v. 31, p. 201-243.

Simkin, T., 1981, Volcanoes of the world: Stroudsburg, Pa., Hutchinson Ross, 232 p.

Sivard, R. L., 1981, World energy survey: Leesburg, Va., World Priorities, 44 p.

Skinner, B. J., 1969, Earth resources: Englewood Cliffs, N.J., Prentice-Hall, 149 p.

Slosson, J. E., 1969, The role of engineering geology in urban planning: Colorado Geological Survey Special Publication, n. 1, p. 8-15.

Solley, W. B., Chase, E. B., and Mann, W. B., IV, 1983, Estimated use of water in the United States in 1980: U.S. Geological Survey Circular 1001, 56 p.

Stallings, J. H., 1975, Soil conservation: Englewood Cliffs, N.J., Prentice-Hall, 575 p.

Starr, C., and Whipple, C., 1980, Risks of risk decisions: Science, v. 208, p. 1114-1119.

State of California v. Sierra Club, 49 U.S.L.W. 4441 (April 28, 1981), reversing 610 F. 2d 581 (9th Cir. 1979) 400 F. Supp. 610 (N.D. Cal. 1975).

Steinbeck, J., 1939, The grapes of wrath: New York, Viking Press, 619 p.

Steinbrugge, K. V., 1971, San Fernando earthquake February 9, 1971: San Francisco, Pacific Fire Rating Bureau, 93 p.

Stephens, J. C., 1958, Subsidence of organic soils in the Florida Everglades: Soil Science Society of America Proceedings, v. 20, p. 77-80.

Stirewalt, G. L., and Tilford, N. R., eds., 1981, Geologic and hydrologic factors in the disposal of hazardous wastes in the southeastern United States: Association of Engineering Geologists Bulletin, v. 18, n. 3, p. 225-275.

Strobel v. Kerr Salt Co., 164 N.Y. 303, 58 N.E. 142 (1900).

Sugisaki, R., 1981, Deep-seated gas emission induced by the Earth-tide: A basic observation for geochemical earthquake prediction: Science, v. 212, p. 1264-1265.

Sun, M., 1983, China faces environmental challenge: Science, v. 221, p. 1271-1272.

Supreme Court of New South Wales, no date, Coal mining under stored water: Report of the Commissioner, New South Wales, Australia, 118 p.

Swanston, D. N., and Swanson, F. J., 1976, Timber harvesting, mass erosion, and steepland forest geomorphology in the Pacific Northwest, *in* Coates, D. R., ed., Geomorphology and engineering: Stroudsburg, Pa., Dowden, Hutchinson & Ross, Inc., p. 199-221.

Swinzow, G. K., 1982, The Alaska Good Friday earthquake of 1964: Cold Regions Research Engineering Laboratory Report 82-1, 26 p.

Tartar v. Spring Creek Water and Mining Co., 5 Cal. 395 (1855).

Tennessee Valley Authority v. Hill, 98 S. Ct. 2279 (1978).

Thomas, W. L., Jr., ed., 1956, Man's role in changing the face of the Earth: Chicago, University Chicago Press, 1193 p.

Toy, T. J., 1982, Accelerated erosion: Process, problems, and prognosis: Geology, v. 10, p. 524-529.

Ulrich, B. and Pankrath, eds., 1983, Effects of accumulation of air pollutants in forest ecosystems: Boston, Reidel, 389 p.

UNESCO, 1961, Symposium on salinity problems in the arid zone, Teheran, 1958: Salinity problems in the arid zones: Proceedings, Paris, UNESCO, 1961, Arid Zone Research, v. 14, 395 p.

United States v. Bishop Processing Company, 287 F. Supp. 624 D. Md. (1968).

United States v. Gerlach Livestock Co., 339 U.S., 723, 738 (1950).

U.S. Bureau of Land Management, 1975, Range condition Report, *prepared for the* Senate Committee on Appropriations: Washington, D.C., U.S. Government Printing Office, var. pages.

U.S. Corps of Engineers, 1964, Land against the sea: U.S. Army Coastal Engineering Center Miscellaneous Paper 4-64, 43 p.

U.S. Corps of Engineers, 1971, Shore management guidelines: National Shoreline Study: Department of Army, Washington, D.C., U.S. Government Printing Office, 56 p.

U.S. Corps of Engineers, 1971, Shore protection guidelines: National Shoreline Study, Department of Army: Washington, D.C., U.S. Government Printing Office, 59 p.

U.S. Department of Agriculture, 1976, Erosion and sediment control guidelines for developing areas in Texas: Soil Conservation Service (USDA), Temple, Tex., var. pages.

U.S. Department of Agriculture, 1983, Floods and landslides ravage national forests: Intermountain Reporter, Forest Service (USDA), Ogden, Utah, 5 p.

U.S. Department of Energy and U.S. Geological Survey, 1979, Earth Science Technical Plan for mined geologic disposal of radioactive waste: U.S. Department of Commerce, TID-29018 (Draft), 151 p.

U.S. Department of Interior, 1967, Surface mining and our environment: Washington, D.C., U.S. Government Printing Office, 124 p.

U.S. Department of Interior, 1983, Mineral revenues: The 1982 report on receipts from federal and Indian leases: Washington, D.C., U.S. Government Printing Office, 68 p.

U.S. Environmental Protection Agency, 1976, A manual of laws, regulations, and institutions for control of ground water pollution: Washington, D.C., EPA, var. pages.

U.S. Geological Survey, 1975, Estimates of undiscovered petroleum resources: A perspective: Washington, D.C., U.S. Government Printing Office, 23 p.

U.S. Geological Survey, 1975, Man against volcano: The eruption on Heimaey, Vestmann Islands, Iceland: USGS Inf-75-22: Washington, D.C., U.S. Government Printing Office, 15 p.

U.S. Geological Survey, 1981, Goals, strategies, priorites and tasks of a national landslide hazard-reduction program: U.S. Geological Survey Open-File Report 81-987, 91 p.

U.S. Geological Survey, 1982, Goals and tasks of the landslide part of a ground-failure hazards reduction program: U.S. Geological Survey Circular 880, 49 p.

U.S. Water Resources Council, 1976, A unified national program for flood plain management: Washington, D.C., U.S. Water Resources Council, var. pages.

U.S. Water Resources Council, 1978, The Nation's water resources 1975-2000: Second Water Assessment: Washington, D.C., U.S. Government Printing Office, 86 p.

Vermont Yankee Nuclear Power Corporation v. Natural Resources Defense Council, 98 S. Ct. 1197 (1978).

Vogt, W., 1948, Road to survival: New York, W. Sloane Associates, 355 p.

Voight, B., 1978, Rockslides and avalanches: Part 1: Natural phenomena: New York, Elsevier, var. pages.

Voight, B., 1980, Rockslides and avalanches: Part 2: Engineering sites: New York, Elsevier, var. pages.

Waggoner, E. B., 1982, The engineering geologist: An interesting hybrid: Association of Engineering Geologists Bulletin, v. 19, n. 2, p. 143-150.

Wakita, H., et al., 1980, Hydrogen release: New indicator of fault activity: Science, v. 210, p. 188-190.

Waldrip, D. B., and Ruhe, R. V., Solid waste disposal by land burial in southern Indiana: Bloomington, Indiana, Water Resources Research Center Technical Report 45, 110 p.

Walker, G. P. L. and McBroome, L. A., 1983, Mount St. Helens 1980 and Mount Pelee 1902: Flow or surge: Geology, v. 11, p. 751-754.

Walton, T. L., and Purpura, J. A., 1977, Beach nourishment along the southeast Atlantic and Gulf coasts: Shore and Beach, July, p. 10-18.

Ward, R., 1978, Floods: A geographical perspective: New York, John Wiley & Sons, 244 p.

White, D. F., and William, D. L., 1975, Assessment of geothermal resources of the United States: U.S. Geological Survey Circular 726, 155 p.

White, G. F., 1964, Choice of adjustment to floods: University of Chicago, Department of Geography Research Paper 93, 150 p.

White, G. F., ed., 1974, Natural hazards: New York, Oxford University Press, 288 p.

White, G. F., 1980, Environment: Science, v. 209, p. 183-190.

Whyatt, T., 1982, Niagara River: A case study: Amicus Journal, v. 3, n. 4, p. 38-40.

Williams, G. P., and Guy, H. P., 1971, Debris avalanches: A geomorphic hazard, *in* Coates, D. R., Environmental geomorphology: Binghamton, State University of New York, Publications in Geomorphology, p. 25-46.

Williams, G. P., and Guy, H. P., 1973, Erosional and depositional aspects of Hurricane Camille in Virginia, 1969: U.S. Geological Survey Professsional Paper 804, 80 p.

William Spangle and Associates, 1980, Land use planning after earthquakes: Portola Valley, Calif., William Spangle and Associates, Inc., var. pages.

Wilshire, H. G., 1980, Human causes of accelerated wind erosion in California's deserts, *in* Coates, D. R., and Vitek, J. D., eds., Thresholds in geomorphology: London, George Allen & Unwin, p. 415-434.

Wilshire, H. G., and Nakata, J. K., 1976, Off-road vehicle effects on California Mojave Desert: California Geology, v. 29, n. 6, p. 123-132.

Wittfogel, K. A., 1956, The hydraulic civilizations, *in* Thomas, F., ed., Man's role in changing the face of the earth: Chicago, University Chicago Press, p. 152-164.

Woodhouse, W. W., Jr., and Hanes, R. E., 1967, Dune stabilization with vegetation on the outer banks of North Carolina: U.S. Army Coastal Engineering Research Center Technical Memoir 22, 45 p.

Woodwell, G. M., et al., 1983, Global deforestation: Contribution to atmospheric carbon dioxide: Science, v. 222, p. 1081-1086.

Zabel and Russell v. U.S. Army Corps of Engineers, 276 F.2d 764, 5th Cir., U.S. Sup. Ct. (1971).

Zaruba, Q., and Mencl, V., 1969, Landslides and their control: Amsterdam, Elsevier, 205 p.

# INDEX



*Page numbers in italic indicate figures.*